How Not To Be Wrong

JORDAN ELLENBERG

How Not To Be Wrong

The Hidden Maths of Everyday Life

ALLEN LANE
an imprint of
PENGUIN BOOKS

ALLEN LANE

Published by the Penguin Group
Penguin Books Ltd, 80 Strand, London WC2R ORL, England
Penguin Group (USA) Inc., 375 Hudson Street, New York, New York 10014, USA
Penguin Group (Canada), 90 Eglinton Avenue East, Suite 700, Toronto, Ontario, Canada M4P 2Y3
(a division of Pearson Penguin Canada Inc.)
Penguin Ireland, 25 St Stephen's Green, Dublin 2, Ireland (a division of Penguin Books Ltd)
Penguin Group (Australia), 707 Collins Street, Melbourne, Victoria 3008, Australia
(a division of Pearson Australia Group Pty Ltd)
Penguin Books India Pvt Ltd, 11 Community Centre, Panchsheel Park, New Delhi – 110 017, India
Penguin Group (NZ), 67 Apollo Drive, Rosedale, Auckland 0632, New Zealand
(a division of Pearson New Zealand Ltd)
Penguin Books (South Africa) (Pty) Ltd, Block D, Rosebank Office Park, 181 Jan Smuts Avenue,
Parktown North, Gauteng 2193, South Africa

Penguin Books Ltd, Registered Offices: 80 Strand, London WC2R ORL, England

www.penguin.com

First published in the United States of America by The Penguin Press, a member of The Penguin Group (USA) LLC 2014
First published in Great Britain by Allen Lane 2014

002

Copyright © Jordan Ellenberg, 2014

Grateful acknowledgment is made for permission to reprint
excerpts from the following copyrighted works:
"Soonest Mended" from *The Double Dream of Spring* by John Ashbery.
Copyright © 1966, 1970 by John Ashbery. Reprinted by permission of
Georges Borchardt, Inc., on behalf of the author.
"Sitting on a Fence," words and music by Ian Cullimore and Paul Heaton.
Copyright © 1986 Universal Music Publishing Ltd. and Universal / Island Music
Ltd. All rights in the United States and Canada controlled and administered by
Universal Polygram International Publishing, Inc. All rights reserved. Used by
permission. Reprinted by permission of Hal Leonard Corporation.

Printed in Great Britain by Clays Ltd, St Ives plc

A CIP catalogue record for this book is available from the British Library

ISBN: 978-1-846-14678-7

www.greenpenguin.co.uk

for Tanya

"What is best in mathematics deserves not merely to be learnt as a task, but to be assimilated as a part of daily thought, and brought again and again before the mind with ever-renewed encouragement."

BERTRAND RUSSELL, "The Study of Mathematics" (1902)

CONTENTS

PART V
Existence

WHEN AM I GOING
TO USE THIS?

R ight now, in a classroom somewhere in the world, a student is
mouthing off to her math teacher. The teacher has just asked
her to spend a substantial portion of her weekend computing a
list of thirty definite integrals.

There are other things the student would rather do. There is, in fact,
hardly anything she would *not* rather do. She knows this quite clearly, because she spent a substantial portion of the previous weekend computing
a different—but not *very* different—list of thirty definite integrals. She
doesn't see the point, and she tells her teacher so. And at some point in
this conversation, the student is going to ask the question the teacher
fears most:

"When am I going to use this?"

Now the math teacher is probably going to say something like:

"I know this seems dull to you, but remember, you don't know what
career you'll choose—you may not see the relevance now, but you might
go into a field where it'll be really important that you know how to compute definite integrals quickly and correctly by hand."

This answer is seldom satisfying to the student. That's because it's
a lie. And the teacher and the student both know it's a lie. The number
of adults who will ever make use of the integral of $(1 - 3x + 4x^2)^{-2}$ dx, or

the formula for the cosine of 3θ, or synthetic division of polynomials, can be counted on a few thousand hands.

The lie is not very satisfying to the teacher, either. I should know: in my many years as a math professor I've asked many hundreds of college students to compute lists of definite integrals.

Fortunately, there's a better answer. It goes something like this:

"Mathematics is not just a sequence of computations to be carried out by rote until your patience or stamina runs out—although it might seem that way from what you've been taught in courses called *mathematics*. Those integrals are to mathematics as weight training and calisthenics are to soccer. If you want to play soccer—I mean, *really play*, at a competitive level—you've got to do a lot of boring, repetitive, apparently pointless drills. Do professional players ever *use* those drills? Well, you won't see anybody on the field curling a weight or zigzagging between traffic cones. But you do see players using the strength, speed, insight, and flexibility they built up by doing those drills, week after tedious week. Learning those drills is part of learning soccer.

"If you want to play soccer for a living, or even make the varsity team, you're going to be spending lots of boring weekends on the practice field. There's no other way. But now here's the good news. If the drills are too much for you to take, you can still play for fun, with friends. You can enjoy the thrill of making a slick pass between defenders or scoring from distance just as much as a pro athlete does. You'll be healthier and happier than you would be if you sat home watching the professionals on TV.

"Mathematics is pretty much the same. You may not be aiming for a mathematically oriented career. That's fine—most people aren't. But you can still do math. You probably already *are* doing math, even if you don't call it that. Math is woven into the way we reason. And math makes you better at things. Knowing mathematics is like wearing a pair of X-ray specs that reveal hidden structures underneath the messy and chaotic surface of the world. Math is a science of not being wrong about things, its techniques and habits hammered out by centuries of hard work and argument. With the tools of mathematics in hand, you can understand the world in a deeper, sounder, and more meaningful way. All you need is a coach, or even just a book, to teach you the rules and some basic tactics. I will be your coach. I will show you how."

For reasons of time, this is seldom what I actually say in the classroom. But in a book, there's room to stretch out a little more. I hope to back up the grand claims I just made by showing you that the problems we think about every day—problems of politics, of medicine, of commerce, of theology—are shot through with mathematics. Understanding this gives you access to insights accessible by no other means.

Even if I did give my student the full inspirational speech, she might—if she is really sharp—remain unconvinced.

"That sounds good, Professor," she'll say. "But it's pretty abstract. You say that with mathematics at your disposal you can get things right you'd otherwise get wrong. But what kind of things? Give me an *actual example*."

And at that point I would tell her the story of Abraham Wald and the missing bullet holes.

ABRAHAM WALD AND THE MISSING BULLET HOLES

This story, like many World War II stories, starts with the Nazis hounding a Jew out of Europe and ends with the Nazis regretting it. Abraham Wald was born in 1902 in what was then the city of Klausenburg in what was then the Austro-Hungarian Empire. By the time Wald was a teenager, one World War was in the books and his hometown had become Cluj, Romania. He was the grandson of a rabbi and the son of a kosher baker, but the younger Wald was a mathematician almost from the start. His talent for the subject was quickly recognized, and he was admitted to study mathematics at the University of Vienna, where he was drawn to subjects abstract and recondite even by the standards of pure mathematics: set theory and metric spaces.

But when Wald's studies were completed, it was the mid-1930s, Austria was deep in economic distress, and there was no possibility that a foreigner could be hired as a professor in Vienna. Wald was rescued by a job offer from Oskar Morgenstern. Morgenstern would later immigrate to the United States and help invent game theory, but in 1933 he was the director of the Austrian Institute for Economic Research, and he hired Wald at a small salary to do mathematical odd jobs. That turned out to be a good move for Wald: his experience in economics got him a fellowship offer at the Cowles Commission, an economic institute then located in Colorado Springs. Despite the ever-worsening political situation, Wald was reluctant to take a step that would lead him away from pure mathematics for good. But then the Nazis conquered Austria, making Wald's decision substantially easier. After just a few months in Colorado, he was offered a professorship of statistics at Columbia; he packed up once again and moved to New York.

And that was where he fought the war.

The Statistical Research Group (SRG), where Wald spent much of World War II, was a classified program that yoked the assembled might of American statisticians to the war effort—something like the Manhattan Project, except the weapons being developed were equations, not explosives. And the SRG was actually *in* Manhattan, at 401 West 118th Street in Morningside Heights, just a block away from Columbia University. The building now houses Columbia faculty apartments and some doctor's offices, but in 1943 it was the buzzing, sparking nerve center of wartime math. At the Applied Mathematics Group–Columbia, dozens of young women bent over Marchant desktop calculators were calculating formulas for the optimal curve a fighter should trace out through the air in order to keep an enemy plane in its gunsights. In another apartment, a team of researchers from Princeton was developing protocols for strategic bombing. And Columbia's wing of the atom bomb project was right next door.

But the SRG was the most high-powered, and ultimately the most influential, of any of these groups. The atmosphere combined the intellectual openness and intensity of an academic department with the shared sense of purpose that comes only with high stakes. "When we made recommendations," W. Allen Wallis, the director, wrote, "frequently things happened. Fighter planes entered combat with their machine guns loaded according to Jack Wolfowitz's* recommendations about mixing

* Paul's dad.

types of ammunition, and maybe the pilots came back or maybe they didn't. Navy planes launched rockets whose propellants had been accepted by Abe Girshick's sampling-inspection plans, and maybe the rockets exploded and destroyed our own planes and pilots or maybe they destroyed the target."

The mathematical talent at hand was equal to the gravity of the task. In Wallis's words, the SRG was "the most extraordinary group of statisticians ever organized, taking into account both number and quality." Frederick Mosteller, who would later found Harvard's statistics department, was there. So was Leonard Jimmie Savage, the pioneer of decision theory and great advocate of the field that came to be called Bayesian statistics.* Norbert Wiener, the MIT mathematician and the creator of cybernetics, dropped by from time to time. This was a group where Milton Friedman, the future Nobelist in economics, was often the fourth-smartest person in the room.

The *smartest* person in the room was usually Abraham Wald. Wald had been Allen Wallis's teacher at Columbia, and functioned as a kind of mathematical eminence to the group. Still an "enemy alien," he was not technically allowed to see the classified reports he was producing; the joke around SRG was that the secretaries were required to pull each sheet of notepaper out of his hands as soon as he was finished writing on it. Wald was, in some ways, an unlikely participant. His inclination, as it always had been, was toward abstraction, and away from direct applications. But his motivation to use his talents against the Axis was obvious. And when you needed to turn a vague idea into solid mathematics, Wald was the person you wanted at your side.

So here's the question. You don't want your planes to get shot down by enemy fighters, so you armor them. But armor makes the plane heavier, and heavier planes are less maneuverable and use more fuel. Armoring the planes too much is a problem; armoring the planes too little is a problem. Somewhere in between there's an optimum. The reason you

* Savage was almost totally blind, able to see only out of one corner of one eye, and at one point spent six months living only on pemmican in order to prove a point about Arctic exploration. Just thought that was worth mentioning.

have a team of mathematicians socked away in an apartment in New York City is to figure out where that optimum is.

The military came to the SRG with some data they thought might be useful. When American planes came back from engagements over Europe, they were covered in bullet holes. But the damage wasn't uniformly distributed across the aircraft. There were more bullet holes in the fuselage, not so many in the engines.

Section of plane	Bullet holes per square foot
Engine	1.11
Fuselage	1.73
Fuel system	1.55
Rest of the plane	1.8

The officers saw an opportunity for efficiency; you can get the same protection with less armor if you concentrate the armor on the places with the greatest need, where the planes are getting hit the most. But exactly how much more armor belonged on those parts of the plane? That was the answer they came to Wald for. It wasn't the answer they got.

The armor, said Wald, doesn't go where the bullet holes are. It goes where the bullet holes *aren't:* on the engines.

Wald's insight was simply to ask: where are the missing holes? The ones that would have been all over the engine casing, if the damage had been spread equally all over the plane? Wald was pretty sure he knew. The missing bullet holes were on the missing planes. The reason planes were coming back with fewer hits to the engine is that planes that got hit in the engine weren't coming back. Whereas the large number of planes returning to base with a thoroughly Swiss-cheesed fuselage is pretty strong evidence that hits to the fuselage can (and therefore should) be tolerated. If you go the recovery room at the hospital, you'll see a lot more people with bullet holes in their legs than people with bullet holes in their chests. But that's not because people don't get shot in the chest; it's because the people who get shot in the chest don't recover.

Here's an old mathematician's trick that makes the picture perfectly clear: *set some variables to zero*. In this case, the variable to tweak is the probability that a plane that takes a hit to the engine manages to stay in the air. Setting that probability to zero means a single shot to the engine is guaranteed to bring the plane down. What would the data look like then? You'd have planes coming back with bullet holes all over the wings, the fuselage, the nose—but none at all on the engine. The military analyst has two options for explaining this: either the German bullets just happen to hit every part of the plane but one, or the engine is a point of total vulnerability. Both stories explain the data, but the latter makes a lot more sense. The armor goes where the bullet holes aren't.

Wald's recommendations were quickly put into effect, and were still being used by the navy and the air force through the wars in Korea and Vietnam. I can't tell you exactly how many American planes they saved, though the data-slinging descendants of the SRG inside today's military no doubt have a pretty good idea. One thing the American defense establishment has traditionally understood very well is that countries don't win wars just by being braver than the other side, or freer, or slightly preferred by God. The winners are usually the guys who get 5% fewer of their planes shot down, or use 5% less fuel, or get 5% more nutrition into their infantry at 95% of the cost. That's not the stuff war movies are made of, but it's the stuff wars are made of. And there's math every step of the way.

Why did Wald see what the officers, who had vastly more knowledge and understanding of aerial combat, couldn't? It comes back to his math-trained habits of thought. A mathematician is always asking, "What assumptions are you making? And are they justified?" This can be annoying. But it can also be very productive. In this case, the officers were making an assumption unwittingly: that the planes that came back were a random sample of all the planes. If that were true, you could draw conclusions about the distribution of bullet holes on all the planes by examining the distribution of bullet holes on only the surviving planes. Once you recognize that you've been making that hypothesis, it only takes a moment to realize it's dead wrong; there's no reason at all to expect the planes to have an equal likelihood of survival no matter where they get hit. In a piece of mathematical lingo we'll come back to in chapter 15, the rate of survival and the location of the bullet holes are *correlated*.

Wald's other advantage was his tendency toward abstraction. Wolfowitz, who had studied under Wald at Columbia, wrote that the problems he favored were "all of the most abstract sort," and that he was "always ready to talk about mathematics, but uninterested in popularization and special applications."

Wald's personality made it hard for him to focus his attention on applied problems, it's true. The details of planes and guns were, to his eye, so much upholstery—he peered right through to the mathematical struts and nails holding the story together. Sometimes that approach can lead you to ignore features of the problem that really matter. But it also lets you see the common skeleton shared by problems that look very different on the surface. Thus you have meaningful experience even in areas where you appear to have none.

To a mathematician, the structure underlying the bullet hole problem is a phenomenon called *survivorship bias*. It arises again and again, in all kinds of contexts. And once you're familiar with it, as Wald was, you're primed to notice it wherever it's hiding.

Like mutual funds. Judging the performance of funds is an area where you don't want to be wrong, even by a little bit. A shift of 1% in annual growth might be the difference between a valuable financial asset and a dog. The funds in Morningstar's Large Blend category, whose mutual funds invest in big companies that roughly represent the S&P 500, look like the former kind. The funds in this class grew an average of 178.4% between 1995 and 2004: a healthy 10.8% per year.* Sounds like you'd do well, if you had cash on hand, to invest in those funds, no?

Well, no. A 2006 study by Savant Capital shone a somewhat colder light on those numbers. Think again about how Morningstar generates its number. It's 2004, you take all the funds classified as Large Blend, and you see how much they grew over the last ten years.

But something's missing: *the funds that aren't there*. Mutual funds don't live forever. Some flourish, some die. The ones that die are, by and large, the ones that don't make money. So judging a decade's worth of

---

* To be fair, the S&P 500 index itself did even better, gaining 212.5% over the same period.

mutual funds by the ones that still exist at the end of the ten years is like judging our pilots' evasive maneuvers by counting the bullet holes in the planes that come back. What would it mean if we never found more than one bullet hole per plane? Not that our pilots are brilliant at dodging enemy fire, but that the planes that got hit twice went down in flames.

The Savant study found that if you included the performance of the dead funds together with the surviving ones, the rate of return dropped down to 134.5%, a much more ordinary 8.9% per year. More recent research backed that up: a comprehensive 2011 study in the *Review of Finance* covering nearly 5,000 funds found that the excess return rate of the 2,641 survivors is about 20% higher than the same figure recomputed to include the funds that didn't make it. The size of the survivorship effect might have surprised investors, but it probably wouldn't have surprised Abraham Wald.

MATHEMATICS IS THE EXTENSION OF COMMON SENSE BY OTHER MEANS

At this point my teenaged interlocutor is going to stop me and ask, quite reasonably: Where's the math? Wald was a mathematician, that's true, and it can't be denied that his solution to the problem of the bullet holes was ingenious, but what's mathematical about it? There was no trig identity to be seen, no integral or inequality or formula.

First of all: Wald did use formulas. I told the story without them, because this is just the introduction. When you write a book explaining human reproduction to preteens, the introduction stops short of the really hydraulic stuff about how babies get inside Mommy's tummy. Instead, you start with something more like "Everything in nature changes; trees lose their leaves in winter only to bloom again in spring; the humble caterpillar enters its chrysalis and emerges as a magnificent butterfly. You are part of nature too, and . . ."

That's the part of the book we're in now.

But we're all adults here. Turning off the soft focus for a second, here's what a sample page of Wald's actual report looks like:

lower bound to the Q_i could be obtained. The assumption here is that the decrease from q_i to q_{i+1} lies between definite limits. Therefore, both an upper and lower bound for the Q_i can be obtained.

We assume that

$$\lambda_1 q_i \leq q_{i+1} \leq \lambda_2 q_i \; ,$$

where $\lambda_1 < \lambda_2 < 1$ and such that the expression

$$\sum_{j=1}^{n} \frac{a_j}{\lambda_1^{\frac{j(j-1)}{2}}} < 1 - a_o \qquad (A)$$

is satisfied.

The exact solution is tedious but close approximations to the upper and lower bounds to the Q_i for $i < n$ can be obtained by the following procedure. The set of hypothetical data used is

$$a_o = .780 \qquad a_3 = .010$$
$$a_1 = .070 \qquad a_4 = .005$$
$$a_2 = .040 \qquad a_5 = .005$$
$$\lambda_1 = .80 \qquad \lambda_2 = .90$$

Condition A is satisfied, since by substitution

$$.07 + \frac{.04}{.8} + \frac{.01}{(.8)^3} + \frac{.005}{(.8)^6} + \frac{.005}{(.8)^{10}} = .20529 \; ,$$

which is less than

$$1 - a_o = .22 \; .$$

THE LOWER LIMIT OF Q_i

The first step is to solve equation 66. This involves the solution of the following four equations for positive roots g_o, g_1, g_2, g_3.

I hope that wasn't too shocking.

Still, the real *idea* behind Wald's insight doesn't require any of the formalism above. We've already explained it, using no mathematical notation of any kind. So my student's question stands. What makes that math? Isn't it just common sense?

Yes. Mathematics *is* common sense. On some basic level, this is clear.

How can you explain to someone why adding seven things to five things yields the same result as adding five things to seven? You can't: that fact is baked into our way of thinking about combining things together. Mathematicians like to give names to the phenomena our common sense describes: instead of saying, "*This* thing added to *that* thing is the *same* thing as *that* thing added to *this* thing," we say, "Addition is commutative." Or, because we like our symbols, we write:

For any choice of a and b, a + b = b + a.

Despite the official-looking formula, we are talking about a fact in-stinctively understood by every child.

Multiplication is a slightly different story. The formula looks pretty similar:

For any choice of a and b, a × b = b × a.

The mind, presented with this statement, does not say "no duh" quite as instantly as it does for addition. Is it "common sense" that two sets of six things amount to the same as six sets of two?

Maybe not; but it can *become* common sense. Here's my earliest math-ematical memory. I'm lying on the floor in my parents' house, my cheek pressed against the shag rug, looking at the stereo. Very probably I am listening to side two of the Beatles' Blue Album. Maybe I'm six. This is the seventies, and therefore the stereo is encased in a pressed wood panel, which has a rectangular array of airholes punched into the side. Eight holes across, six holes up and down. So I'm lying there, looking at the airholes. The six rows of holes. The eight columns of holes. By focus-ing my gaze in and out I could make my mind flip back and forth be-tween seeing the rows and seeing the columns. Six rows with eight holes each. Eight columns with six holes each.

And then I had it—eight groups of six were the same as six groups of eight. Not because it was a rule I'd been told, but because it could not be any other way. The number of holes in the panel was the number of holes in the panel, no matter which way you counted them.

MY PARENTS' STEREO,
1977

We tend to teach mathematics as a long list of rules. You learn them in order and you have to obey them, because if you don't obey them you get a C-. *This is not mathematics.* Mathematics is the study of things that come out a certain way because there is no other way they could possibly be.

Now let's be fair: not everything in mathematics can be made as perfectly transparent to our intuition as addition and multiplication. You can't do calculus by common sense. But calculus is still *derived* from our common sense—Newton took our physical intuition about objects moving in straight lines, formalized it, and then built on top of that formal structure a universal mathematical description of motion. Once you have Newton's theory in hand, you can apply it to problems that would make your head spin if you had no equations to help you. In the same way, we have built-in mental systems for assessing the likelihood of an uncertain outcome. But those systems are pretty weak and unreliable, especially when it comes to events of extreme rarity. That's when we shore up our intuition with a few sturdy, well-placed theorems and techniques, and make out of it a mathematical theory of probability.

The specialized language in which mathematicians converse with each other is a magnificent tool for conveying complex ideas precisely and swiftly. But its foreignness can create among outsiders the impression of a sphere of thought totally alien to ordinary thinking. That's exactly wrong.

Math is like an atomic-powered prosthesis that you attach to your common sense, vastly multiplying its reach and strength. Despite the power of mathematics, and despite its sometimes forbidding notation and abstraction, the actual mental work involved is little different from the way we think about more down-to-earth problems. I find it helpful to keep in mind an image of Iron Man punching a hole through a brick wall. On the one hand, the actual wall-breaking force is being supplied, not by Tony Stark's muscles, but by a series of exquisitely synchronized servomechanisms powered by a compact beta particle generator. On the other hand, from Tony Stark's point of view, what he is doing is punching a wall, exactly as he would without the armor. Only much, much harder.

To paraphrase Clausewitz: Mathematics is the extension of common sense by other means.

Without the rigorous structure that math provides, common sense can lead you astray. That's what happened to the officers who wanted to armor the parts of the planes that were already strong enough. But formal mathematics without common sense—without the constant interplay between abstract reasoning and our intuitions about quantity, time, space, motion, behavior, and uncertainty—would just be a sterile exercise in rule-following and bookkeeping. In other words, math would actually be what the peevish calculus student believes it to be.

That's a real danger. John von Neumann, in his 1947 essay "The Mathematician," warned:

> As a mathematical discipline travels far from its empirical source, or still more, if it is a second and third generation only indirectly inspired by ideas coming from "reality" it is beset with very grave dangers. It becomes more and more purely aestheticizing, more and more purely *l'art pour l'art*. This need not be bad, if the field is surrounded by correlated subjects, which still have closer empirical connections, or if the discipline is under the influence of men with an exceptionally well-developed taste. But there is a grave danger that the subject will develop along the line of least resistance, that the stream, so far from its source, will separate into a multitude of insignificant branches, and that the discipline will become a disorga-

nized mass of details and complexities. In other words, at a great distance from its empirical source, or after much "abstract" inbreeding, a mathematical subject is in danger of degeneration.[*]

WHAT KINDS OF MATHEMATICS WILL APPEAR IN THIS BOOK?

If your acquaintance with mathematics comes entirely from school, you have been told a story that is very limited, and in some important ways false. School mathematics is largely made up of a sequence of facts and rules, facts which are certain, rules which come from a higher authority and cannot be questioned. It treats mathematical matters as completely settled.

Mathematics is not settled. Even concerning the basic objects of study, like numbers and geometric figures, our ignorance is much greater than our knowledge. And the things we do know were arrived at only after massive effort, contention, and confusion. All this sweat and tumult is carefully screened off in your textbook.

There are facts and there are facts, of course. There has never been much controversy about whether $1 + 2 = 3$. The question of *how and whether we can truly prove* that $1 + 2 = 3$, which wobbles uneasily between mathematics and philosophy, is another story—we return to that at the end of the book. But that the computation is correct is a plain truth. The tumult lies elsewhere. We'll come within sight of it several times.

Mathematical facts can be simple or complicated, and they can be shallow or profound. This divides the mathematical universe into four quadrants:

---

[*] Von Neumann's view of the nature of math is solid, but it's fair to feel a bit queasy about his characterization of mathematics carried out for purely aesthetic ends as "degenerate." Von Neumann is writing this just ten years after the *entartene Kunst* ("degenerate art") exhibition in Hitler's Berlin, whose point was that "*l'art pour l'art*" was the sort of thing that Jews and Communists liked, and was designed to undercut the healthy "realist" art required by a vigorous Teutonic state. Under the circumstances, one feels a little defensive toward mathematics that serves no apparent purpose. A writer with different political commitments than my own would, at this point, bring up von Neumann's energetic work on the development and delivery of nuclear weapons.

Basic arithmetic facts, like $1 + 2 = 3$, are simple and shallow. So are basic identities like $\sin(2x) = 2 \sin x \cos x$ or the quadratic formula: they might be slightly harder to convince yourself of than $1 + 2 = 3$, but in the end they don't have much conceptual heft.

Moving over to complicated/shallow, you have the problem of multiplying two ten-digit numbers, or the computation of an intricate definite integral, or, given a couple of years of graduate school, the trace of Frobenius on a modular form of conductor 2377. It's conceivable you might, for some reason, need to know the answer to such a problem, and it's undeniable that it would be somewhere between annoying and impossible to work it out by hand; or, as in the case of the modular form, it might take some serious schooling even to understand what's being asked for. But knowing those answers doesn't really enrich your knowledge about the world.

The complicated/profound quadrant is where professional mathematicians like me try to spend most of our time. That's where the celebrity theorems and conjectures live: the Riemann Hypothesis, Fermat's Last Theorem,* the Poincaré Conjecture, P vs. NP, Gödel's Theorem . . . Each

---

* Which, among pros, is now called Wiles's Theorem, since Andrew Wiles proved it (with a critical assist from Richard Taylor) and Fermat did not. But the traditional name will probably never be dislodged.

one of these theorems involves ideas of deep meaning, fundamental importance, mind-blowing beauty, and brutal technicality, and each of them is the protagonist of books of its own.

But not this book. This book is going to hang out in the upper left quadrant: simple and profound. The mathematical ideas we want to address are ones that can be engaged with directly and profitably, whether your mathematical training stops at pre-algebra or extends much further. And they are not "mere facts," like a simple statement of arithmetic—they are principles, whose application extends far beyond the things you're used to thinking of as mathematical. They are the go-to tools on the utility belt, and used properly they will help you not be wrong.

Pure mathematics can be a kind of convent, a quiet place safely cut off from the pernicious influences of the world's messiness and inconsistency. I grew up inside those walls. Other math kids I knew were tempted by applications to physics, or genomics, or the black art of hedge fund management, but I wanted no such *rumspringa*.* As a graduate student, I dedicated myself to number theory, what Gauss called "the queen of mathematics," the purest of the pure subjects, the sealed garden at the center of the convent, where we contemplated the same questions about numbers and equations that troubled the Greeks and have gotten hardly less vexing in the twenty-five hundred years since.

At first I worked on number theory with a classical flavor, proving facts about sums of fourth powers of whole numbers that I could, if pressed, explain to my family at Thanksgiving, even if I couldn't explain how I proved what I proved. But before long I got enticed into even more abstract realms, investigating problems where the basic actors— "residually modular Galois representations," "cohomology of moduli schemes," "dynamical systems on homogeneous spaces," things like that—were impossible to talk about outside the archipelago of seminar halls and faculty lounges that stretches from Oxford to Princeton to Kyoto to Paris to Madison, Wisconsin, where I'm a professor now. When I tell you this stuff is thrilling, and meaningful, and beautiful, and that

* To be honest, I did spend some part of my early twenties thinking I might want to be a Serious Literary Novelist. I even finished a Serious Literary Novel, called *The Grasshopper King*, and got it published. But in the process I discovered that every day I devoted to Serious Literary Novel-writing was a day half spent moping around wishing I were working on math problems.

I'll never get tired of thinking about it, you may just have to believe me, because it takes a long education just to get to the point where the objects of study rear into view.

But something funny happened. The more abstract and distant from lived experience my research got, the more I started to notice how much math was going on in the world outside the walls. Not Galois representations or cohomology, but ideas that were simpler, older, and just as deep—the northwest quadrant of the conceptual foursquare. I started writing articles for magazines and newspapers about the way the world looked through a mathematical lens, and I found, to my surprise, that even people who said they hated math were willing to read them. It was a kind of math teaching, but very different from what we do in a classroom.

What it has in common with the classroom is that the reader gets asked to do some work. Back to von Neumann on "The Mathematician":

"It is harder to understand the mechanism of an airplane, and the theories of the forces which lift and which propel it, than merely to ride in it, to be elevated and transported by it—or even to steer it. It is exceptional that one should be able to acquire the understanding of a process without having previously acquired a deep familiarity with running it, with using it, before one has assimilated it in an instinctive and empirical way."

In other words: it is pretty hard to *understand* mathematics without *doing* some mathematics. There's no royal road to geometry, as Euclid told Ptolemy, or maybe, depending on your source, as Menaechmus told Alexander the Great. (Let's face it, famous old maxims attributed to ancient scientists are probably made up, but they're no less instructive for that.)

This will not be the kind of book where I make grand, vague gestures at great monuments of mathematics, and instruct you in the proper manner of admiring them from a great distance. We are here to get our hands a little dirty. We'll compute some things. There will be a few formulas and equations, when I need them to make a point. No formal math beyond arithmetic will be required, though lots of math way beyond arithmetic will be explained. I'll draw some crude graphs and charts. We'll encounter some topics from school math, outside their usual habitat; we'll see how trigonometric functions describe the extent to which two variables are related to each other, what calculus has to say about the relationship between linear and nonlinear phenomena, and how the quadratic formula serves as a cognitive model for scientific inquiry. And we'll also run into some of the mathematics that usually gets put off to college or beyond, like the crisis in set theory, which appears here as a kind of metaphor for Supreme Court jurisprudence and baseball umpiring; recent developments in analytic number theory, which demonstrate the interplay between structure and randomness; and information theory and combinatorial designs, which help explain how a group of MIT undergrads won millions of dollars by understanding the guts of the Massachusetts state lottery.

There will be occasional gossip about mathematicians of note, and a certain amount of philosophical speculation. There will even be a proof or two. But there will be no homework, and there will be no test.

PART I

· · · · ·

Linearity

Includes: the Laffer curve, calculus explained in one page, the Law of Large Numbers, assorted terrorism analogies, "Everyone in America will be overweight by 2048," why South Dakota has more brain cancer than North Dakota, the ghosts of departed quantities, the habit of definition

ONE

LESS LIKE SWEDEN

A few years ago, in the heat of the battle over the Affordable Care
Act, Daniel J. Mitchell of the libertarian Cato Institute posted
a blog entry with the provocative title: "Why Is Obama Trying
to Make America More Like Sweden when Swedes Are Trying to Be
Less Like Sweden?"

Good question! When you put it that way, it does seem pretty
perverse. Why, Mr. President, are we swimming against the current of
history, while social welfare states around the world—even rich little
Sweden!—are cutting back on expensive benefits and high taxes? "If
Swedes have learned from their mistakes and are now trying to reduce
the size and scope of government," Mitchell writes, "why are American
politicians determined to repeat those mistakes?"

Answering this question will require an extremely scientific chart.
Here's what the world looks like to the Cato Institute:

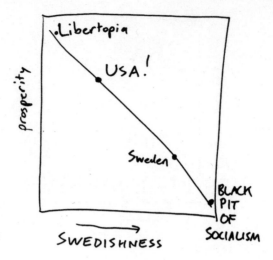

The x-axis represents Swedishness,* and the y-axis is some measure of prosperity. Don't worry about exactly how we're quantifying these things. The point is just this: according to the chart, the more Swedish you are, the worse off your country is. The Swedes, no fools, have figured this out and are launching their northwestward climb toward free-market prosperity. But Obama's sliding in the wrong direction.

Let me draw the same picture from the point of view of people whose economic views are closer to President Obama's than to those of the Cato Institute. See the next image.

This picture gives very different advice about how Swedish we should be. Where do we find peak prosperity? At a point more Swedish than America, but less Swedish than Sweden. If this picture is right, it makes perfect sense for Obama to beef up our welfare state while the Swedes trim theirs down.

The difference between the two pictures is the difference between linearity and nonlinearity, one of the central distinctions in mathematics. The Cato curve is a line;† the non-Cato curve, the one with the hump in

---

* Here "Swedishness" refers to "quantity of social services and taxation," not to other features of Sweden such as "ready availability of herring in dozens of different sauces," a condition to which all nations should obviously aspire.
† Or a line segment, if you must. I won't make a big deal of this distinction.

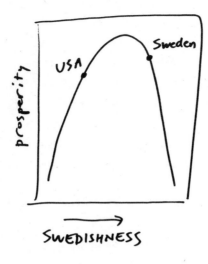

the middle, is not. A line is one kind of curve, but not the only kind, and lines enjoy all kinds of special properties that curves in general may not. The highest point on a line segment—the maximum prosperity, in this example—has to be on one end or the other. That's just how lines are. If lowering taxes is good for prosperity, then lowering taxes even more is even better. And if Sweden wants to de-Swede, so should we. Of course, an anti-Cato think tank might posit that the line slopes in the other direction, going southwest to northeast. And if that's what the line looks like, then no amount of social spending is too much. The optimal policy is Maximum Swede.

Usually, when someone announces they're a "nonlinear thinker" they're about to apologize for losing something you lent them. But nonlinearity is a real thing! And in this context, thinking nonlinearly is crucial, because not all curves are lines. A moment of reflection will tell you that the real curves of economics look like the second picture, not the first. They're nonlinear. Mitchell's reasoning is an example of *false linearity*—he's assuming, without coming right out and saying so, that the course of prosperity is described by the line segment in the first picture, in which case Sweden stripping down its social infrastructure means we should do the same.

But as long as you believe there's such a thing as too much welfare state and such a thing as too little, you know the linear picture is wrong. Some principle more complicated than "More government bad, less government good" is in effect. The generals who consulted Abraham Wald faced the same kind of situation: too little armor meant planes got shot down, too much meant the planes couldn't fly. It's not a question of whether adding more armor is good or bad; it could be either, depending on how heavily armored the planes are to start with. If there's an optimal answer, it's somewhere in the middle, and deviating from it in either direction is bad news.

Nonlinear thinking means *which way you should go depends on where you already are.*

This insight isn't new. Already in Roman times we find Horace's famous remark *"Est modus in rebus, sunt certi denique fines, quos ultra citraque nequit consistere rectum"* ("There is a proper measure in things. There are, finally, certain boundaries short of and beyond which what is right cannot exist"). And further back still, in the *Nicomachean Ethics*, Aristotle observes that eating either too much or too little is troubling to the constitution. The optimum is somewhere in between; because the relation between eating and health isn't linear, but curved, with bad outcomes on both ends.

SOMETHING-DOO ECONOMICS

The irony is that economic conservatives like the folks at Cato used to understand this better than anybody. That second picture I drew up there? The extremely scientific one with the hump in the middle? I am not the first person to draw it. It's called the *Laffer curve*, and it's played a central role in Republican economics for almost forty years. By the middle of the Reagan administration, the curve had become such a commonplace of economic discourse that Ben Stein ad-libbed it into his famous soul-killing lecture in *Ferris Bueller's Day Off*:

Anyone know what this is? Class? Anyone? . . . Anyone? Anyone seen this before? The Laffer curve. Anyone know what this says? It says that at this point on the revenue curve, you will get exactly the same amount of revenue as at this point. This is very controversial. Does anyone know what Vice President Bush called this in 1980? Anyone? Something-doo economics. "Voodoo" economics.

The legend of the Laffer curve goes like this: Arthur Laffer, then an economics professor at the University of Chicago, had dinner one night in 1974 with Dick Cheney, Donald Rumsfeld, and *Wall Street Journal* editor Jude Wanniski at an upscale hotel restaurant in Washington, DC. They were tussling over President Ford's tax plan, and eventually, as intellectuals do when the tussling gets heavy, Laffer commandeered a napkin* and drew a picture. The picture looked like this:

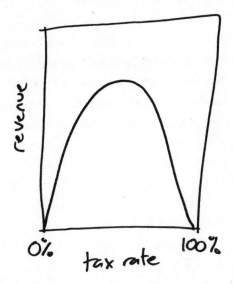

The horizontal axis here is level of taxation, and the vertical axis represents the amount of revenue the government takes in from taxpayers. On the left edge of the graph, the tax rate is 0%; in that case, by definition, the government gets no tax revenue. On the right, the tax rate is 100%; whatever income you have, whether from a business you run or a salary you're paid, goes straight into Uncle Sam's bag.

---

* Laffer disputes the napkin portion of the story, recalling that the restaurant had classy cloth napkins that he would never have vandalized with an economic doodle.

Which is empty. Because if the government vacuums up every cent of the wage you're paid to show up and teach school, or sell hardware, or middle-manage, why bother doing it? Over on the right edge of the graph, people don't work at all. Or, if they work, they do so in informal economic niches where the tax collector's hand can't reach. The government's revenue is zero once again.

In the intermediate range in the middle of the curve, where the government charges us somewhere between none of our income and all of it—in other words, in the real world—the government does take in some amount of revenue.

That means the curve recording the relationship between tax rate and government revenue cannot be a straight line. If it were, revenue would be maximized at either the left or right edge of the graph; but it's zero both places. If the current income tax is really close to zero, so that you're on the left-hand side of the graph, then raising taxes increases the amount of money the government has available to fund services and programs, just as you might intuitively expect. But if the rate is close to 100%, raising taxes actually *decreases* government revenue. If you're to the right of the Laffer peak, and you want to decrease the deficit without cutting spending, there's a simple and politically peachy solution: lower the tax rate, and thereby increase the amount of taxes you take in. *Which way you should go depends on where you are.*

So where are we? That's where things get sticky. In 1974, the top income tax rate was 70%, and the idea that America was on the right-hand downslope of the Laffer curve held a certain appeal—especially for the few people lucky enough to pay tax at that rate, which only applied to income beyond the first $200,000.* And the Laffer curve had a potent advocate in Wanniski, who brought his theory into the public consciousness in a 1978 book rather self-assuredly titled *The Way the World Works.*† Wanniski was a true believer, with the right mix of zeal and political canniness to get people to listen to an idea considered fringy even by tax-cut advocates. He was untroubled by being called a nut. "Now, what does 'nut' mean?" he asked an interviewer. "Thomas Edison was a nut, Leibniz

---

* Somewhere between a half million and a million dollars a year in today's income.
† Like I'm one to talk.

was a nut, Galileo was a nut, so forth and so on. Everybody who comes with a new idea to the conventional wisdom, comes with an idea that's so far outside the mainstream, that's considered nutty."

(Aside: it's important to point out here that people with out-of-the-mainstream ideas who compare themselves to Edison and Galileo are *never actually right*. I get letters with this kind of language at least once a month, usually from people who have "proofs" of mathematical statements that have been known for hundreds of years to be false. I can guarantee you Einstein did not go around telling people, "Look, I know this theory of general relativity sounds wacky, but that's what they said about Galileo!")

The Laffer curve, with its compact visual representation and its agreeably counterintuitive sting, turned out to be an easy sell for politicians with a preexisting hunger for tax cuts. As economist Hal Varian put it, "You can explain it to a Congressman in six minutes and he can talk about it for six months." Wanniski became an advisor first to Jack Kemp, then to Ronald Reagan, whose experiences as a wealthy movie star in the 1940s formed the template for his view of the economy four decades later. His budget director, David Stockman, recalls:

> "I came into the Big Money making pictures during World War II," [Reagan] would always say. At that time the wartime income surtax hit 90 percent. "You could only make four pictures and then you were in the top bracket," he would continue. "So we all quit working after about four pictures and went off to the country." High tax rates caused less work. Low tax rates caused more. His experience proved it.

These days it's hard to find a reputable economist who thinks we're on the downslope of the Laffer curve. Maybe that's not surprising, considering top incomes are currently taxed at just 35%, a rate that would have seemed absurdly low for most of the twentieth century. But even in Reagan's day, we were probably on the left-hand side of the curve. Greg Mankiw, an economist at Harvard and a Republican who chaired the Council of Economic Advisors under the second President Bush, writes in his microeconomics textbook:

Subsequent history failed to confirm Laffer's conjecture that lower tax rates would raise tax revenue. When Reagan cut taxes after he was elected, the result was less tax revenue, not more. Revenue from personal income taxes (per person, adjusted for inflation) fell by 9 percent from 1980 to 1984, even though average income (per person, adjusted for inflation) grew by 4 percent over this period. Yet once the policy was in place, it was hard to reverse.

Some sympathy for the supply-siders is now in order. First of all, maximizing government revenue needn't be the goal of tax policy. Milton Friedman, whom we last met during World War II doing classified military work for the Statistical Research Group, went on to become a Nobel-winning economist and advisor to presidents, and a powerful advocate for low taxes and libertarian philosophy. Friedman's famous slogan on taxation is "I am in favor of cutting taxes under any circumstances and for any excuse, for any reason, whenever it's possible." He didn't think we should be aiming for the top of the Laffer curve, where government tax revenue is as high as it can be. For Friedman, money obtained by the government would eventually be money spent by the government, and that money, he felt, was more often spent badly than well.

More moderate supply-side thinkers, like Mankiw, argue that lower taxes can increase the motivation to work hard and launch businesses, leading eventually to a bigger, stronger economy, even if the immediate effect of the tax cut is decreased government revenue and bigger deficits. An economist with more redistributionist sympathies would observe that this cuts both ways; maybe the government's diminished ability to spend means it constructs less infrastructure, regulates fraud less stringently, and generally does less of the work that enables free enterprise to thrive.

Mankiw also points out that the very richest people—the ones who'd been paying 70% on the top tranche of their income—*did* contribute more tax revenue after Reagan's tax cuts.* That leads to the somewhat vexing possibility that the way to maximize government revenue is to

---

* Whether the increased tax receipts are because the rich started working harder once less encumbered by income tax, as supply-side theory predicts, is more difficult to say for certain.

jack up taxes on the middle class, who have no choice but to keep on working, while slashing rates on the rich; those guys have enough stock-piled wealth to make credible threats to withhold or offshore their eco-nomic activity, should their government charge them a rate they deem too high. If that story's right, a lot of liberals will uncomfortably climb in the boat with Milton Friedman: maybe maximizing tax revenue isn't so great after all.

Mankiw's final assessment is a rather polite, "Laffer's argument is not completely without merit." I would give Laffer more credit than that! His drawing made the fundamental and incontrovertible mathematical point that the relationship between taxation and revenue is necessarily nonlinear. It doesn't, of course, have to be a single smooth hill like the one Laffer sketched; it could look like a trapezoid

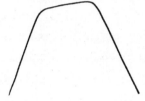

or a dromedary's back

or a wildly oscillating free-for-all*

* Or, more likely still, it might not be a single curve at all, as Martin Gardner illustrated by means of the snarly "Neo-Laffer curve" in his acid assessment of supply-side theory, "The Laffer Curve."

but if it slopes upward in one place, it has to slope downward somewhere else. There is such a thing as being too Swedish. That's a statement no economist would disagree with. It's also, as Laffer himself pointed out, something that was understood by many social scientists before him. But to most people, it's not at all obvious—at least, not until you see the picture on the napkin. Laffer understood perfectly well that his curve didn't have the power to tell you whether or not any given economy at any given time was overtaxed or not. That's why he didn't draw any numbers on the picture. Questioned during congressional testimony about the precise location of the optimal tax rate, he conceded, "I cannot measure it frankly, but I can tell you what the characteristics of it are; yes, sir." All the Laffer curve says is that lower taxes could, under some circumstances, increase tax revenue; but figuring out what those circumstances are requires deep, difficult, empirical work, the kind of work that doesn't fit on a napkin.

There's nothing wrong with the Laffer curve—only with the uses people put it to. Wanniski and the politicans who followed his panpipe fell prey to the oldest false syllogism in the book:

> It *could* be the case that lowering taxes will increase government revenue;
>
> I *want it* to be the case that lowering taxes will increase government revenue;
>
> Therefore, *it is* the case that lowering taxes will increase government revenue.

TWO

STRAIGHT LOCALLY, CURVED GLOBALLY

Y ou might not have thought you needed a professional mathematician to tell you that not all curves are straight lines. But linear reasoning is everywhere. You're doing it every time you say
that if something is good to have, having more of it is even better. Political shouters rely on it: "You support military action against Iran? I guess
you'd like to launch a *ground invasion* of every country that *looks at us
funny!*" Or, on the other hand, "Engagement with Iran? You probably
also think *Adolf Hitler* was just *misunderstood.*"

Why is this kind of reasoning so popular, when a moment's thought
reveals its wrongness? Why would anyone think, even for a second, that
all curves are straight lines, when they're obviously not?

One reason is that, in a sense, they are. That story starts with Archimedes.

EXHAUSTION

What's the area of the following circle?

In the modern world, that's a problem so standard you could put it on
the SAT. The area of a circle is πr^2, and in this case the radius r is 1, so

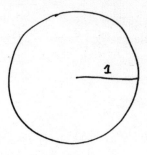

the area is π. But two thousand years ago this was a vexing open question, important enough to draw the attention of Archimedes.

Why was it so hard? For one thing, the Greeks didn't really think of π as a number, as we do. The numbers they understood were whole numbers, numbers that counted things: 1, 2, 3, 4 . . . But the first great success of Greek geometry—the Pythagorean Theorem*—turned out to be the ruin of their number system.

Here's a picture:

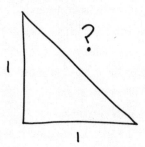

The Pythagorean Theorem tells you that the square of the *hypotenuse*—the side drawn diagonally here, the one that doesn't touch the right angle—is the sum of the squares of the other two sides, or *legs*. In this picture, that says the square of the hypotenuse is $1^2 + 1^2 = 1 + 1 = 2$. In particular, the hypotenuse is longer than 1 and shorter than 2 (as you can check with your eyeballs, no theorem required). That the length is not a

---

* By the way, we don't know who first proved the Pythagorean Theorem, but scholars are almost certain it was not Pythagoras himself. In fact, beyond the bare fact, attested by contemporaries, that a learned man by that name lived and gained fame in the sixth century BCE, we know almost nothing about Pythagoras. The main accounts of his life and work date from almost eight hundred years after his death. By that time, Pythagoras the real person had been completely replaced by Pythagoras the myth, a kind of summing up in an individual of the philosophy of the scholars who called themselves Pythagoreans.

whole number was not, in itself, a problem for the Greeks. Maybe we just measured everything in the wrong units. If we choose our unit of length to make the legs 5 units long, you can check with a ruler that the hypotenuse is just about 7 units long. Just about—but a bit too long. For the square of the hypotenuse is

$$5^2 + 5^2 = 25 + 25 = 50$$

and if the hypotenuse were 7, its square would be $7 \times 7 = 49$.

Or if you make the legs 12 units long, the hypotenuse is almost exactly 17 units, but is tantalizingly too short, because $12^2 + 12^2$ is 288, a smidgen less than 17^2, which is 289.

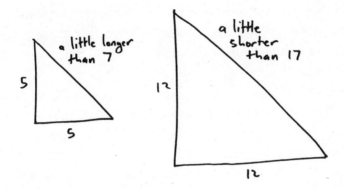

And at some point around the fifth century BCE, a member of the Pythagorean school made a shocking discovery: there was *no way* to measure the isosceles right triangle so that the length of each side was a whole number. Modern people would say "the square root of 2 is irrational"— that is, it is not the ratio of any two whole numbers. But the Pythagoreans would not have said that. How could they? Their notion of quantity was built on the idea of proportions between whole numbers. To them, the length of that hypotenuse had been revealed to be *not a number at all*.

This caused a fuss. The Pythagoreans, you have to remember, were extremely weird. Their philosophy was a chunky stew of things we'd now call mathematics, things we'd now call religion, and things we'd now call mental illness. They believed that odd numbers were good and even numbers evil; that a planet identical to our own, the Antichthon, lay on the other side of the sun; and that it was wrong to eat beans, by some accounts because they were the repository of dead people's souls. Pythagoras himself was said to have had the ability to talk to cattle (he told them not to eat beans) and to have been one of the very few ancient Greeks to wear pants.

The mathematics of the Pythagoreans was inseparably bound up with their ideology. The story (probably not really true, but it gives the right impression of the Pythagorean style) is that the Pythagorean who discovered the irrationality of the square root of 2 was a man named Hippasus, whose reward for proving such a nauseating theorem was to be tossed into the sea by his colleagues, to his death.

But you can't drown a theorem. The Pythagoreans' successors, like Euclid and Archimedes, understood that you had to roll up your sleeves and measure things, even if this brought you outside the pleasant walled garden of the whole numbers. No one knew whether the area of a circle could be expressed using whole numbers alone.* But wheels must be built and silos filled;† so the measurement must be done.

The original idea comes from Eudoxus of Cnidus; Euclid included it as book 12 of the elements. But it was Archimedes who really brought the project to its full fruition. Today we call his approach *the method of exhaustion*. And it starts like this.

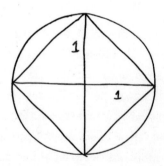

---

* It can't, in fact, but nobody figured out how to prove this until the eighteenth century.
† Actually, silos weren't round until the early twentieth century, when a University of Wisconsin professor, H.W. King, invented the now-ubiquitous cylindrical design in order to solve the problem of spoilage in the corners.

The square in the picture is called the *inscribed square*; each of its corners just touches the circle, but it doesn't extend beyond the circle's boundary. Why do this? Because circles are mysterious and intimidating, and squares are easy. If you have before you a square whose side has length X, its area is X times X—indeed, that's why we call the operation of multiplying a number by itself squaring! A basic rule of mathematical life: if the universe hands you a hard problem, try to solve an easier one instead, and hope the simple version is close enough to the original problem that the universe doesn't object.

The inscribed square breaks up into four triangles, each of which is none other than the isosceles triangle we just drew.* So the square's area is four times the area of the triangle. That triangle, in turn, is what you get when you take a 1 x 1 square and cut it diagonally in half like a tuna fish sandwich.

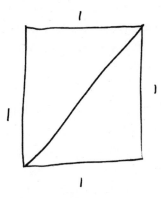

The area of the tuna fish sandwich is $1 \times 1 = 1$, so the area of each triangular half-sandwich is 1/2, and the area of the inscribed square is 4 times 1/2, or 2.

By the way, suppose you *don't* know the Pythagorean Theorem. Guess what—you do now! Or at least you know what it has to say about this particular right triangle. Because the right triangle that makes up the lower half of the tuna fish sandwich is exactly the same as the one that is the northwest quarter of the inscribed square. And its hypotenuse is the

---

* Rather, each of the four pieces can be obtained from the original isosceles right triangle by sliding and rotating it around the plane; we take as given that these manipulations don't change a figure's area.

inscribed square's side. So when you square the hypotenuse, you get the area of the inscribed square, which is 2. That is, the hypotenuse is that number which, when squared, yields 2; or, in the usual more concise lingo, the square root of 2.

The inscribed square is entirely contained within the circle. If its area is 2, the area of the circle must be *at least* 2.

Now we draw another square.

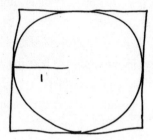

This one is called the *circumscribed* square; it, too, touches the circle at just four points. But this square contains the circle. Its sides have length 2, so its area is 4; and so we know the area of the circle is at most 4.

To have shown that pi is between 2 and 4 is perhaps not so impressive. But Archimedes is just getting started. Take the four corners of your inscribed square and mark new points on the circle halfway between each adjacent pair of corners. Now you've got eight equally spaced points, and when you connect those, you get an inscribed octagon, or, in technical language, a "stop sign":

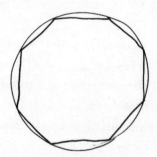

Computing the area of the inscribed octagon is a bit harder, and I'll spare you the trigonometry. The important thing is that it's about straight lines and angles, not curves, and so it was doable with the methods available to Archimedes. And the area is twice the square root of 2, which is about 2.83.

You can play the same game with the circumscribed octagon

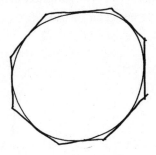

whose area is $8(\sqrt{2} - 1)$, a little over 3.31.

So the area of the circle is trapped in between 2.83 and 3.31.

Why stop there? You can stick points in between the corners of the octagon (whether inscribed or circumscribed) to make a 16-gon; after some more trigonometric figuring, that tells you that the area of the circle is in between 3.06 and 3.18. Do it again, to make a 32-gon; and again, and again, and pretty soon you have something that looks like this:

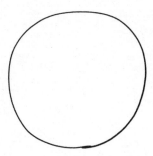

Wait, isn't that just the circle? Of course not! It's a regular polygon with 65,536 sides. Couldn't you tell?

The great insight of Eudoxus and Archimedes was that *it doesn't matter* whether it's a circle or a polygon with very many very short sides. The two areas will be close enough for any purpose you might have in mind. The area of the little fringe between the circle and the polygon has been "exhausted" by our relentless iteration. The circle has a curve to it, that's true. But every tiny little piece of it can be well approximated by a perfectly straight line, just as the tiny little patch of the earth's surface we stand on is well approximated by a perfectly flat plane.*

The slogan to keep in mind: straight locally, curved globally.

Or think of it like this. You are streaking downward toward the circle as from a great height. At first you can see the whole thing:

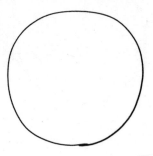

Then just one segment of arc:

And a still smaller segment:

* At least if, like me, you live in the midwestern United States.

Until, zooming in, and zooming in, what you see is pretty much indistinguishable from a line. An ant on the circle, aware only of his own tiny immediate surroundings, would think he was on a straight line, just as a person on the surface of the earth (unless she is clever enough to watch objects crest the horizon as they approach from afar) feels like she's standing on a plane.

THE PAGE WHERE I TEACH YOU CALCULUS

I will now teach you calculus. Ready? The idea, for which we have Isaac Newton to thank, is that there's nothing special about a perfect circle. *Every* smooth curve, when you zoom in enough, looks just like a line. Doesn't matter how winding or snarled it is—just that it doesn't have any sharp corners.

When you fire a missile, its path looks like this:

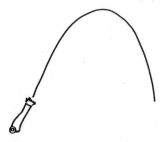

The missile goes up, then down, in a parabolic arc. Gravity makes all motion curve toward the earth; that's among the fundamental facts of our physical life. But if we zoom in on a very short segment, the curve starts to look like this:

And then like this:

Just like the circle, the missile's path looks to the naked eye like a straight line, progressing upward at an angle. The deviation from straightness caused by gravity is too small to see—but it's still there, of course. Zooming in to an even smaller region of the curve makes the curve even more like a straight line. Closer and straighter, closer and straighter . . .

Now here's the conceptual leap. Newton said, look, let's go all the way. Reduce your field of view until it's *infinitesimal*—so small that it's smaller than any size you can name, but not zero. You're studying the missile's arc, not over a very short time interval, but at a single moment. What was *almost* a line becomes *exactly* a line. And the slope of this line is what Newton called the *fluxion*, and what we'd now call the *derivative*.

That's a kind of jump Archimedes wasn't willing to make. He understood that polygons with shorter sides got closer and closer to the circle—but he would never have said that the circle actually *was* a polygon with infinitely many infinitely short sides.

Some of Newton's contemporaries, too, were reluctant to go along for the ride. The most famous objector was George Berkeley, who denounced Newton's infinitesimals in a tone of high mockery sadly absent from current mathematical literature: "And what are these fluxions? The velocities of evanescent increments. And what are these same evanescent increments? They are neither finite quantities, nor quantities infinitely small, nor yet nothing. May we not call them the ghosts of departed quantities?"

And yet calculus *works*. If you swing a rock in a loop around your head and suddenly release it, it'll shoot off along a linear trajectory at constant speed,* exactly in the direction that calculus says the rock is moving at the precise moment you let go. That's yet another Newtonian insight; objects in motion tend to proceed in a straight-line path, unless some other force intercedes to nudge the object one way or the other. That's one reason linear thinking comes so naturally to us: our intuition about time and motion is formed by the phenomena we observe in the world. Even before Newton codified his laws, something in us knew that things like to move in straight lines, unless given a reason to do otherwise.

EVANESCENT INCREMENTS AND UNNECESSARY PERPLEXITIES

Newton's critics had a point; his construction of the derivative didn't amount to what we'd call rigorous mathematics nowadays. The problem is the notion of the infinitely small, which was a slightly embarrassing sticking point for mathematicians for thousands of years. The trouble started with Zeno, a fifth-century-BCE Greek philosopher of the Eleatic school who specialized in asking innocent-seeming questions about the physical world that inevitably blossomed into huge philosophical brouhahas.

His most famous paradox goes like this. I decide to walk to the ice cream store. Now certainly I can't get to the ice cream store until I've gone halfway there. And once I've gone halfway, I can't get to the store until I've gone half the distance that remains. Having done so, I still have to cover half the remaining distance. And so on, and so on. I may get closer and closer to the ice cream store—but no matter how many steps of this process I undergo, I never actually *reach* the ice cream store. I am always some tiny but nonzero distance away from my two scoops with jimmies. Thus, Zeno concludes, to walk to the ice cream store is impossible. The argument works just as well for any destination: it's equally impossible to walk across the street, or to take a single step, or to wave your hand. All motion is ruled out.

---

* Apart from the effects of gravity, air resistance, etc., etc. But on a short timescale, the linear approximation is good enough.

Diogenes the Cynic was said to have refuted Zeno's argument by standing up and walking across the room. Which is a pretty good argument that motion is actually possible; so something must be wrong with Zeno's argument. But where's the mistake?

Break down the trip to the store numerically. First you go halfway. Then you go half of the remaining distance, which is 1/4 of the total distance, and you've got 1/4 left to go. So half of what's left is 1/8, then 1/16, then 1/32. Your progress toward the store looks like this:

$$1/2 + 1/4 + 1/8 + 1/16 + 1/32 + \ldots$$

If you add up ten terms of this sequence you get about 0.999. If you add up twenty terms it's more like 0.999999. In other words, you are getting really, really, really close to the store. But no matter how many terms you add, you never get to 1.

Zeno's paradox is much like another conundrum: is the repeating decimal 0.99999. . . . . . equal to 1?

I have seen people come nearly to blows over this question.* It's hotly disputed on websites ranging from World of Warcraft fan pages to Ayn Rand forums. Our natural feeling about Zeno is "of course you eventually get your ice cream." But in this case, intuition points the other way. Most people, if you press them, say 0.9999 . . . doesn't equal 1. It doesn't *look* like 1, that's for sure. It looks smaller. But not much smaller! Like Zeno's hungry ice cream lover, it gets closer and closer to its goal, but never, it seems, quite makes it there.

And yet, math teachers everywhere, myself included, will tell them, "No, it's 1."

How do I convince someone to come over to my side? One good trick is to argue as follows. Everyone knows that

$$0.33333. . . . . = 1/3.$$

Multiply both sides by 3 and you'll see

---

* Admittedly, these particular people were teenagers at a summer math camp.

$0.99999. \ldots = 3/3 = 1.$

If that doesn't sway you, try multiplying $0.99999 \ldots$ by 10, which is just a matter of moving the decimal point one spot to the right.

$10 \times (0.99999 \ldots) = 9.99999. \ldots$

Now subtract the vexing decimal from both sides:

$10 \times (0.99999 \ldots) - 1 \times (0.99999 \ldots)$
$= 9.99999 \ldots - 0.99999. \ldots .$

The left-hand side of the equation is just $9 \times (0.99999 \ldots)$, because 10 times something minus that something is 9 times the aforementioned thing. And over on the right-hand side, we have managed to cancel out the terrible infinite decimal, and are left with a simple 9. So we end up with

$9 \times (0.99999 \ldots) = 9.$

If 9 times something is 9, that something just has to be 1—doesn't it?

These arguments are often enough to win people over. But let's be honest: they lack something. They don't really address the anxious uncertainty induced by the claim $0.99999 \ldots = 1$; instead, they represent a kind of algebraic intimidation. "You believe that 1/3 is 0.3 repeating—don't you? *Don't you?*"

Or worse: maybe you bought my argument based on multiplication by 10. But how about this one? What is

$1 + 2 + 4 + 8 + 16 + \ldots ?$

Here the "..." means "carry on the sum forever, adding twice as much each time." Surely such a sum must be infinite! But an argument much like the apparently correct one concerning $0.9999 \ldots$ seems to suggest otherwise. Multiply the sum above by 2 and you get

$$2 \times (1 + 2 + 4 + 8 + 16 + \ldots) = 2 + 4 + 8 + 16 + \ldots$$

which looks a lot like the original sum; indeed, it is just the original sum $(1 + 2 + 4 + 8 + 16 + \ldots)$ with the 1 lopped off the beginning, which means that $2 \times (1 + 2 + 4 + 8 + 16 + \ldots)$ is 1 less than $(1 + 2 + 4 + 8 + 16 + \ldots)$. In other words,

$$2 \times (1 + 2 + 4 + 8 + 16 + \ldots) - 1 \times (1 + 2 + 4 + 8 + 16 + \ldots)$$
$$= -1.$$

But the left-hand side simplifies to the very sum we started with, and we're left with

$$1 + 2 + 4 + 8 + 16 + \ldots = -1.$$

Is *that* what you want to believe?* That adding bigger and bigger numbers, ad infinitum, flops you over into negativeland?

More craziness: What is the value of the infinite sum

$$1 - 1 + 1 - 1 + 1 - 1 + \ldots$$

One might first observe that the sum is

$$(1 - 1) + (1 - 1) + (1 - 1) + \ldots = 0 + 0 + 0 + \ldots$$

and argue that the sum of a bunch of zeroes, even infinitely many, has to be 0. On the other hand, $1 - 1 + 1$ is the same thing as $1 - (1 - 1)$, because the negative of a negative is a positive; applying this fact again and again, we can rewrite the sum as

$$1 - (1 - 1) - (1 - 1) - (1 - 1) \ldots = 1 - 0 - 0 - 0 \ldots$$

which seems to demand, in the same way, that the sum is equal to 1!

---

* So as not to leave you hanging: there is a context, that of *2-adic numbers*, in which this crazy-looking argument is completely correct. More on this in the endnotes, for number theory enthusiasts.

So which is it, 0 or 1? Or is it somehow 0 half the time and 1 half the time? It seems to depend where you stop—but infinite sums never stop!

Don't decide yet, because it gets worse. Suppose T is the value of our mystery sum:

$$T = 1 - 1 + 1 - 1 + 1 - 1 + \ldots$$

Taking the negative of both sides gives you

$$-T = -1 + 1 - 1 + 1 \ldots$$

But the sum on the right-hand side is precisely what you get if you take the original sum defining T and lop off that first 1, thus subtracting 1; in other words,

$$-T = -1 + 1 - 1 + 1 \ldots = T - 1.$$

So $-T = T - 1$, an equation concerning T which is satisfied only when T is equal to 1/2. Can a sum of infinitely many whole numbers somehow magically become a fraction? If you say no, you have the right to be at least a little suspicious of slick arguments like this one. But note that some people said yes, including the Italian mathematician/priest Guido Grandi, after whom the series $1 - 1 + 1 - 1 + 1 - 1 + \ldots$ is usually named; in a 1703 paper, he argued that the sum of the series is 1/2, and moreover that this miraculous conclusion represented the creation of the universe from nothing. (Don't worry, I don't follow that last step either.) Other leading mathematicians of the time, like Leibniz and Euler, were on board with Grandi's strange computation, if not his interpretation.

But in fact, the answer to the 0.999 . . . riddle (and to Zeno's paradox, and to Grandi's series) lies a little deeper. You don't have to give in to my algebraic strong-arming. You might, for instance, insist that 0.999 . . . is not equal to 1, but rather 1 minus some tiny infinitesimal number. And, for that matter, you might further insist that 0.333 . . . is not *exactly* equal to 1/3, but also falls short by an infinitesimal quantity. This point of view requires some stamina to push through to completion, but it can be done. I once had a calculus student named Brian who, unhappy with

the classroom definitions, worked out a fair chunk of the theory by himself, referring to his infinitesimal quantities as "Brian numbers."

Brian was not actually the first to get there. There's a whole field of
mathematics that specializes in contemplating numbers of this kind,
called *nonstandard analysis*. The theory, developed by Abraham Robinson in the mid-twentieth century, finally made sense of the "evanescent
increments" that Berkeley found so ridiculous. The price you have to pay
(or, from another point of view, the reward you get to reap) is a profusion
of novel kinds of numbers; not only infinitely small ones, but infinitely
large ones, a huge spray of them in all shapes and sizes.*

As it happened, Brian was in luck—I had a colleague at Princeton,
Edward Nelson, who was an expert in nonstandard analysis. I set up a
meeting for the two of them so Brian could learn more about it. The
meeting, Ed told me later, didn't go well. As soon as Ed made it clear that
infinitesimal quantities were not in fact going to be called *Brian numbers*,
Brian lost all interest.

(Moral lesson: people who go into mathematics for fame and glory
don't stay in mathematics for long.)

But we're no closer to settling our dispute. What is 0.999 . . . , *really?*
Is it 1? Or is it some number infinitesimally less than 1, a crazy kind of
number that hadn't even been discovered a hundred years ago?

The right answer is to unask the question. What is 0.999. . . . , *really?*
It appears to refer to a kind of sum:

.9 + .09 + .009 + .0009 + . . .

But what does that mean? That pesky ellipsis is the real problem.
There can be no controversy about what it means to add up two, or three,
or a hundred numbers. This is just mathematical notation for a physical
process we understand very well: take a hundred heaps of stuff, mush
them together, see how much you have. But infinitely many? That's a different story. In the real world, you can never have infinitely many heaps.

---

* The *surreal numbers*, developed by John Conway, are especially charming and weird examples,
as their name suggests; they are strange hybrids between numbers and games of strategy and their
depths have not yet been fully explored. The book *Winning Ways*, by Berlekamp, Conway, and
Guy, is a good place to learn about these exotic numbers, and lots more about the rich mathematics of game playing besides.

What's the numerical value of an infinite sum? It doesn't have one—*until we give it one.* That was the great innovation of Augustin-Louis Cauchy, who introduced the notion of *limit* into calculus in the 1820s.*

The British number theorist G. H. Hardy, in his 1949 book *Divergent Series*, explains it best:

> It does not occur to a modern mathematician that a collection of mathematical symbols should have a "meaning" until one has been assigned to it by definition. It was not a triviality even to the greatest mathematicians of the eighteenth century. They had not the habit of definition: it was not natural to them to say, in so many words, "by X we *mean* Y." . . . It is broadly true to say that mathematicians before Cauchy asked not, "How shall we *define* $1 - 1 + 1 - 1 + \ldots$" but "What *is* $1 - 1 + 1 - 1 + \ldots$?" and that this habit of mind led them into unnecessary perplexities and controversies which were often really verbal.

This is not just loosey-goosey mathematical relativism. Just because we *can* assign whatever meaning we like to a string of mathematical symbols doesn't mean we should. In math, as in life, there are good choices and there are bad ones. In the mathematical context, the good choices are the ones that settle unnecessary perplexities without creating new ones.

The sum $.9 + .09 + .009 + \ldots$ gets closer and closer to 1 the more terms you add. And it never gets any farther away. No matter how tight a cordon we draw around the number 1, the sum will eventually, after some finite number of steps, penetrate it, and never leave. Under those circumstances, Cauchy said, we should simply *define* the value of the infinite sum to be 1. And then he worked very hard to prove that committing oneself to his definition didn't cause horrible contradictions to pop up elsewhere. By the time this labor was done, he'd constructed a framework that made Newton's calculus completely rigorous. When we say a curve looks locally like a straight line at a certain angle, we now

---

* Like all mathematical breakthroughs, Cauchy's theory of limits had precursors—for instance, Cauchy's definition was very much in the spirit of d'Alembert's bounds for the error terms of binomial series. But there's no question that Cauchy was the turning point; after him, analysis is modern.

mean more or less this: as you zoom in tighter and tighter, the curve re-sembles the given line more and more closely. In Cauchy's formulation, there's no need to mention infinitely small numbers, or anything else that would make a skeptic blanch.

Of course there is a cost. The reason the 0.999 . . . problem is diffi-cult is that it brings our intuitions into conflict. We would like the sum of an infinite series to play nicely with arithmetic manipulations like the ones we carried out on the previous pages, and this seems to demand that the sum equal 1. On the other hand, we would like each number to be represented by a unique string of decimal digits, which conflicts with the claim that the same number can be called either 1 or 0.999 . . . , as we like. We can't hold on to both of these desires at once; one must be discarded. In Cauchy's approach, which has amply proved its worth in the two centuries since he invented it, it's the uniqueness of the decimal expansion that goes out the window. We're untroubled by the fact that the English language sometimes uses two different strings of letters (i.e., two words) to refer synonymously to the same thing in the world; in the same way, it's not so bad that two different strings of digits can refer to the same number.

As for Grandi's 1 − 1 + 1 − 1 + . . . , it is one of the series outside the reach of Cauchy's theory: that is, one of the *divergent series* that formed the subject of Hardy's book. The Norwegian mathematician Niels Hen-rik Abel, an early fan of Cauchy's approach, wrote in 1828, "Divergent series are the invention of the devil, and it is shameful to base on them any demonstration whatsoever."[*] Hardy's view, which is our view today, is more forgiving; there are some divergent series to which we ought to assign values and some to which we ought not, and some to which we ought or ought not depending on the context in which the series arises. Modern mathematicians would say that if we are to assign the Grandi series a value, it should be 1/2, because, as it turns out, all interesting theories of infinite sums either give it the value 1/2 or decline, like Cau-chy's theory, to give it any value at all.[†]

To write Cauchy's definitions down precisely takes a bit more work.

* Ironic, considering Grandi's original theological application of his divergent series!
† In the famous words of Lindsay Lohan, "The limit does not exist!"

This was especially true for Cauchy himself, who had not quite phrased the ideas in their clean, modern form.* (In mathematics, you very seldom get the clearest account of an idea from the person who invented it.) Cauchy was an unwavering conservative and a royalist, but in his mathematics he was proudly revolutionary and a scourge to academic authority. Once he understood how to do things without the dangerous infinitesimals, he unilaterally rewrote his syllabus at the École Polytechnique to reflect his new ideas. This enraged everyone around him: his mystified students, who had signed up for freshman calculus, not a seminar on cutting-edge pure mathematics; his colleagues, who felt that the engineering students at the École had no need for Cauchy's level of rigor; and the administrators, whose commands to stick to the official course outline he completely ignored. The École imposed a new curriculum from above that emphasized the traditional infinitesimal approach to calculus, and placed note takers in Cauchy's classroom to make sure he complied. Cauchy did not comply. Cauchy was not interested in the needs of engineers. Cauchy was interested in the truth.

It's hard to defend Cauchy's stance on pedagogical grounds. But I'm sympathetic with him anyway. One of the great joys of mathematics is the incontrovertible feeling that you've understood something the right way, all the way down to the bottom; it's a feeling I haven't experienced in any other sphere of mental life. And when you know how to do something the right way, it's hard—for some stubborn people, impossible—to make yourself explain it the wrong way.

* If you've ever taken a math course that uses epsilons and deltas, you've seen the descendants of Cauchy's formal definitions.

THREE

EVERYONE IS OBESE

The stand-up comic Eugene Mirman tells this joke about statistics. He says he likes to tell people, "I read that 100% of Americans were Asian."

"But Eugene," his confused companion protests, "*you're* not Asian."

And the punch line, delivered with magnificent self-assurance: "I read that I was!"

I thought of Mirman's joke when I encountered a paper in the journal *Obesity* whose title posed the discomfiting question: "Will all Americans become overweight or obese?" As if the rhetorical question weren't enough, the article supplies an answer: "Yes—by 2048."

In 2048 I'll be seventy-seven years old, and I hope not to be overweight. But I read I would be!

The *Obesity* paper got plenty of press, as you might imagine. ABC News warned of an "obesity apocalypse." The *Long Beach Press-Telegram* went with the simple headline "We're Getting Fatter." The study's results resonated with the latest manifestation of the fevered, ever-shifting anxiety with which Americans have always contemplated our national moral status. Before I was born, boys grew long hair and thus we were bound to get whipped by the Communists. When I was a kid, we played arcade games too much, which left us doomed to be outcompeted by the industrious Japanese. Now, we eat too much fast food, and we're all going to die weak and immobile, surrounded by empty chicken buckets, puddled into the couches from which we long ago became unable to hoist ourselves. The paper certified this anxiety as a fact proved by science.

I have some good news. We're not all going to be overweight in the year 2048. Why? Because not every curve is a line.

But every curve, as we just learned from Newton, is pretty close to a line. That's the idea that drives *linear regression*, the statistical technique that is to social science as the screwdriver is to home repair. It's the one tool you're pretty much definitely going to use, whatever the task. Every time you read in the newspaper that people with more cousins are happier, or that countries that have more Burger Kings have looser morals, or that halving your intake of niacin doubles your risk of athlete's foot, or that every extra $10,000 of income makes you 3% more likely to vote Republican,* you're encountering the result of a linear regression.

Here's how it works. You have two things you want to relate; let's say, the cost of tuition at a university and the average SAT score of its incoming students. You might think schools with higher SATs are likely to be pricier; but a look at the data tells you that's not a universal law. Elon University, just outside Burlington, North Carolina, has an average combined math and verbal score of 1217, and charges $20,441 tuition a year. Nearby Guilford College, in Greensboro, is a bit pricier at $23,420, but entering first-years there averaged only 1131 on the SAT.

Still, if you look at a whole list of schools—say, the thirty-one private universities that reported their tuition and scores to the North Carolina Career Resource Network in 2007—you see a clear trend.

Each dot on the plot represents one of the colleges. Those two dots way up in the upper right-hand corner, with sky-high SAT scores and prices to match? Those are Wake Forest and Davidson. The lonely dot near the bottom, the only private school on the list with tuition under $10K, is Cabarrus College of Health Sciences.

---

* More details on these studies can be found in the *Journal of Stuff I Totally Made Up in Order to Illustrate My Point*.

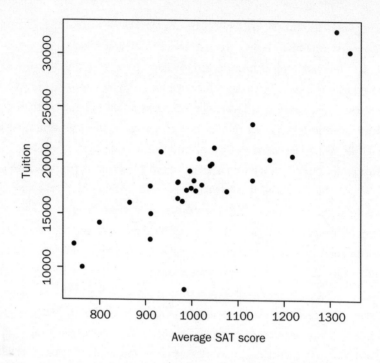

The picture shows clearly that schools with higher scores have higher prices, by and large. But how *much* higher? That's where linear regression enters the picture. The points in the picture above are obviously not on a line. But you can see that they're not far off. You could probably draw a straight line freehand that cuts pretty much through the middle of this cloud of points. Linear regression takes the guesswork out, finding the line that comes closest* to passing through all the points. For the North Carolina colleges, it looks like the following figure.

The line in the picture has a slope of about 28. That means: if tuition were actually completely determined by SAT scores according to the line I drew on the chart, each extra point of SAT would correspond to an

---

* "Closest," in this context, is measured as follows: if you replace the actual tuition at each school by the estimate the line suggests, and then you compute the difference between the estimated and actual tuition for each school, and then you square each of these numbers, and you add all those squares up, you get some kind of total measure of the extent to which the line misses the points, and you choose the line that makes this measure as small as possible. This business of summing up squares smells like Pythagoras, and indeed the underlying geometry of linear regression is no more than Pythagoras's theorem transposed and upgraded to a much-higher-dimensional setting; but that story requires more algebra than I want to deploy in this space. See the discussion of correlation and trigonometry in chapter 15 for a little more in this vein, though.

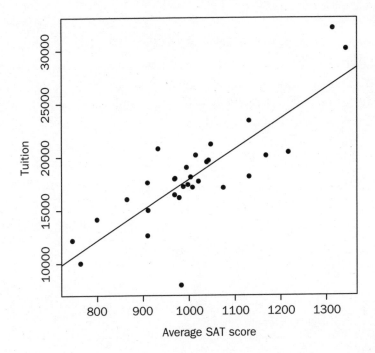

extra $28 in tuition. If you can raise the average SAT score of your in-coming first-years by 50 points on average, you can charge $1,400 more in tuition. (Or, from the parent's point of view, your kid improving 100 points is going to cost you an extra $2,800 a year. That test-prep course was more expensive than you thought!)

Linear regression is a marvelous tool, versatile, scalable, and as easy to execute as clicking a button on your spreadsheet. You can use it for data sets involving two variables, like the ones I've drawn here, but it works just as well for three variables, or a thousand. Whenever you want to understand which variables drive which other variables, and in which direction, it's the first thing you reach for. And it works on any data set at all.

That's a weakness as well as a strength. You can do linear regression without thinking about whether the phenomenon you're modeling is actually close to linear. *But you shouldn't.* I said linear regression was like a screwdriver, and that's true; but in another sense, it's more like a table saw. If you use it without paying careful attention to what you're doing, the results can be gruesome.

Take, for instance, the missile we fired off in the last chapter. Perhaps you were not the one who fired the missile at all. Perhaps you are, instead, the missile's intended recipient. As such, you have a keen interest in analyzing the missile's path as accurately as possible.

Maybe you have plotted the vertical position of the missile at five points in time, and it looks like this:

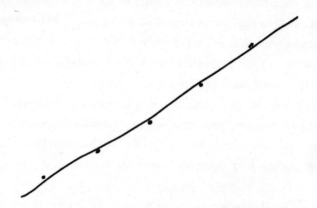

Now you do a quick linear regression, and you get great results. There's a line that passes almost exactly through the points you plotted:

(This is where your hand starts to creep, unthinkingly, toward the table saw's keening blade.)

Your line gives a very precise model for the missile's motion: for every minute that passes, the missile increases its altitude by some fixed amount: say, 400 meters. After an hour it's 24 km above the earth's surface. When does it come down? It never comes down! An upward sloping line just keeps on sloping upward. That's what lines do.

(Blood, gristle, screams.)

Not every curve is a line. And the curve of a missile's flight is *most emphatically* not a line; it's a parabola. Just like Archimedes's circle, it looks like a line close up; and that's why the linear regression will do a great job telling you where the missile is five seconds after the last time you tracked it. But an hour later? Forget it. Your model says the missile is in the lower stratosphere, when, in fact, it is probably approaching your house.

The most vivid warning I know against thoughtless linear extrapolation was set down not by a statistician but by Mark Twain, in *Life on the Mississippi*:

The Mississippi between Cairo and New Orleans was twelve hundred and fifteen miles long one hundred and seventy-six years ago. It was eleven hundred and eighty after the cut-off of 1722. It was one thousand and forty after the American Bend cut-off. It has lost sixty-seven miles since. Consequently its length is only nine hundred and seventy-three miles at present. . . . In the space of one hundred and seventy-six years the Lower Mississippi has shortened itself two hundred and forty-two miles. This is an average of a trifle over one mile and a third per year. Therefore, any calm person, who is not blind or idiotic, can see that in the Old Oolitic Silurian Period, just a million years ago next November, the Lower Mississippi River was upward of one million three hundred thousand miles long, and stuck out over the Gulf of Mexico like a fishing-rod. And by the same token any person can see that seven hundred and forty-two years from now the Lower Mississippi will be only a mile and three-quarters long, and Cairo and New Orleans will have joined their streets together, and be plodding comfortably along under a single mayor and a mutual board of aldermen. There is something fascinating about science. One gets such wholesale returns of conjecture out of such a trifling investment of fact.

ASIDE: HOW TO GET PARTIAL CREDIT
ON MY CALCULUS EXAM

The methods of calculus are a lot like linear regression: they're purely mechanical, your calculator can carry them out, and it is very dangerous to use them inattentively. On a calculus exam you might be asked to compute the weight of water left in a jug after you punch some kind of hole and let some kind of flow take place for some amount of time, blah blah blah. It's easy to make arithmetic mistakes when doing a problem like this under time pressure. And sometimes that leads to a student arriving at a ridiculous result, like a jug of water whose weight is −4 grams.

If a student arrives at −4 grams and writes, in a desperate, hurried hand, "I screwed up somewhere, but I can't find my mistake," I give them half credit.

If they just write "−4g" at the bottom of the page and circle it, they get zero—even if the entire derivation was correct apart from a single misplaced digit somewhere halfway down the page.

Working an integral or performing a linear regression is something a computer can do quite effectively. Understanding whether the result makes sense—or deciding whether the method is the right one to use in the first place—requires a guiding human hand. When we teach mathematics we are supposed to be explaining how to be that guide. A math course that fails to do so is essentially training the student to be a very slow, buggy version of Microsoft Excel.

And let's be frank: that really is what many of our math courses are doing. To make a long, contentious story short (but still contentious), the teaching of mathematics to children has for decades now been the arena of the so-called math wars. On one side, you have teachers who favor an emphasis on memorization, fluency, traditional algorithms, and exact answers; on the other, teachers who think math teaching should be about learning meaning, developing ways of thinking, guided discovery, and approximation. Sometimes the first approach is called *traditional* and the second *reform*, although the supposedly nontraditional discovery approach has been around in some form for decades, and whether "reform" truly counts as a reform is exactly what's up for debate. *Fierce* debate. At a math dinner party it's okay to bring up politics or religion, but start an argument about math pedagogy and it's likely to end with somebody storming out in either a traditionalist or reformist huff.

I don't count myself in either camp. I can't go along with those reformists who want to throw out memorization of the multiplication table. When doing any serious mathematical thinking, you're going to have to multiply 6 by 8 sometimes, and if you have to reach for your calculator each time you do that, you'll never achieve the kind of mental flow that actual thinking requires. You can't write a sonnet if you have to look up the spelling of each word as you go.

Some reformists go so far as to say that the classical algorithms (like "add two multidigit numbers by stacking one atop the other and carrying the one when necessary") should be taken out of the classroom, lest they interfere with the students' process of discovering the properties of mathematical objects on their own.*

That seems like a terrible idea to me: these algorithms are useful tools that people worked hard to make, and there's no reason we should have to start completely from scratch.

On the other hand, there are algorithms I think we can safely discard in the modern world. We don't need to teach students how to extract square roots by hand, or in their head (though the latter skill, I can tell you from long personal experience, makes a great party trick in sufficiently nerdy circles). Calculators are also useful tools that people worked hard to make—we should use them, too, when the situation demands! I don't even care whether my students can divide 430 by 12 using long division—though I *do* care that their number sense is sufficiently developed to reckon mentally that the answer's a little more than 35.

The danger of overemphasizing algorithms and precise computations is that algorithms and precise computations are easy to assess. If we set-

* It's a little reminiscent of Orson Scott Card's short story "Unaccompanied Sonata," which is about a musical prodigy who is carefully kept alone and ignorant of all other music in the world so his originality won't be compromised, but then a guy sneaks in and plays him some Bach, and of course the music police can tell what happened, and the prodigy ends up getting banished from music, and later I think his hands get cut off and he's blinded or something, because Orson Scott Card has this weird ingrown thing about punishment and mortification of the flesh, but anyway, the point is, don't try to keep young musicians from hearing Bach, because Bach is great.

tle on a vision of mathematics that consists of "getting the answer right" and no more, and test for that, we run the risk of creating students who test very well but know no mathematics at all. This might be satisfying to those whose incentives are driven by test scores foremost and only, but it is not satisfying to me.

Of course it's no better (in fact, it's substantially worse) to pass along a population of students who've developed some wispy sense of mathematical meaning but can't work examples swiftly and correctly. A math teacher's least favorite thing to hear from a student is "I get the concept, but I couldn't do the problems." Though the student doesn't know it, this is shorthand for "I don't get the concept." The ideas of mathematics can sound abstract, but they make sense only in reference to concrete computations. William Carlos Williams put it crisply: *no ideas but in things*.

Nowhere is the battle more starkly defined than in plane geometry. Here is the last redoubt of the teaching of proofs, the bedrock practice of mathematics. By many professional mathematicians it is considered a sort of last stand of "real math." But it's not clear to what extent we're really teaching the beauty, power, and surprise of proof when we teach geometry. It's easy for the course to become an exercise in repetition as arid as a list of thirty definite integrals. The situation is so dire that the Fields Medalist David Mumford has suggested that we might dispense with plane geometry entirely and replace it with a first course in programming. A computer program, after all, has much in common with a geometric proof: both require the student to put together several very simple components from a small bag of options, one after the other, so that the sequence as a whole accomplishes some meaningful task.

I'm not as radical as that. In fact, I'm not radical at all. Dissatisfying as it may be to partisans, I think we have to teach a mathematics that values precise answers but also intelligent approximation, that demands the ability to deploy existing algorithms fluently but also the horse sense to work things out on the fly, that mixes rigidity with a sense of play. If we don't, we're not really teaching mathematics at all.

It's a tall order—but it's what the best math teachers are doing, anyway, while the math wars rage among the administrators overhead.

BACK TO THE OBESITY APOCALYPSE

So what percentage of Americans are going to be overweight in 2048? You can guess by now how Youfa Wang and his *Obesity* coauthors generated their projection. The National Health and Nutrition Examination Study, or NHANES, tracks the health data of a large, representative sample of Americans, covering everything from hearing loss to sexually transmitted infections. In particular, it gives very good data for the proportion of Americans who are overweight, which for present purposes is defined as having a body-mass index of 25 or higher.* There's no question that the prevalence of overweight has increased in recent decades. In the early 1970s, just under half of Americans had a BMI that high. By the early 1990s that figure had risen to almost 60%, and by 2008 almost three-quarters of the U.S. population was overweight.

You can plot the prevalence of obesity against time just as we did with the missile's vertical progress:

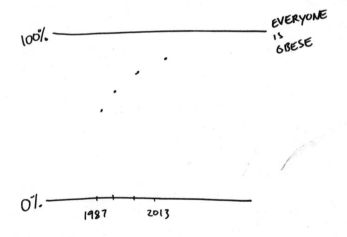

And you can generate a linear regression, which will look something like this:

---

* In the research literature, "overweight" means "BMI at least 25 but less than 30" and "obese" means "BMI 30 or above," but I'll refer to both groups together as "overweight" to avoid having to type "overweight or obese" umpteen times.

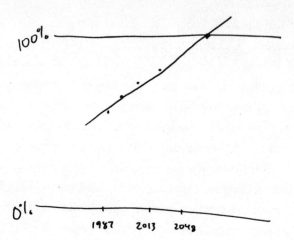

In 2048, the line crosses 100%. And that's why Wang writes that all Americans will be overweight in 2048, if current trends continue.

But current trends will not continue. They can't! If they did, by 2060, a whopping 109% of Americans would be overweight.

In reality, the graph of an increasing proportion bends toward 100%, like this:

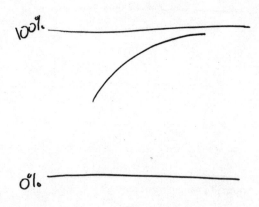

That's not an ironclad law, like the gravity that bends the missile's path into a parabola, but it's as close as you're going to get in medicine. The higher the proportion of overweight people, the fewer skinny malinkies are left to convert, and the more slowly the proportion increases toward 100%. In fact, the curve probably goes horizontal at some point below 100%. The thin we have always with us! And indeed, just four years later, the NHANES survey showed that the upward march of over-weight prevalence had already begun to slow.

But the *Obesity* paper conceals a worse crime against mathematics and common sense. Linear regression is easy to do—and once you've done one, it's cake to do more. So Wang and company broke down their data by ethnic group and sex. Black men, for instance, were less likely to be overweight than the average American; and, more important, their rate of overweight was growing only half as quickly. If we superimpose the proportion of overweight black men on the proportion of overweight Americans overall, together with the linear regressions Wang and company worked out, we get a picture that looks like this.

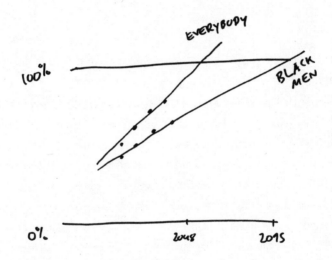

Nice work, black men! Not until 2095 will all of you be overweight. In 2048, only 80% of you will be.

See the problem? If *all* Americans are supposed to be overweight in 2048, where are those one in five future black men without a weight problem supposed to be? Offshore?

That basic contradiction goes unmentioned in the paper. It's the epi-demiological equivalent of saying there are −4 grams of water left in the bucket. Zero credit.

FOUR

HOW MUCH IS THAT IN DEAD AMERICANS?

How bad is the conflict in the Middle East? Counterterrorism specialist Daniel Byman of Georgetown University lays down some cold, hard numbers in *Foreign Affairs*: "The Israeli military reports that from the start of the second intifada [in 2000] through the end of October 2005, Palestinians killed 1,074 Israelis and wounded 7,520—astounding figures for such a small country, the proportional equivalent of more than 50,000 dead and 300,000 wounded for the United States." This kind of computation has become commonplace in discussions of the region. In December 2001 the U.S. House of Representatives declared that the 26 people killed by a series of attacks in Israel were "the equivalent, on a proportional basis, of 1,200 American deaths." Newt Gingrich in 2006: "Remember that when Israel loses eight people, because of the difference in population, it's the equivalent of losing almost 500 Americans." Not to be outdone, Ahmed Moor wrote in the *Los Angeles Times*: "When Israel killed 1,400 Palestinians in Gaza—proportionally equivalent to 300,000 Americans—in Operation Cast Lead, incoming President Obama stayed mum."

The rhetoric of proportion isn't reserved for the Holy Land. In 1988, Gerald Caplan wrote in the *Toronto Star*, "Some 45,000 Nicaraguans on both sides of the struggle have been killed, wounded or kidnapped in the past eight years; in perspective, that's the equivalent of 300,000 Canadians or 3 million Americans." Robert McNamara, the Vietnam-era secretary of defense, said in 1997 that the nearly 4 million Vietnamese deaths during the war were "equivalent to 27 million Americans." Any time a lot of people in a small country come to a bad end, editorialists get out their slide rules and start figuring: how much is that in dead Americans?

Here's how you generate these numbers. The 1,074 Israelis killed by terrorists amount to about 0.015% of the Israeli population (which between 2000 and 2005 ranged from about 6 to 7 million). So the pundits are reckoning that the death of 0.015% of the much larger United States population, which indeed comes to about 50,000, would have roughly the same impact here.

This is lineocentrism in its purest form. According to the argument by proportion, you can find the equivalent of 1,074 Israelis anywhere around the globe via the graph below:

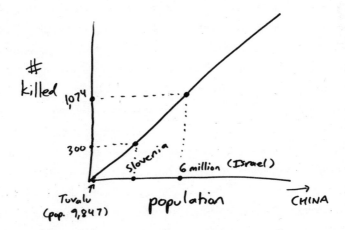

The 1,074 Israeli victims are equivalent to 7,700 Spaniards or 223,000 Chinese, but only 300 Slovenes and either one or two Tuvaluans.

Eventually (or perhaps immediately?) this reasoning starts to break down. When there are two men left in the bar at closing time, and one of them coldcocks the other, it is not equivalent in context to 150 million Americans getting simultaneously punched in the face.

Or: when 11% of the population of Rwanda was wiped out in 1994,

all agree that it was among the worst crimes of the century. But we don't
describe the bloodshed there by saying, "In the context of 1940s Europe,
it was nine times as bad as the Holocaust." And to do so would set teeth
rightly on edge.

An important rule of mathematical hygiene: when you're field-testing
a mathematical method, try computing the same thing several different
ways. If you get several different answers, something's wrong with your
method.

For example: the 2004 bombings at the Atocha train station in Madrid killed almost 200 people. What would be an equivalently deadly
bombing at Grand Central Station?

The United States has almost seven times the population of Spain.
So if you think of 200 people as 0.0004% of the Spanish population, you
find that an equivalent attack would kill 1,300 people in the United
States. On the other hand, 200 people is 0.006% of the population of
Madrid; scaling up to New York City, which is two and a half times as
large, gives you 463 victims. Or should we compare the province of Madrid with the state of New York? That gives you something closer to
600. This multiplicity of conclusions should be a red flag. Something is
fishy with the method of proportions.

One can't, of course, reject proportions entirely. Proportions matter!
If you want to know which parts of America have the biggest brain cancer problem, it doesn't make much sense to look at the states with the
most deaths from brain cancer: those are California, Texas, New York,
and Florida, which have the most brain cancer because they have the
most people. Stephen Pinker makes a similar point in his recent best
seller *The Better Angels of Our Nature*, which argues that the world has
steadily grown less violent throughout human history. The twentieth
century gets a bad rap because of the vast numbers of people caught in
the gears of great-power politics. But the Nazis, the Soviets, the Communist Party of China, and the colonial overlords were actually not particularly effective slaughterers on a proportional basis, Pinker argues—
there are just so many more people to kill nowadays! These days we
don't spare much grief for antique bloodlettings like the Thirty Years'
War. But that war took place in a smaller world, and by Pinker's estimate

killed one out of every hundred people on Earth. To do that now would mean wiping out 70 million people, more than the number who died in both world wars together.

So it's better to study rates: deaths as a proportion of total population. For instance, instead of counting raw numbers of brain cancer deaths by state, we can compute the proportion of each state's population that dies of brain cancer each year. That makes for a very different leaderboard. South Dakota takes the unwelcome first prize, with 5.7 brain cancer deaths per 100,000 people per year, well above the national rate of 3.4. South Dakota is followed on the list by Nebraska, Alaska, Delaware, and Maine. These are the places to avoid if you don't want to get brain cancer, it seems. So where should you move? Scrolling down to the bottom of the list, you find Wyoming, Vermont, North Dakota, Hawaii, and the District of Columbia.

Now this is strange. Why should South Dakota be brain cancer central and North Dakota nearly tumor free? Why would you be safe in Vermont but imperiled in Maine?

The answer: South Dakota isn't necessarily causing brain cancer, and North Dakota isn't necessarily preventing it. The five states at the top have something in common, and the five states at the bottom do, too. And it's the *same* thing: hardly anyone lives there. Of the nine states (and one District) that finished at the top and bottom, the biggest is Nebraska, which is currently locked with West Virginia in a close struggle to be the 37th most populous state. Living in a small state, apparently, makes it either much more or much less likely you'll get brain cancer.

Since that makes no sense, we'd better seek another explanation.

To see what's going on, let's play an imaginary game. The game is called *who's the best at flipping coins*. It's pretty simple. You flip a bunch of coins and whoever gets the most heads wins. To make this a little more interesting, though, not everybody has the same number of coins. Some people—Team Small—have only ten coins, while the members of Team Big have a hundred each.

If we score by absolute number of heads, one thing's for almost sure—the winner of this game is going to come from Team Big. The typical Big player is going to get around 50 heads, a figure none of the

Smalls can possibly match. Even if Team Small has a hundred members, the high scorer among them is likely to get an 8 or 9.[*]

That doesn't seem fair! Team Big has got a massive built-in advantage. So here's a better idea. Instead of scoring by raw number, let's score by proportion. That should put the two teams on a fairer footing.

But it doesn't. As I said, if there are a hundred Smalls, at least one is likely to get 8 heads. So that person's score is going to be at least 80%. And the Bigs? None of the Bigs is going to get 80% heads. It's physically possible, of course. But it's not going to happen. In fact, you'd need about two *billion* players on the Big team before you'd get a reasonable chance of seeing any outcome that lopsided. This ought to fit your intuition about probability. The more coins you throw, the more likely you are to be close to 50-50.

You can try it yourself! I did, and here's what happened. Repeatedly flipping 10 coins at a time to simulate Small players, I got a sequence of head counts that looked like this:

4, 4, 5, 6, 5, 4, 3, 3, 4, 5, 5, 9, 3, 5, 7, 4, 5, 7, 7, 9 . . .

With a hundred coins, like the Bigs, I got:

46, 54, 48, 45, 45, 52, 49, 47, 58, 40, 57, 46, 46, 51, 52, 51, 50, 60, 43, 45 . . .

And with a thousand:

486, 501, 489, 472, 537, 474, 508, 510, 478, 508, 493, 511, 489, 510, 530, 490, 503, 462, 500, 494 . . .

Okay, to be honest, I didn't flip a thousand coins. I asked my computer to simulate coin flips. Who has time to flip a thousand coins?

One person who did was J. E. Kerrich, a mathematician from South Africa who made an ill-advised visit to Europe in 1939. His semester

---

[*] I'm not going to do these computations on the page, but if you want to check my work, the key term is "binomial theorem."

abroad quickly turned into an unscheduled stint in an internment camp in Denmark. Where a less statistically minded prisoner might have passed the time by scratching the days on the cell wall, Kerrich flipped a coin, 10,000 times in all, keeping track of the number of heads as he went. His results looked like this:

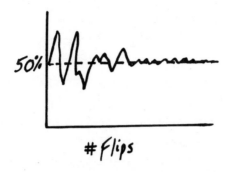

As you can see, the fraction of heads converges inexorably toward 50% as you flip more and more coins, as if squeezed by an invisible vise. You can see the same effect in the simulations. The proportions of heads in the first group of tries, the Smalls, range from 30% to 90%. With a hundred flips at a time, the range narrows: just 40% to 60%. And with a thousand flips, the range of proportions is only 46.2% to 53.7%. Something is pushing those numbers closer and closer to 50%. That something is the cold, strong hand of the Law of Large Numbers. I won't state that theorem precisely (though it is stunningly handsome!), but you can think of it as saying the following: the more coins you flip, the more and more extravagantly unlikely it is that you'll get 80% heads. In fact, if you flip enough coins, there's only the barest chance of getting as many as 51%! Observing a highly unbalanced result in ten flips is unremarkable; getting the same proportional imbalance in a hundred flips would be so startling as to make you wonder whether someone has mucked with your coins.

The understanding that the results of an experiment tend to settle down to a fixed average when the experiment is repeated again and again is not new. In fact, it's almost as old as the mathematical study of chance itself; an informal form of the principle was asserted in the sixteenth century by Girolamo Cardano, though it was not until the early 1800s that Siméon-Denis Poisson came up with the pithy name *"la loi des grands nombres"* to describe it.

THE GENDARME'S HAT

By the early eighteenth century, Jakob Bernoulli had worked out a precise statement and mathematical proof of the Law of Large Numbers. It was now no longer an observation, but a theorem.

And the theorem tells you that the Big–Small game isn't fair. The Law of Large Numbers will always push the Big players' scores toward 50%, while those of the Smalls are apt to vary much more widely. But it would be nuts to conclude that the Small team is "better" at flipping heads, even though that team wins every game. For if you average the proportion of heads flipped by *all* the Small players, not just the top scorer, they'll likely be at just about 50%, same as the Bigs. And if we look for the player with the fewest heads instead of the most, Team Small suddenly looks bad at getting heads: it's very likely one of their players will have only 20% heads, and none of the Big players will ever score that badly. Scoring by raw number of heads gives the Big team an insuperable advantage; but using percentages slants the game just as badly in favor of the Smalls. The smaller the number of coins—what we'd call in statistics the *sample size*—the greater the variation in the proportion of heads.

It's the very same effect that makes political polls less reliable when fewer voters are polled. And it's the same, too, for brain cancer. Small states have small sample sizes—they are thin reeds whipped around by the winds of chance, while the big states are grand old oaks that barely bend. Measuring the absolute number of brain cancer deaths is biased toward the big states; but measuring the highest *rates*—or the lowest ones!—puts the smallest states in the lead. That's how South Dakota can have one of the highest rates of brain cancer death while North Dakota claims one of the lowest. It's not because Mount Rushmore or Wall Drug is somehow toxic to the brain; it's because smaller populations are inherently more variable.

That's a mathematical fact you already know, even if you don't know you already know it. Who's the most accurate shooter in the NBA? A month into the 2011–12 season, five players were locked in a tie for the highest shooting percentage in the league: Armon Johnson, DeAndre Liggins, Ryan Reid, Hasheem Thabeet, and Ronny Turiaf.

Who?

That's the point. These were not the five best shooters in the NBA. These were people who barely ever played. Armon Johnson, for instance, appeared in one game for the Portland Trail Blazers. He took one shot. He made it. The five guys on the list took thirteen shots between them and hit them all. Small samples are more variable, so the leading shooter in the NBA is always going to be somebody who's only taken a handful of shots and who got lucky every time. You would never declare that Armon Johnson was a more accurate shooter than the highest-ranking full-time player on the list, Tyson Chandler of the Knicks, who made 141 out of 202 shots over the same time period.* (Any doubt on this point can be put to rest by looking at Johnson's 2010–11 season, when he shot a steadfastly ordinary 45.5% from the field.) That's why the standard leaderboard doesn't show guys like Armon Johnson. Instead, the NBA restricts the rankings to players who've reached a certain threshold of playing time; otherwise, part-time nobodies with their small sample sizes would dominate the list.

But not every ranking system has the quantitative savvy to make allowances for the Law of Large Numbers. The state of North Carolina, like many others in this age of educational accountability, instituted incentive programs for schools that do well on standardized tests. Each school is rated on the average improvement of student test scores from one spring to the next; the top twenty-five schools in the state on this measure get a banner to hang in the gym and bragging rights over the surrounding towns.

Who wins this kind of contest? The top scorer in 1999, with a 91.5 "performance composite score," was C. C. Wright Elementary in North Wilkesboro. That school was on the small side, with 418 students in a state where elementary schools average almost 500 kids. Not far behind Wright were Kingswood Elementary, with a score of 90.9, and Riverside

---

* And yes, shooting percentage is as much a function of which shots you choose to take as your intrinsic skill at hitting the basket; the big man whose shots are mostly layups and dunks starts with a big advantage. But that's orthogonal to the point we're making here.

Elementary, with 90.4. Kingswood had just 315 students, and tiny Riverside, in the Appalachian town of Newland, had only 161.

In fact, the small schools cleaned up on North Carolina's measure in general. A study by Thomas Kane and Douglas Staiger found that 28% of the smallest schools in the state made the top twenty-five at some point in the seven-year window they studied; among all schools, only 7% ever got the banner in the gym.

It sounds like small schools, where teachers really know the students and their families and have time to deliver individualized instruction, are better at raising test scores.

But maybe I should mention that the title of Kane and Staiger's paper is "The Promise and Pitfalls of Using Imprecise School Accountability Measures." And that smaller schools did not show any tendency, on average, to have significantly higher scores on the tests. And that the schools that were assigned state "assistance teams" (read: that got a dressing-down from state officials for low test scores) were *also* predominantly smaller schools.

In other words, as far as we know, Riverside Elementary is no more one of the top elementary schools in North Carolina than Armon Johnson is the sharpest shooter in the league. The reason small schools dominate the top twenty-five isn't because small schools are better, but because small schools have more variable test scores. A few child prodigies or a few third-grade slackers can swing a small school's average wildly; in a large school, the effect of a few extreme scores will simply dissolve into the big average, hardly budging the overall number.

So how are we supposed to know which school is best, or which state is most cancer-prone, if taking simple averages doesn't work? If you're an executive managing a lot of teams, how can you accurately assess performance when the smaller teams are more likely to predominate at both the top and bottom tier of your rankings?

There is, unfortunately, no easy answer. If a tiny state like South Dakota experiences a rash of brain cancer, you might presume that the spike is in large measure due to luck, and you might estimate that the rate of brain cancer in the future is likely to be closer to the overall national number. You could accomplish this by taking some kind of weighted average of the South Dakota rate with the national rate. But how to weight the two numbers? That's a bit of an art, involving a fair amount of technical labor I'll spare you here.

One relevant fact was first observed by Abraham de Moivre, an early contributor to the modern theory of probability. De Moivre's 1756 book *The Doctrine of Chances* was one of the key texts on the subject. (Even then, the popularization of mathematical advances was a vigorous industry; Edmond Hoyle, whose authority in matters of card games was so great that people still use the phrase "according to Hoyle," wrote a book to help gamblers master the new theory, called *An Essay Towards Making the Doctrine of Chances Easy to those who Understand Vulgar Arithmetic only, to which is added some useful tables on annuities.*)

De Moivre wasn't satisfied with the Law of Large Numbers, which said that in the long run the proportion of heads in a sequence of flips gets closer and closer to 50%. He wanted to know *how much* closer. To understand what he found, let's go back and look at those coin flip counts again. But now, instead of listing the total number of heads, we're going to record the *difference* between the number of heads actually flipped and the number of heads you might expect, 50% of the flips. In other words, we're measuring how far off we are from perfect head-tail parity.

For the ten-coin trials, you get:

1, 1, 0, 1, 0, 1, 2, 2, 1, 0, 0, 4, 2, 0, 2, 1, 0, 2, 2, 4 . . .

For the hundred-coin trials:

4, 4, 2, 5, 2, 1, 3, 8, 10, 7, 4, 4, 1, 2, 1, 0, 10, 7, 5 . . .

And for the thousand-coin trials:

14, 1, 11, 28, 37, 26, 8, 10, 22, 8, 7, 11, 11, 10, 30, 10, 3, 38, 0, 6 . . .

You can see that the discrepancies from 50-50 get bigger in absolute terms as the number of coin flips grows, even though (as the Law of Large Numbers demands) they're getting smaller as a proportion of the number of flips. De Moivre's insight is that the size of the typical discrepancy* is governed by the *square root* of the number of coins you toss. Toss a hundred times as many coins as before and the typical discrepancy grows by a factor of 10—at least, in absolute terms. As a proportion of the total number of tosses, the discrepancy *shrinks* as the number of coins grows, because the square root of the number of coins grows much more slowly than does the number of coins itself. The thousand-coin flippers sometimes miss an even distribution by as many as 38 heads; but as a proportion of total throws, that's only 3.8% away from 50-50.

De Moivre's observation is the same one that underlies the computation of the standard error in a political poll. If you want to make the error bar half as big, you need to survey four times as many people. And if you want to know how impressed to be by a good run of heads, you can ask how many square roots away from 50% it is. The square root of 100 is 10. So when I got 60 heads in 100 tries, that was exactly one square root away from 50-50. The square root of 1,000 is about 31; so when I got 538 heads in 1,000 tries, I did something even more surprising, even though I got only 53.8% heads in the latter case and 60% heads in the former.

But de Moivre wasn't done. He found that the discrepancies from 50-50, in the long run, always tend to form themselves into a perfect bell curve, or, as we call it in the biz, the normal distribution. (Statistics pioneer Francis Ysidro Edgeworth proposed that the curve be called the *gendarme's hat*, and I have to say I'm sorry this didn't catch on.)

The bell curve/gendarme's hat is tall in the middle and very flat near the edges, which is to say that the farther a discrepancy is from zero, the less likely it is to be encountered. And this can be precisely quantified. If you flip N coins, the chance that you'll end up being off by at most the square root of N from 50% heads is about 95.45%. The square root of 1,000 is about 31; indeed, eighteen of our twenty big thousand-coin trials above, or 90%, were within 31 heads of 500. If I kept playing the game, the fraction of times I ended up somewhere between 469 and 531 heads would get closer and closer to that 95.45% figure.[†]

---

* Experts will note that I am carefully avoiding the phrase "standard deviation." Non-experts who wish to go deeper should look the term up.
† To be precise, it's a little less, more like 95.37%, since 31 is not *quite* the square root of 1,000 but a little smaller.

LE
GENDARME

It feels like something is *making* it happen. Indeed, de Moivre himself might have felt this way. By many accounts, he viewed the regularities in the behavior of repeated coin flips (or any other experiment subject to chance) as the work of God's hand itself, which turned the short-term irregularities of coins, dice, and human life into predictable long-term behavior, governed by immutable laws and decipherable formulae.

It's dangerous to feel this way. Because if you think somebody's transcendental hand—God, Lady Luck, Lakshmi, doesn't matter—is pushing the coins to come up half heads, you start to believe in the so-called law of averages: five heads in a row and the next one's almost sure to land tails. Have three sons, and a daughter is surely up next. After all, didn't de Moivre tell us that extreme outcomes, like four straight sons, are highly unlikely? He did, and they are. But *if you've already had three sons*, a fourth son is not so unlikely at all. In fact, you're just as likely to have a son as a first-time parent.

This seems at first to be in conflict with the Law of Large Numbers, which ought to be pushing your brood to be split half and half between boys and girls.* But the conflict is an illusion. It's easier to see what's going on with the coins. I might start flipping and get 10 heads in a row.

---

* Actually, closer to 51.5% boys and 48.5% girls, but who's counting?

What happens next? Well, one thing that might happen is you'd start to suspect something was funny about the coin. We'll return to that issue in part II, but for now let's assume the coin is fair. So the law demands that the proportion of heads must approach 50% as I flip the coin more and more times.

Common sense suggests that, at this point, tails must be slightly more likely, in order to correct the existing imbalance.

But common sense says much more insistently that the coin can't remember what happened the first ten times I flipped it!

I won't keep you in suspense—the second common sense is right. The law of averages is not very well named, because laws should be true, and this one is false. Coins have no memory. So the next coin you flip has a 50-50 chance of coming up heads, the same as any other. The way the overall proportion settles down to 50% isn't that fate favors tails to compensate for the heads that have already landed; it's that those first ten flips become less and less important the more flips we make. If I flip the coin a thousand more times, and get about half heads, then the proportion of heads in the first 1,010 flips is also going to be close to 50%. *That's* how the Law of Large Numbers works: not by balancing out what's already happened, but by diluting what's already happened with new data, until the past is so proportionally negligible that it can safely be forgotten.

SURVIVORS

What applies to coins and test scores applies to massacres and genocides, too. If you rate your bloodshed by proportion of national population eliminated, the worst offenses will tend to be concentrated in the smallest countries. Matthew White, author of the agreeably morbid *Great Big Book of Horrible Things*, ranked the bloodlettings of the twentieth century in this order, and found that the top three were the massacre of the Herero of Namibia by their German colonists, the slaughter of Cambodians by Pol Pot, and King Leopold's war in the Congo. Hitler, Stalin, Mao, and the big populations they decimated don't make the list.

This bias toward less populous nations presents a problem—where is our mathematically certified rule for figuring out precisely how much distress to experience when we read about the deaths of people in Israel, Palestine, Nicaragua, or Spain?

Here's a rule of thumb that makes sense to me: if the magnitude of a disaster is so great that it feels right to talk about "survivors," then it makes sense to measure the death toll as a proportion of total population. When you talk about a survivor of the Rwandan genocide, you could be talking about any Tutsi living in Rwanda; so it makes sense to say that the genocide wiped out 75% of the Tutsi population. And you might be justified to say that a catastrophe that killed 75% of the population of Switzerland was the "Swiss equivalent" of what befell the Tutsi.

But it would be absurd to call someone in Seattle a "survivor" of the World Trade Center attack. So it's probably not useful to think of deaths at the World Trade Center as a proportion of all Americans. Only about one in a hundred thousand Americans, or 0.001%, died at the World Trade Center that day. That number is too close to zero for your intuition to grasp hold of it; you have no feeling for what that proportion means. And so it's dicey to say that the Swiss equivalent to the World Trade Center attacks would be a mass murder that killed 0.001% of the Swiss, or eighty people.

So how are we supposed to rank atrocities, if not by absolute numbers and not by proportion? Some comparisons are clear. The Rwanda genocide was worse than 9/11 and 9/11 was worse than Columbine and Columbine was worse than one person getting killed in a drunk-driving accident. Others, separated by vast differences in time and space, are harder to compare. Was the Thirty Years' War really more deadly than World War I? How does the horrifyingly rapid Rwanda genocide stack up against the long, brutal war between Iran and Iraq?

Most mathematicians would say that, in the end, the disasters and atrocities of history form what we call a partially ordered set. That's a fancy way of saying that some pairs of disasters can be meaningfully compared, and others cannot. This isn't because we don't have accurate enough death counts, or firm enough opinions as to the relative merits of being annihilated by a bomb versus dying of war-induced famine. It's because the question of whether one war was worse than another is fundamentally unlike the question of whether one number is bigger than another. The latter question always has an answer. The former does not. And if you want to imagine what it means for twenty-six people to be killed by terrorist bombings, imagine twenty-six people killed by terrorist bombings—not halfway across the world, but in your own city. That computation is mathematically and morally unimpeachable, and no calculator is required.

FIVE

MORE PIE THAN PLATE

Proportions can be misleading even in simpler, seemingly less am-
biguous cases.

A recent working paper by economists Michael Spence and
Sandile Hlatshwayo painted a striking picture of job growth in the United
States. It's traditional and pleasant to think of America as an industrial
colossus, whose factories run furiously night and day producing the goods
the world demands. Contemporary reality is rather different. Between
1990 and 2008, the U.S. economy gained a net 27.3 million jobs. Of
those, 26.7 million, or 98%, came from the "nontradable sector": the part
of the economy including things like government, health care, retail, and
food service, which can't be outsourced and which don't produce goods
to be shipped overseas.

That number tells a powerful story about recent American industrial
history, and it was widely repeated, from *The Economist* to Bill Clinton's
latest book. But you have to be careful about what it means. Ninety-eight
percent is really, really close to 100%. So does the study say that growth
is as concentrated in the nontradable part of the economy as it could possibly be? That's what it sounds like—but that's not quite right. Jobs in
the tradable sector grew by a mere 620,000 between 1990 and 2008,
that's true. But it could have been worse—they could have declined!

That's what happened between 2000 and 2008; the tradable sector lost about 3 million jobs, while the nontradable sector added 7 million. So the nontradable sector accounted for 7 million jobs out of the total gain of 4 million, or 175%!

The slogan to live by here is:

Don't talk about percentages of numbers when the numbers might be negative.

This may seem overly cautious. Negative numbers are numbers, and as such they can be multiplied and divided like any others. But even this is not as trivial as it first appears. To our mathematical predecessors, it wasn't even clear that negative numbers were numbers at all—they do not, after all, represent quantities in exactly the same way positive numbers do. I can have seven apples in my hand, but not negative seven. The great sixteenth-century algebraists, like Cardano and François Viète, argued furiously about whether a negative times a negative equaled a positive; or rather, they understood that consistency seemed to demand that this be so, but there was real division about whether this had been proved factual or was only a notational expedient. Cardano, when an equation he was studying had a negative number among its solutions, had the habit of calling the offending solution *ficta*, or fake.

The arguments of Italian Renaissance mathematicians can at times seem as recondite and irrelevant to us as their theology. But they weren't wrong that there's something about the combination of negative quantities and arithmetic operations like percentage that short-circuits one's intuition. When you disobey the slogan I gave you, all sorts of weird incongruities start to bubble up.

For example, say I run a coffee shop. People, sad to say, are not buying my coffee; last month I lost $500 on that part of my business. Fortunately, I had the prescience to install a pastry case and a CD rack, and those two operations made a $750 profit each.

In all, I made $1000 this month, and 75% of that amount came from my pastry case. Which sounds like the pastry case is what's really moving my business right now; almost all my profit is croissant-driven. Except that it's just as correct to say that 75% of my profits came from the CD rack. And imagine if I'd lost $1000 more on coffee—then my total profits would be zero, infinity percent of which would be coming from

MORE PIE THAN PLATE

pastry!* "Seventy-five percent" sounds like it means "almost all," but when you're dealing with numbers that could be either positive or negative, like profits, it might mean something very different.

This problem never arises when you study numbers that are constrained to be *positive*, like expenses, revenues, or populations. If 75% of Americans think Paul McCartney was the cutest Beatle, then it's not possible that another 75% give the nod to Ringo Starr; he, George,† and John have to split the remaining 25% between them.

You can see this phenomenon in the jobs data, too. Spence and Hlatshwayo might have pointed out that about 600,000 jobs were created in finance and insurance; that's almost 100% of the total jobs created by the tradable sector as a whole. They didn't point that out, because they weren't trying to trick you into believing that no other part of the economy was growing over that time span. As you might remember, there was at least one other part of the U.S. economy that added a lot of jobs between 1990 and today: the sector classified as "computer systems design and related services," which tripled its job numbers, adding more than a million jobs all by itself. The total jobs added by finance and computers were way over the 620,000 jobs added by the tradable sector as a whole; those gains were balanced out by big losses in manufacturing. The combination of positive and negative allows you, if you're not careful, to tell a fake story, in which the whole work of job creation in the tradable sector was done by the financial industry.

One can't object very much to what Spence and Hlatshwayo wrote. It's true, the total job growth in an aggregate of hundreds of industries *can* be negative, but in a normal economic context over a reasonably long time interval, it's extremely likely to be positive. The population keeps growing, after all, and, absent total disaster, that tends to drag the absolute number of jobs along with it.

But other percentage flingers are not so careful. In June 2011, the Republican Party of Wisconsin issued a news release touting the job-

* Safety warning: never divide by zero unless a licensed mathematician is present.
† Actual cutest Beatle.

creating record of Governor Scott Walker. It had been another weak
month for the U.S. economy as a whole, which added only eighteen
thousand jobs nationally. But the state employment numbers looked
much better: a net increase of ninety-five hundred jobs. "Today," the
statement read, "we learned that over 50 percent of U.S. job growth in
June came from our state." The talking point was picked up and distributed by GOP politicians, like Representative Jim Sensenbrenner, who
told an audience in a Milwaukee suburb, "The labor report that came
out last week had an anemic eighteen thousand created in this country,
but half of them came here in Wisconsin. Something we are doing here
must be working."

This is a perfect example of the soup you get into when you start
reporting percentages of numbers, like net job gains, that might be either
positive or negative. Wisconsin added ninety-five hundred jobs, which is
good; but neighboring Minnesota, under Democratic governor Mark
Dayton, added more than thirteen thousand in the same month. Texas,
California, Michigan, and Massachusetts also outpaced Wisconsin's job
gains. Wisconsin had a good month, that's true—but it didn't contribute
as many jobs as the rest of the country put together, as the Republican
messaging suggested. In fact, what was going on is that job losses in other
states almost exactly balanced out the jobs created in places like Wisconsin, Massachusetts, and Texas. That's how Wisconsin's governor could
claim his state accounted for half the nation's job growth, and Minnesota's governor, if he'd cared to, could have said that his own state was responsible for 70% of it, and they could both, in this technically correct
but fundamentally misleading way, be right.

Or take a recent *New York Times* op-ed by Steven Rattner, which
used the work of economists Thomas Piketty and Emmanuel Saez to
argue that the current economic recovery is unequally distributed among
Americans:

New statistics show an ever-more-startling* divergence between the
fortunes of the wealthy and everybody else—and the desperate need

---

* Math pedantry: in order to claim some phenomenon is "ever-more-startling," you have to do
more than show that it is startling; you have to show that its startlingness is *increasing*. This issue
is not addressed in the body of the op-ed.

to address this wrenching problem. Even in a country that sometimes seems inured to income inequality, these takeaways are truly
stunning.

In 2010, as the nation continued to recover from the recession, a
dizzying 93 percent of the additional income created in the country
that year, compared to 2009—$288 billion—went to the top 1 per-
cent of taxpayers, those with at least $352,000 in income. . . . The
bottom 99 percent received a microscopic $80 increase in pay per
person in 2010, after adjusting for inflation. The top 1 percent,
whose average income is $1,019,089, had an 11.6 percent increase in
income.

The article comes packaged with a handsome infographic that breaks
the income gains up even further: 37% to the ultrarich members of
the top 0.01%, with 56% to the rest of the top 1%, leaving a meager 7%
for the remaining 99% of the population. You can make a little pie chart:

Now let's slice the pie one more time, and ask about the people who
are in the top 10%, but not the top 1%. Here you've got the family doctors, the non-elite lawyers, the engineers, and the upper-middle manag-
ers. How big is their slice? You can get this from Piketty and Saez's data,
which they've helpfully put online. And you find something curious.
This group of Americans had an average income of about $159,000 in
2009, which increased to a little over $161,000 in 2010. That's a modest
gain compared to what the richest percentile racked up, but it still accounts for 17% of the total income gained between 2010 and 2011.

Try to fit a 17% slice of the pie in with the 93% share held by the one-percenters and you find you've got more pie than plate.

93% and 17% add up to more than 100%; how does this make sense? It makes sense because the bottom 90% actually had *lower* average income in 2011 than they did in 2010, recovery or no recovery. Negative numbers in the mix make percentages act wonky.

Looking at the Piketty-Saez data for different years, you see the same pattern again and again. In 1992, 131% of the national gains in income were accrued by the top 1% of earners! That's certainly an impressive figure, but one which clearly indicates that the percentage doesn't mean quite what you're used to it meaning. You can't put 131% in a pie chart. Between 1982 and 1983, as another recession retreated into memory, 91% of the national income gain went to the 10%-but-not-1% group. Does that mean that the recovery was captured by the reasonably wealthy professionals, leaving the middle class and the very rich behind? Nope—the top 1% saw a healthy increase that year too, accounting for 63% of the national income gain all by themselves. What was really going on then, as now, was that the bottom 90% continued to lose ground while the situation brightened for everybody else.

None of which is to deny that morning in America comes a little earlier in the day for the richest Americans than it does for the middle class. But it does put a slightly different spin on the story. It's not that the 1% are benefitting while the rest of America languishes. The people in the top 10% but not the top 1%—a group that includes, not to put too fine a point on it, many readers of the *New York Times* opinion page—are doing fine too, capturing more than twice as much as the 7% share that the pie chart appears to allow them. It's the *other* 90% of the country whose tunnel still looks dark at the end.

Even when the numbers involved *happen* to be positive, there's room for spinners to tell a misleading story about percentages. In April 2012, Mitt Romney's presidential campaign, facing poor poll numbers among women voters, released a statement asserting, "The Obama administration has brought hard times to American women. Under President Obama, more women have struggled to find work than at any other time in recorded history. Women account for 92.3% of all jobs lost under Obama."

That statement is, in a manner of speaking, correct. According to the Bureau of Labor Statistics, total employment in January 2009 was 133,561,000, and in March 2012, just 132,821,000: a net loss of 740,000 jobs. Among women, the numbers were 66,122,000 and 65,439,000; so 683,000 fewer women were employed in March 2012 than in January 2009, when Obama took office. Divide the second number by the first and you get the 92% figure. It's almost as if President Obama had been going around ordering businesses to fire all the women.

But no. Those numbers are *net* job losses. We have no idea how many jobs were created and how many destroyed over the three-year period; only that the difference of those two numbers is 740,000. The net job loss is positive sometimes, and negative other times, which is why taking percentages of it is a dangerous business. Just imagine what would have happened if the Romney campaign had started their count one month later, in February 2009.* At that point, another brutal month into the recession, total employment was down to 132,837,000. Between then and March 2012, the economy suffered a net loss of just 16,000 jobs. Among women alone, the jobs lost were 484,000 (balanced, of course, by a corresponding gain for men). What a missed opportunity for the Romney campaign—if they'd started their reckoning in February, the first full month of the Obama presidency, they could have pointed out that women accounted for over *3,000%* of all jobs lost on Obama's watch!

But that would have signaled to any but the thickest voters that this percentage was somehow not the right measure.

What actually happened to men and women in the workforce between Obama's inauguration and March 2012? Two things. Between January 2009 and February 2010, employment plunged for both men and women as the recession and its aftermath took their toll.

January 2009–February 2010:
Net job loss for men: 2,971,000
Net job loss for women: 1,546,000

---

* The analysis here is indebted to that of Glenn Kessler, who wrote about the Romney ad in the April 10, 2012, edition of the *Washington Post*.

And then, post-recession, the employment picture started slowly improving:

> February 2010–March 2012:
> Net job gain for men: 2,714,000
> Net job gain for women: 863,000

During the steep decline, men took it on the chin, suffering almost twice as many job losses as women. And in the recovery, men account for 75% of the jobs gained. When you add both periods together, the men's figures happen to cancel out almost exactly, leaving them with about as many jobs at the end as the beginning. But the idea that the current economic period has been almost exclusively bad for women is badly misguided.

The *Washington Post* graded the Romney campaign's 92.3% figure as "true but false." That classification drew mockery by Romney supporters, but I think it's just right, and has something deep to say about the use of numbers in politics. There's no question about the accuracy of the number. You divide the net jobs lost by women by the net jobs lost, and you get 92.3%.

But that makes the claim "true" only in a very weak sense. It's as if the Obama campaign had released a statement saying, "Mitt Romney has never denied allegations that for years he's operated a bicontinental cocaine-trafficking ring in Colombia and Salt Lake City."

That statement is also 100% true! But it's designed to create a false impression. So "true but false" is a pretty fair assessment. It's the right answer to the wrong question. Which makes it worse, in a way, than a plain miscalculation. It's easy to think of the quantitative analysis of policy as something you do with a calculator. But the calculator only enters once you've figured out what calculation you want to do.

I blame word problems. They give a badly wrong impression of the relation between mathematics and reality. "Bobby has three hundred marbles and gives 30% of them to Jenny. He gives half as many to Jimmy as he gave to Jenny. How many does he have left?" That *looks* like it's about the real world, but it's just an arithmetic problem in a not very convincing disguise. The word problem has nothing to do with marbles.

It might as well just say: type "300 − (0.30 × 300) − (0.30 × 300)/2 ="
into your calculator and copy down the answer!

But real-world questions aren't like word problems. A real-world
problem is something like "Has the recession and its aftermath been especially bad for women in the workforce, and if so, to what extent is this
the result of Obama administration policies?" Your calculator doesn't
have a button for this. Because in order to give a sensible answer, you need
to know more than just numbers. What shape do the job-loss curves for
men and women have in a typical recession? Was this recession notably
different in that respect? What kind of jobs are disproportionately held
by women, and what decisions has Obama made that affect that sector of
the economy? It's only after you've started to formulate these questions
that you take out the calculator. But at that point the real mental work is
already finished. Dividing one number by another is mere computation;
figuring out *what* you should divide by *what* is mathematics.

PART II

· · · · ·

Inference

Includes: hidden messages in the Torah, the dangers of wiggle room, null hypothesis significance testing, B. F. Skinner vs. William Shakespeare, "Turbo Sexophonic Delight," the clumpiness of prime numbers, torturing the data until it confesses, the right way to teach creationism in public schools

SIX

THE BALTIMORE
STOCKBROKER AND
THE BIBLE CODE

Peole use mathematics to get a handle on problems ranging from the everyday ("How long should I expect to wait for the next bus?") to the cosmic ("What did the universe look like three trillionths of a second after the Big Bang?").

But there's a realm of questions out beyond cosmic, questions about The Meaning and Origin of It All, questions you might think mathematics could have no purchase on.

Never underestimate the territorial ambitions of mathematics! You want to know about God? There are mathematicians on the case.

The idea that earthly humans can learn about the divine world by rational observation is a very old one, as old, according to the twelfth-century Jewish scholar Maimonides, as monotheism itself. Maimonides's central work, the *Mishneh Torah*, gives this account of Abraham's revelation:

> After Abraham was weaned, while still an infant, his mind began to reflect. By day and by night he was thinking and wondering: "How is it possible that this [celestial] sphere should continuously be guiding the world and have no one to guide it and cause it to turn round;

for it cannot be that it turns round of itself?" . . . His mind was busily working and reflecting until he had attained the way of truth, apprehended the correct line of thought, and knew that there is one God, that He guides the celestial sphere and created everything, and that among all that exist, there is no god besides Him. . . . He then began to proclaim to the whole world with great power and to instruct the people that the entire universe had but one Creator and that Him it was right to worship. . . . When the people flocked to him and questioned him regarding his assertions, he would instruct each one according to his capacity till he had brought him to the way of truth, and thus thousands and tens of thousands joined him.

This vision of religious belief is extremely congenial to the mathematical mind. You believe in God not because you were touched by an angel, not because your heart opened up one day and let the sunshine in, and certainly not because of something your parents told you, but because God is a thing that *must be*, as surely as 8 times 6 must be the same as 6 times 8.

Nowadays, the Abrahamic argument—just *look* at everything, how could it all be so awesome if there weren't a designer behind it?—has been judged wanting, at least in most scientific circles. But then again, now we have microscopes and telescopes and computers. We are not restricted to gaping at the moon from our cribs. We have data, lots of data, and we have the tools to mess with it.

The favorite data set of the rabbinical scholar is the Torah, which is, after all, a sequentially arranged string of characters drawn from a finite alphabet, which we attempt faithfully to transmit without error from synagogue to synagogue. Despite being written on parchment, it's the original digital signal.

And when a group of researchers at the Hebrew University in Jerusalem started analyzing that signal, in the mid-1990s, they found something very strange; or, depending on your theological perspective, not strange at all. The researchers came from different disciplines: Eliyahu Rips was a senior professor of mathematics, a well-known group theorist; Yoav Rosenberg a graduate student in computer science; and Doron Witztum a former student with a master's degree in physics. But all shared a taste for the strand of Torah study that searches for esoteric texts hidden beneath the stories, genealogies, and admonitions that make up the Torah's surface. Their tool of choice was the "equidistant letter sequence," henceforth ELS, a string of text obtained by plucking characters from the Torah at regular intervals. For example, in the phrase

DON YOUR BRACES ASKEW

you can read every fifth letter, starting from the first, to get

DON **Y**OUR **B**RACES **A**SKEW

so the ELS would be **DUCK**, whether as warning or waterfowl identification to be determined from context.

Most ELSs don't spell words; if I make an ELS out of every third letter in the sentence you're reading, I get gibberish like **MTSOSLO . . .** , which is more typical. Still, the Torah is a long document, and if you look for patterns, you'll find them.

As a mode of religious inquiry, this seems strange at first. Is the God of the Old Testament really the kind of deity who signals his presence by showing up in a word search? In the Torah, when God wants you to know he's there, you *know*—ninety-year-old women get pregnant, bushes catch fire and talk, dinner falls from the sky.

Still, Rips, Witztum, and Rosenberg were not the first to look for messages concealed in the ELSs of the Torah. There's some sporadic precedent among the classical rabbis, but the method was really pioneered in the twentieth century by Michael Dov Weissmandl, a rabbi in Slovakia who spent World War II trying, largely in vain, to raise enough money from the West to buy respite for Slovakia's Jews from bribable German officials. Weissmandl found several interesting ELSs in the Torah. Most famously, he observed that starting from a certain "mem" (the Hebrew letter that sounds like "m") in the Torah, and counting forward in steps of 50 letters, you found the sequence *"mem shin nun hay,"* which spells out the Hebrew word *Mishneh*, the first word of the title of

Maimonides's Torah commentary. Now you skip forward 613 letters (why 613? because that's the exact number of commandments in the Torah, please try to keep up) and start counting every 50th letter again. You find that the letters spell out *Torah*—in other words, that the title of Maimonides's book is recorded in ELS form in the Torah, a document set down more than a thousand years before his birth.

Like I said, the Torah is a long document—by one count, it has 304,805 letters in all. So it's not clear what to make, if anything, from patterns like the one Weissmandl found—there are lots of ways to slice and dice the Torah, and inevitably some of them are going to spell out words.

Witztum, Rips, and Rosenberg, mathematically as well as religiously trained, set themselves a more systematic task. They chose thirty-two notable rabbis from the whole span of modern Jewish history, from Avraham HaMalach to The Yaabez. In Hebrew, numbers can be recorded in alphabetic characters, so the birth and death dates of the rabbis provided more letter sequences to play with. So the question is: Do the names of the rabbis appear in equidistant letter sequences unusually close to their birth and death dates?

Or, more provocatively: did the Torah know the future?

Witztum and his colleagues tested this hypothesis in a clever way. First they searched the book of Genesis for ELSs spelling out the rabbis' names and dates, and computed how close in the text the sequences yielding the names were to the ones yielding the corresponding dates. Then they shuffled the thirty-two dates, so that each one was now matched with a random rabbi, and they ran the test again. Then they did the same thing a million times.* If there were no relation in the Torah's text between the names of the rabbis and the corresponding dates, you'd expect the true matching between rabbis and dates to do about as well as one of the random shuffles. That's not what they found. The correct association ended up very near the top of the rankings, notching the 453rd highest score among the 1 million contenders.

---

* Which is only a tiny fraction of the possible *permutations* of thirty-two dates, of which there are 263,130,836,933,693,530,167,218,012,160,000,000.

They tried the same thing with other texts: *War and Peace*, the book of Isaiah (part of Scripture, but not the part that God is understood to have written), and a version of Genesis with the letters scrambled up at random. In all these cases, the real rabbinical birthdays stayed in the middle of the pack.

The authors' conclusion, written with characteristic mathematical sobriety: "We conclude that the proximity of ELSs with related meanings in the Book of Genesis is not due to chance."

Despite the quiet language, this was understood to be a startling finding, made more so by the mathematical credentials of the authors, especially Rips. The paper was refereed and published in 1994 in the journal *Statistical Science*, accompanied by an unusual preface by editor Robert E. Kass, who wrote:

> Our referees were baffled: their prior beliefs made them think the Book of Genesis could not possibly contain meaningful references to modern-day individuals, yet when the authors carried out additional analyses and checks the effect persisted. The paper is thus offered to *Statistical Science* readers as a challenging puzzle.

Despite its startling findings, the Witztum paper didn't immediately draw a lot of public attention. All that changed when the American journalist Michael Drosnin got wind of the paper. Drosnin went hunting for ELSs of his own, jettisoning scientific restraint and counting every cluster of sequences he could find as a divine foretelling of future events. In 1997, he published a book, *The Bible Code*, whose cover features a faded, ancient-looking Torah scroll, with circled sequences of letters spelling out the Hebrew words for "Yitzhak Rabin" and "assassin who will assassinate." Drosnin's claims to have warned Rabin of his 1995 assassination a year in advance were a potent advertisement for his book, which also features Torah-certified predictions of the Gulf War and the 1994 collision of Comet Shoemaker-Levy 9 with Jupiter. Witztum, Rips, and Rosenberg denounced Drosnin's ad hoc method, but death and prophecy move units: *The Bible Code* was a best seller. Drosnin appeared on *The Oprah Winfrey Show* and CNN, and had personal audiences with Yasser

Arafat, Shimon Peres, and Clinton chief of staff John Podesta during which he shared his theories about the upcoming End of Days.* Millions saw what looked like mathematical proof that the Bible was the word of God; modern people with a scientific worldview were presented with an unexpected avenue toward accepting religious faith, and many took it. I have it on good assurance that one new father from a secular Jewish family waited until the *Statistical Science* paper was officially accepted before deciding to circumcise his son. (For the kid's sake, I hope the refereeing process was on the speedy side.)

But just as the codes were drawing wide acceptance in public, their foundations were coming under attack in the mathematical world. The controversy was especially bitter among the large community of Orthodox Jewish mathematicians. The Harvard math department, where I was a PhD student at the time, had on the faculty both David Kazhdan, who had expressed a modest openness to the codes, and Shlomo Sternberg, a vocal opponent who thought promotion of the codes made the Orthodox look like dupes and fools. Sternberg launched a broadside in the *Notices of the American Mathematical Society* in which he called the Witztum-Rips-Rosenberg paper "a hoax" and said that Kazhdan and others with similar views "have not only brought shame on themselves, they have disgraced mathematics."

The math department afternoon tea was kind of awkward the day Sternberg's article came out, let me tell you.

Religious scholars, too, were resistant to the lure of the codes. Some, like the leaders of the yeshiva Aish HaTorah, embraced the codes as a means of drawing unobservant Jews back into a more rigorous version of the faith. Others were suspicious of a mechanism that represented a sharp break from conventional Torah study. I heard of one distinguished rabbi who, at the end of a long and traditionally boozy Purim dinner, asked one of his guests, a code adherent, "So tell me, what would you do if you found a code in the Torah that said the Sabbath was supposed to be on Sunday?"

There wouldn't be such a code, the colleague said, because God commanded that the Sabbath is on Saturday.

* Which was supposed to happen in 2006, so, whew, I guess?

The old rabbi didn't give up. "Okay," he said, "but what if there were?"

The young colleague was silent for a time, and finally said, "Then I guess I'd have to think about it."

At this point, the rabbi determined that the codes were to be rejected; for while there is indeed a Jewish tradition, particularly among rabbis with mystical leanings, of carrying out numerical analysis of the letters of the Torah, the process is meant only to aid in understanding and appreciating the holy book. If the method could be used, even in principle, to induce doubt as to the basic laws of the faith, it was about as authentically Jewish as a bacon cheeseburger.

Why did mathematicians reject what seemed plain evidence of the Torah's divine inspiration? To explain, we need to introduce a new character: the Baltimore stockbroker.

THE BALTIMORE STOCKBROKER

Here's a parable. One day, you receive an unsolicited newsletter from a stockbroker in Baltimore, containing a tip that a certain stock is due for a big rise. A week passes, and just as the Baltimore stockbroker predicted, the stock goes up. The next week, you get a new edition of the newsletter, and this time, the tip is about a stock whose price the broker thinks is going to fall. And indeed, the stock craters. Ten weeks go by, each one bringing a new issue of the mysterious newsletter with a new prediction, and each time, the prediction comes true.

On the eleventh week, you get a solicitation to invest money with the Baltimore stockbroker, naturally with a hefty commission to cover the keen view of the market so amply demonstrated by the newsletter's ten-week run of golden picks.

Sounds like a pretty good deal, right? Surely the Baltimore stockbroker is onto something—it seems incredibly unlikely that a complete duffer, with no special knowledge about the market, would get ten up-or-down predictions in a row correct. In fact, you can compute the odds on the nose: if the duffer has a 50% chance of getting each prediction right, then the chance of his getting the first *two* predictions right is half of half, or a quarter, his chance of getting the first three right is half of that quarter, or an eighth, and so on. Continuing this computation, his chance of hitting the mark ten times in a row[*] is

$$(1/2) \times (1/2) \times (1/2) \times (1/2) \times (1/2) \times (1/2) \times (1/2) \times (1/2)$$
$$\times (1/2) \times (1/2) = (1/1024).$$

In other words, the chance that a duffer would do so well is next to nil.

But things look different when you retell the story from the Baltimore stockbroker's point of view. Here's what you didn't see the first time. That first week, you weren't the only person who got the broker's newsletter; he sent out 10,240.[†] But the newsletters weren't all the same. Half of them were like yours, predicting a rise in the stock. The others predicted exactly the opposite. The 5,120 people who got a dud prediction from the stockbroker never heard from him again. But you, and the 5,119 other people who got your version of the newsletter, get another tip next week. Of those 5,120 newsletters, half say what yours said and half say the opposite. And after that week, there are still 2,560 people who've received two correct predictions in a row.

And so on.

After the tenth week, there are going to be ten lucky (?) people who've gotten ten straight winning picks from the Baltimore stockbroker— *no matter what* the stock market does. The broker might be an eagle-eyed observer of the market, or he might pick stocks by slapping chicken guts against the wall and reading the stains—either way, there are ten newsletter recipients out there to whom he looks like a genius. Ten people from whom he can expect to collect substantial fees. Ten people for whom past performance is going to be no guarantee of future results.

I've often heard the Baltimore stockbroker parable told as a true story, but I couldn't locate any evidence that it's ever really happened.

---

[*] There's a useful principle, the *product rule*, hiding in this computation. If the chance of foo happening is p, and the chance of bar happening is q, and if foo and bar are independent—that is, foo happening doesn't make bar any more or less likely—then the chance of both foo *and* bar happening is p × q.

[†] This story certainly dates back to the days when this process would have involved reproducing and stapling ten thousand physical documents, but is even more realistic now that this kind of mass mailing can be carried out electronically at essentially zero expense.

The closest thing I found was a 2008 reality TV show—reality TV being where we go for parables nowadays—in which British magician Derren Brown pulled off a similar stunt, mailing various horse-racing picks to thousands of Britons with the result of eventually convincing a single person that he'd devised a foolproof prediction system. (Brown, who likes dispelling mystical claims more than he does promoting them, exposed the mechanism of the trick at the end of the show, probably doing more for math education in the UK than a dozen sober BBC specials.)

But if you tweak the game, making it less clearly fraudulent but leaving unchanged the potential to mislead, you find the Baltimore stockbroker is alive and well in the financial industry. When a company launches a mutual fund, they often maintain the fund in-house for some time before opening it to the public, a practice called *incubation*. The life of an incubated fund is not as warm and safe as the name might suggest. Typically, companies incubate lots of funds at once, experimenting with numerous investment strategies and allocations. The funds jostle and compete in the womb. Some show handsome returns, and are quickly made available to the public, with extensive documentation of their earnings so far. But the runts of the litter are mercy-killed, often without any public notice that they ever existed.

Now it might be that the mutual funds that make it out of the incubator did so because they actually represented smarter investments. The companies selling the mutual funds may even believe that. Who doesn't, when a gamble goes right, think their own smarts and know-how are in some way due the credit? But the data suggests the opposite: the incubator funds, once the public gets their hands on them, don't maintain their excellent prenatal performance, instead offering roughly the same returns as the median fund.

What does this mean for you, if you're fortunate enough to have some money to invest? It means you're best off resisting the lure of the hot new fund that made 10% over the last twelve months. Better to follow the deeply unsexy advice you're probably sick of hearing, the "eat your vegetables and take the stairs" of financial planning: instead of hunting for a magic system or an advisor with a golden touch, put your money in a big dull low-fee index fund and forget about it. When you sink your savings into the incubated fund with the eye-popping returns, you're like the newsletter getter who invests his life savings with the Baltimore stockbroker; you've been swayed by the impressive results, but you don't know *how many chances* the broker had to get those results.

It's a lot like playing Scrabble with my eight-year-old son. If he's unsatisfied with the letters he pulls from the bag, he dumps them back in and draws again, repeating this process until he gets letters he likes. In his view this is perfectly fair; after all, he's closing his eyes, so he has no way of knowing what letters he's going to draw! But if you give yourself enough chances, you'll eventually come across that Z you're waiting for. And it's not because you're lucky; it's because you're cheating.

The Baltimore stockbroker con works because, like all good magic tricks, it doesn't try to fool you outright. That is, it doesn't try to tell you something false—rather, it tells you something true from which you're likely to draw incorrect conclusions. It really *is* improbable that ten stock picks in a row would come out the right way, or that a magician who bet on six horse races would get the winner right every time, or that a mutual fund would beat the market by 10%. The mistake is in being surprised by this encounter with the improbable. The universe is big, and if you're sufficiently attuned to amazingly improbable occurrences, you'll find them. *Improbable things happen a lot.*

It's massively improbable to get hit by a lightning bolt, or to win the lottery; but these things happen to people all the time, because there are a lot of people in the world, and a lot of them buy lottery tickets, or go golfing in a thunderstorm, or both. Most coincidences lose their snap when viewed from the appropriate distance. On July 9, 2007, the North Carolina Cash 5 lottery numbers came up 4, 21, 23, 34, 39. Two days later, the same five numbers came up again. That seems highly unlikely, and it seems that way because it is. The chance of those two lottery draws matching by pure chance was tiny, less than two in a million. But that's not the relevant question, if you're deciding how impressed to be. After all, the Cash 5 game had already been going on for almost a year, offering many opportunities for coincidence; it turns out the chance *some* three-day period would have seen two identical Cash 5 draws was a much less miraculous one in a thousand. And Cash 5 isn't the only game in town. There are hundreds of five-number lottery games running all over the country, and have been for years; when you put them all together, it's not at all surprising that you get a coincidence like two identical draws in three days. That doesn't make each individual coincidence any less improbable. But here comes the chorus again: *improbable things happen a lot.*

Aristotle, as usual, was here first: despite lacking any formal notion of probability, he was able to understand that "it is probable that improbable things will happen. Granted this, one might argue that *what is improbable is probable.*"

Once you've truly absorbed this fundamental truth, the Baltimore stockbroker has no power over you. That the stockbroker handed you ten straight good stock picks is very unlikely; that he handed *somebody* such a good run of picks, given ten thousand chances, is not even remotely surprising. In the British statistician R. A. Fisher's famous formulation, "the 'one chance in a million' will undoubtedly occur, with no less and no more than its appropriate frequency, however surprised we may be that it should occur to *us*."

WIGGLE ROOM AND THE NAMES OF THE RABBIS

The Bible decoders didn't write ten thousand versions of their paper and send them to ten thousand statistical journals. So it's hard to see, at first, how their story resembles the Baltimore stockbroker con.

But when mathematicians took up the "challenge" Kass had posed in his journal preface, looking for some explanation other than "God did it" for the Bible code results, they found the matter wasn't as simple as Witztum and company had made it seem. The pace was set by Brendan McKay, an Australian computer scientist, and Dror Bar-Natan, an Israeli mathematician then at Hebrew University. They made the critical point that medieval rabbis didn't have passports or birth certificates granting them official names. They were referred to by appellations, and different authors might denote the same rabbi in different ways. If Dwayne "The Rock" Johnson were a famous rabbi, for example, would you look for a prediction of his birth in the Torah under Dwayne Johnson, The Rock, Dwayne "The Rock" Johnson, D.T.R. Johnson, or all of these?

This ambiguity creates some wiggle room for code hunters. Consider

Rabbi Avraham ben Dov Ber Friedman, an eighteenth-century Hasidic mystic who lived and worked in the shtetl of Fastov, in the Ukraine. Witztum, Rips, and Rosenberg use "Rabbi Avraham" and "HaMalach" ("the angel") as appellations. But why, McKay and Bar-Natan ask, do they use "HaMalach" alone but not "Rabbi Avraham HaMalach," a name by which the rebbe was also often known?

McKay and Bar-Natan found that wiggle room in the choices of names led to drastic changes in the quality of the results. They made a different set of choices about the appellations of the rabbis; their choices, according to biblical scholars, make just as much sense as the ones picked by Witztum (one rabbi called the two lists of names "equally appalling.") And they found that with the new list of names, something quite amazing transpired. The Torah no longer seemed to detect the birth and death dates of the rabbinic notables. But the Hebrew edition of *War and Peace* nailed it, identifying the rabbis with their correct dates about as well as the book of Genesis did in the Witztum paper.

What can this mean? Not, I hurry to say, that Leo Tolstoy composed his novel with the names of rabbis concealed therein, designed to be uncovered only once modern Hebrew was developed and classic works of world literature translated into it. Rather, McKay and Bar-Natan are making a potent point about the power of wiggle room. Wiggle room is what the Baltimore stockbroker has when he gives himself plenty of chances to win; wiggle room is what the mutual fund company has when it decides which of its secretly incubating funds are winners and which are trash. Wiggle room is what McKay and Bar-Natan used to work up a list of rabbinical names that jibed well with *War and Peace*. When you're trying to draw reliable inferences from improbable events, wiggle room is the enemy.

In a later paper, McKay and Bar-Natan asked Simcha Emanuel, a Talmud professor then at the University of Tel Aviv, to draw up another list of appellations, this one not designed for compatibility with either the Torah or *War and Peace*. On this list, the Torah did only a little better than chance. (How Tolstoy did is left unreported.)

It is very unlikely that any given set of rabbinic appellations is well matched to birth and death dates in the book of Genesis. But with so many ways of choosing the names, it's not at all improbable that among all the choices there would be *one* that made the Torah look uncannily prescient. Given enough chances, finding codes is a cinch. It's especially easy if you use Michael Drosnin's less scientific approach to code-finding. Drosnin said of code skeptics, "When my critics find a message about the assassination of a prime minister encrypted in *Moby Dick*, I'll believe them." McKay quickly found equidistant letter sequences in Moby Dick referring to the assassination of John F. Kennedy, Indira Gandhi, Leon Trotsky, and, for good measure, Drosnin himself. As I write this, Drosnin remains alive and well despite the prophecy. He is on his third Bible code book, the last of which he advertised by taking out a full-page ad in a December 2010 edition of the *New York Times*, warning President Obama that, according to letter sequences hidden in Scripture, Osama bin Laden might already have a nuclear weapon.

Witztum, Rips, and Rosenberg insist they weren't like the masters of the incubator funds, displaying to the public only the experiments that gave the best possible results; their precise list of names was chosen in advance, they say, before running any tests. And that may well be true. But even if it is, it casts the miraculous success of the Bible codes in a very different light. That the Torah, like *War and Peace*, can successfully be mined for *some* version of the rabbis' names is not surprising. The miracle, if there is one, is that Witztum and his colleagues were moved to choose precisely those versions of the names on which the Torah scores best.

There's one loose end that should trouble you, though. McKay and Bar-Natan made a compelling case that the wiggle room in the design of Witztum's experiment was enough to explain the Bible codes. But the Witztum paper was carried out using standard statistical tests, the same ones scientists use to judge claims about everything from medicines to economic policies. It wouldn't have been accepted in *Statistical Science* otherwise. If the paper passed that test, shouldn't we have accepted its conclusions, however otherworldly they may have seemed? Or, to put it another way: if we now feel comfortable rejecting the conclusions of the Witztum study, what does that say about the reliability of our standard statistical tests?

It says you ought to be a little worried about them. And it turns out that, without any input from the Torah, scientists and statisticians have already been worrying about them for quite some time.

SEVEN

DEAD FISH DON'T
READ MINDS

Because here's the thing: the Bible code kerfuffle is not the only occasion on which the standard statistical tool kit has been used to derive a result that sounds like magic. One of the hottest topics in medical science is functional neuroimaging, which promises to let scientists see your thoughts and feelings flickering across your synapses in real time through ever-more-accurate sensors. At the 2009 Organization for Human Brain Mapping conference in San Francisco, UC Santa Barbara neuroscientist Craig Bennett presented a poster called "Neural correlates of interspecies perspective taking in the post-mortem Atlantic Salmon: An argument for multiple comparisons correction." It takes a second to unwrap the jargony title, but when you do, the poster announces pretty clearly the unusual nature of its results. A dead fish, scanned in an fMRI device, was shown a series of photographs of human beings, and was found to have a surprisingly strong ability to correctly assess the emotions the people in the pictures displayed. That would be impressive enough for a dead person or a live fish—for a dead fish, it's Nobel Prize material!

But the paper, of course, is a deadpan gag. (And a well-executed one: I especially like the "Methods" section, which starts "One mature Atlantic Salmon (*Salmo salar*) participated in the fMRI study. The salmon was approximately 18 inches long, weighed 3.8 lbs, and was not alive at the time of scanning. . . . Foam padding was placed within the head coil as a method of limiting salmon movement during the scan, but proved to be largely unnecessary as subject motion was exceptionally low.") The joke, like all jokes, is a veiled attack: in this case, an attack on sloppy methodology among those neuroimaging researchers who make the mistake of ignoring the fundamental truth that improbable things happen a lot. Neuroscientists divvy up their fMRI scans into tens of thousands of small pieces, called *voxels*, each corresponding to a small region of the brain. When you scan a brain, even a cold dead fish brain, there's a certain amount of random noise coming through on each voxel. It's pretty unlikely that the noise will happen to spike exactly at the moment that you show the fish a snapshot of a person in emotional extremity. But the nervous system is a big place, with tens of thousands of voxels to choose from. The odds that one of those voxels provides data matching up well with the photos is pretty good. That's exactly what Bennett and his collaborators found; in fact, they located two groups of voxels that did an excellent job empathizing with human emotion, one in the salmon's medial brain cavity and the other in the upper spinal column. The point of Bennett's paper is to warn that the standard methods of assessing results, the way we draw our thresholds between a real phenomenon and random static, come under dangerous pressure in this era of massive data sets, effortlessly obtained. We need to think very carefully about whether our standards for evidence are strict enough, if the empathetic salmon makes the cut.

The more chances you give yourself to be surprised, the higher your threshold for surprise had better be. If a random Internet stranger who eliminated all North American grains from his food intake reports that he dropped fifteen pounds and his eczema went away, you shouldn't take that as powerful evidence in favor of the maize-free plan. Somebody's selling a book about that plan, and thousands of people bought that book and tried it, and the odds are very good that, by chance alone, one among them will experience some weight loss and clear skin the next week. And that's the guy who's going to log in as **saygoodbye2corn452** and post his excited testimonial, while the people for whom the diet failed stay silent.

The really surprising result of Bennett's paper isn't that one or two voxels in a dead fish passed a statistical test; it's that a substantial proportion of the neuroimaging articles he surveyed *didn't* use statistical safeguards (known as "multiple comparisons correction") that take into account the ubiquity of the improbable. Without those corrections, scientists are at serious risk of running the Baltimore stockbroker con, not only on their colleagues but on themselves. Getting excited about the fish voxels that matched the photos and ignoring the rest is as potentially dangerous as getting excited about the successful series of stock newsletters while ignoring the many more editions that blew their calls and went in the trash.

REVERSE ENGINEERING, OR, WHY ALGEBRA IS HARD

There are two moments in the course of education where a lot of kids fall off the math train. The first comes in the elementary grades, when fractions are introduced. Until that moment, a number is a *natural number*, one of the figures 0, 1, 2, 3 . . . It is the answer to a question of the form "how many."[*] To go from this notion, so primitive that many animals are said to understand it, to the radically broader idea that a number can mean "what portion of," is a drastic philosophical shift. ("God made the natural numbers," the nineteenth-century algebraist Leopold Kronecker famously said, "and all the rest is the work of man.")

The second dangerous twist in the track is algebra. Why is it so hard? Because, until algebra shows up, you're doing numerical computations in a straightforwardly algorithmic way. You dump some numbers into the addition box, or the multiplication box, or even, in traditionally minded schools, the long-division box, you turn the crank, and you report what comes out the other side.

Algebra is different. It's computation backward. When you're asked to solve

---

[*] There is a long-standing and profoundly unimportant controversy about whether the term "natural number" ought to be defined to include 0 or not. Feel free to pretend I didn't say "0," if you are a die-hard antizeroist.

$$x + 8 = 15$$

you know what came *out* of the addition box (namely, 15) and you're being asked to reverse-engineer what, along with the 8, went in.

In this case, as your seventh-grade math teacher no doubt once told you, you can flip things over to get things right-side round again:

$$x = 15 - 8$$

at which point you can just toss 15 and 8 in the subtraction box (making sure now to keep track of which one you toss first . . .) and find that x must be 7.

But it's not always so easy. You might need to solve a *quadratic equation*, like

$$x^2 - x = 1.$$

Really? (I hear you cry.) *Might* you? Other than that your teacher asked you to, why would you?

Think back to that missile from chapter 2, still traveling furiously toward you:

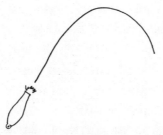

Maybe you know that the missile launched from 100 meters above ground level, with upward velocity of 200 meters per second. If there was no such thing as gravity, the missile would just keep on rising along a straight line in accordance with Newton's laws, getting 200 meters higher each second, and its height after x seconds would be described by the linear function

$$height = 100 + 200x.$$

But there *is* such a thing as gravity, which bends the arc and forces the missile to curve back toward earth. It turns out that the effect of gravity is described by adding a quadratic term:

$$\text{height} = 100 + 200x - 5x^2$$

where the quadratic term is *negative* just because gravity pushes missiles down, not up.

There are a lot of questions you might ask about a missile heading toward you, but one of particularly great import is: When will it land? To answer this is just to answer the question: When will the height of the missile be zero? That is, for what value of x is it the case that

$$100 + 200x - 5x^2 = 0?$$

It is by no means clear how you're supposed to "flip" this equation around and solve for x. But maybe you don't have to. Trial and error is a very powerful weapon. If you plug x = 10 into the above formula, to see how high the missile is after 10 seconds, you get 1,600 meters. Plug in x = 20 and you get 2,100 meters, so it looks like the missile may still be rising. When x = 30, you get 1,600 again: promising; we must be past the peak. At x = 40 the missile is once again just 100 meters above the ground. We could walk forward 10 more seconds, but when we're so close to impact already that's surely overdoing it. If you plug in x = 41 you get −105 meters, which doesn't mean you're predicting the missile has actually begun burrowing under the earth's surface, but rather that impact has already happened, so that your nice, clean model of the missile's motion is, as we say in ballistics, no longer operative.

So if 41 seconds is too long, what about 40.5? That gives −1.25 meters, just a little bit below 0. Turn back the clock a little to 40.4, and you get 19.2m, so impact hasn't happened yet. 40.49? Very close, just 0.8m above the ground. . . .

You can see that by playing the trial and error game, carefully turning the time knob back and forth, you can approximate the time of impact as closely as you like.

But have we "solved" the equation? You're probably hesitant to say you have—after all, even if you keep fine-tuning your guesses until you get the time of impact pinned down to

40.4939015319 . . .

seconds after launch, you don't know *the* answer, but just an *approximation* of the answer. In practice, though, it doesn't help you to time the impact to the millionth of a second, does it? Probably just saying "about 40 seconds" is enough. Try to generate an answer any more precise than that and you're wasting your time, and you'll probably be wrong, besides, because our very simple model of the missile's progress fails to take into account many other factors, like air resistance, the *variation* in air resistance coming from weather, the spin of the missile itself, and so on. These effects may be small, but they're surely big enough to keep you from knowing down to the microsecond when the projectile will show up for its appointment with the ground.

If you want a satisfyingly exact solution, never fear—the quadratic formula is here to help. You may well have memorized this formula once in your life, but unless you have an unusually gifted memory or you are twelve, you don't have it in mind just at the moment. So here it is: if x is a solution to

$$c + bx + ax^2 = 0$$

where a, b, and c are any numbers whatsoever, then

$$x = -\frac{1}{2a}\left(b \pm \sqrt{b^2 - 4ac}\right)$$

In the case of the missile, c = 100, b = 200, and a = −5. So what the quadratic formula has to say about x is that

$$x = \frac{1}{10}\left(200 \pm \sqrt{200^2 + 4 \cdot 5 \cdot 100}\right)$$

Most of the symbols in there are things you could type in your calculator, but there's one funny outlier, the ±. It looks like a plus sign and a minus sign that love each other very much, and this isn't so far off. It indicates that, although we started our mathematical sentence, all confidence, with

$$x =$$

we end up in a state of ambivalence. The ±, something like a blank Scrabble tile, can be read as either a + or a –, as we choose. Each choice we make produces a value of x that makes the equation $100 + 200x - 5x^2$ = 0 hold. There is no single solution to this equation. There are two.

That there are two values of x which satisfy the equation can be made apparent to the eye, even if you long ago forgot the quadratic formula. You can draw a graph of the equation $y = 100 + 200x - 5x^2$ and get a nice upside-down parabola, like this:

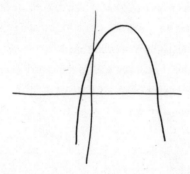

The horizontal line is the x-axis, those points on the plane whose y-coordinate is 0. When the curve $y = 100 + 200x - 5x^2$ meets the x-axis, it must be the case both that y is $100 + 200x - 5x^2$ and that y = 0; so $100 + 200x - 5x^2 = 0$, precisely the equation we were trying to solve, now given geometric form as a question about the intersection between a curve and a horizontal line.

And geometric intuition demands that if such a parabola noses its way above the x-axis at all, it must strike the x-axis in exactly two places, no more, no fewer. In other words, there are two values of x such that $100 + 200x - 5x^2 = 0$.

So what are these two values?

If we choose to read ± as +, we get

$$x = 20 + 2\sqrt{105}$$

which is 40.4939015319 . . . , the same answer we came up with by trial and error. But if we choose −, we get

$$x = 20 - 2\sqrt{105}$$

which is −0.4939015319 . . .

As an answer to our original question, this is somewhat nonsensical. The answer to "When is that missile going to hit me?" can't be "Half a second ago."

Yet this negative value of x is a perfectly good solution to the equation, and when math tells us something we should at least try to listen. What does the negative number mean? Here's one way to understand it. We said the missile was launched from 100 meters off the ground, at a velocity of 200 meters per second. But all we really used was that, at time 0, the missile was traveling upward at that velocity from that position. What if that wasn't actually the launch? Maybe the launch took place, not at time 0, from 100 meters up, but at some earlier time, directly from the ground. What time?

The computation tells us: there are exactly two times when the missile is at ground level. One time is 0.4939 . . . seconds ago. That's when the missile was launched. The other time is 40.4939 . . . seconds from now. That's when the missile lands.

Perhaps it doesn't seem so troubling, especially if you're used to the quadratic formula, to get two answers to the same question. But when you're twelve it represents a real philosophical shift. You've spent six long years in grade school figuring out what *the* answer is, and now, suddenly, there is no such thing.

And those are just quadratic equations! What if you have to solve

$$x^3 + 2x^2 - 11x = 12?$$

This is a *cubic* equation, which is to say it involves x raised to the third power. Fortunately, there *is* a cubic formula that allows you to figure out, by a direct computation, what values of x could have gone in the box to make 12 fall out when you turn the crank. But you didn't learn the cubic formula in school, and the reason you didn't learn it in school is that it's kind of a mess, and wasn't worked out until the late Renaissance, when itinerant algebraists roamed across Italy, engaging each other in fierce public equation-solving battles with money and status on the line. The few people who knew the cubic formula kept it to themselves or wrote it down in cryptic rhymed verse.

Long story. The point is, reverse engineering is hard.

The problem of inference, which is what the Bible coders were wrestling with, is hard because it's exactly this kind of problem. When we are scientists, or Torah scholars, or toddlers gaping at the clouds, we are presented with *observations* and asked to build *theories*—what went into the box to produce the world that we see? Inference is a hard thing, maybe the hardest thing. From the shape of the clouds and the way they move we struggle to go backward, to solve for x, the system that made them.

DEFEATING THE NULL

We've been circling around the fundamental question: How surprised should I be by what I see in the world? This is a book about math, and you must suspect that there's a numerical way to get at this. There is. But it is fraught with danger. We need to talk about *p-values*.

But first we need to talk about improbability, about which we've been unacceptably vague so far. There's a reason for that. There are parts of math, like geometry and arithmetic, that we teach to children and that children, to some extent, teach themselves. Those are the parts that are closest to our native intuition. We are born almost knowing how to count, and how to categorize objects by their location and shape, and the formal, mathematical renditions of these concepts are not so different from the ones we start with.

Probability is different. We certainly have built-in intuition for thinking about uncertain things, but it's much harder to articulate. There's a reason that the mathematical theory of probability came so late in mathematical history, and appears so late in the math curriculum, when it appears at all. When you try to think carefully about what probability *means*, you get a little woozy. When we say, "The probability that a flipped coin will land heads is 1/2," we're invoking the Law of Large Numbers from chapter 4, which says that if you flip the coin many, many times, the proportion of heads will almost inevitably approach 1/2, as if constrained by a narrowing channel. This is what's called the *frequentist view of probability*.

But what can we mean when we say, "The probability that it will rain tomorrow is 20%"? Tomorrow only happens once; it's not an experiment we can repeat like a coin flip again and again. With some effort, we can shoehorn the weather into the frequentist model; maybe we mean that among some large population of days with conditions similar to this one, the following day was rainy 20% of the time. But then you're stuck when asked, "What's the probability that the human race will go extinct in the next thousand years?" This is, almost by definition, an experiment you can't repeat. We use probability even to talk about events that cannot possibly be thought of as subject to chance. What's the probability that consuming olive oil prevents cancer? What's the probability that Shakespeare was the author of Shakespeare's plays? What's the probability that God wrote the Bible and cooked up the earth? It's hard to license talking about these things in the same language we use to assess the outcomes of coin flips and dice rolls. And yet—we find ourselves able to say, of questions like this, "It seems improbable" or "It seems likely." Once we've done so, how can we resist the temptation to ask, *"How* likely?"

It's one thing to ask, another to answer. I can think of no experiment that directly assesses the likelihood that the Man Upstairs actually is Upstairs (or is a Man, for that matter). So we have to do the next best thing—or at least, what traditional statistical practice holds to be the next best thing. (As we'll see, there's controversy on this point.)

We said it was improbable that the names of medieval rabbis are hidden in the letters of the Torah. But is it? Many religious Jews start from the view that everything there is to know is contained, somehow or other, in the Torah's words. If that's the case, the presence of the rabbis' names and birthdays there is not improbable at all; indeed, it's almost required.

You can tell a similar story about the North Carolina lottery. It sounds improbable that an identical set of winning numbers would come up twice in a single week. And that's true, if you agree with the hypothesis that the numbers are drawn from the cage completely at random. But maybe you don't. Maybe you think the randomization system is malfunctioning, and the numbers 4, 21, 23, 34, 39 are more likely to come up than others. Or maybe you think a corrupt lottery official is picking the numbers to match his own favorite ticket. Under either of those hypotheses, the amazing coincidence is not improbable at all. Improbability, as described here, is a *relative* notion, not an absolute one; when we say an outcome is improbable, we are always saying, explicitly or not, that it is improbable under some set of hypotheses we've made about the underlying mechanisms of the world.

Many scientific questions can be boiled down to a simple yes or no: Is something going on, or not? Does a new drug make a dent in the illness it proposes to cure, or does it do nothing? Does a psychological intervention make you happier/peppier/sexier or does it do nothing at all? The "does nothing" scenario is called the *null hypothesis*. That is, the null hypothesis is the hypothesis that the intervention you're studying has no effect. If you're the researcher who developed the new drug, the null hypothesis is the thing that keeps you up at night. Unless you can rule it out, you don't know whether you're on the trail of a medical breakthrough or just barking up the wrong metabolic pathway.

So how do you rule it out? The standard framework, called the *null hypothesis significance test*, was developed in its most commonly used form by R. A. Fisher, the founder of the modern practice of statistics,* in the early twentieth century.

It goes like this. First, you have to run an experiment. You might start with a hundred subjects, then randomly select half to receive your proposed wonder drug while the other half gets a placebo. Your hope, obviously, is that the patients on the drug will be less likely to die than the ones getting the sugar pill.

---

* You might object here that Fisher's methods are *statistics*, not *mathematics*. I am the child of two statisticians and I know that the disciplinary boundary between the two is real. But for our purposes, I'm going to treat statistical thinking as a species of mathematical thinking, and make the case for both.

From here, the protocol might seem simple: if you observe fewer deaths among the drug patients than the placebo patients, declare victory and file a marketing application with the FDA. But that's wrong. It's not enough that the data be consistent with your theory; they have to be *inconsistent* with the negation of your theory, the dreaded null hypothesis. I may assert that I possess telekinetic abilities so powerful that I can drag the sun out from beneath the horizon—if you want proof, just go outside at about five in the morning and see the results of my work! But this kind of evidence is no evidence at all, because, under the null hypothesis that I lack psychic gifts, the sun would come up just the same.

Interpreting the result of a clinical trial requires similar care. Let's make this numerical. Suppose we're in null hypothesis land, where the chance of death is exactly the same (say, 10%) for the fifty patients who got your drug and the fifty who got the placebo. But that doesn't mean that five of the drug patients die and five of the placebo patients die. In fact, the chance that exactly five of the drug patients die is about 18.5%; not very likely, just as it's not very likely that a long series of coin tosses would yield precisely as many heads as tails. In the same way, it's not very likely that exactly the same number of drug patients and placebo patients expire during the course of the trial. I computed:

13.3% chance equally many drug and placebo patients die
43.3% chance fewer placebo patients than drug patients die
43.3% chance fewer drug patients than placebo patients die.

Seeing better results among the drug patients than the placebo patients says very little, since this isn't at all unlikely even under the null hypothesis that your drug doesn't work.

But things are different if the drug patients do a *lot* better. Suppose five of the placebo patients die during the trial, but none of the drug patients do. If the null hypothesis is right, both classes of patients should have a 90% chance of survival. But in that case, it's highly unlikely that all fifty of the drug patients would survive. The first of the drug patients has a 90% chance; now the chance that not only the first but also the second patient survives is 90% of that 90%, or 81%—and if you want the third patient to survive as well, the chance of that happening is only 90%

of that 81%, or 72.9%. Each new patient whose survival you stipulate shaves a little off the chances, and by the end of the process, where you're asking about the probability that all fifty will survive, the slice of probability that remains is pretty slim:

$(0.9) \times (0.9) \times (0.9) \times \ldots$ fifty times! $\ldots \times (0.9) \times (0.9) =$ 0.00515 . . .

Under the null hypothesis, there's only one chance in two hundred of getting results this good. That's much more compelling. If I claim I can make the sun come up with my mind, and it does, you shouldn't be impressed by my powers; but if I claim I can make the sun *not* come up, and it doesn't, then I've demonstrated an outcome very unlikely under the null hypothesis, and you'd best take notice.

So here's the procedure for ruling out the null hypothesis, in executive bullet-point form:

1. Run an experiment.
2. Suppose the null hypothesis is true, and let p be the probability (under that hypothesis) of getting results as extreme as those observed.
3. The number p is called the *p-value*. If it is very small, rejoice; you get to say your results are *statistically significant*. If it is large, concede that the null hypothesis has not been ruled out.

How small is "very small"? There's no principled way to choose a sharp dividing line between what is significant and what is not; but there's a tradition, which starts with Fisher himself and is now widely adhered to, of taking p = 0.05, or 1/20, to be the threshold.

Null hypothesis significance testing is popular because it captures our intuitive way of reasoning about uncertainty. Why do we find the Bible codes compelling, at least at first glance? Because codes like the ones Witztum uncovered are very unlikely under the null hypothesis that the Torah doesn't know the future. The value of p—the likelihood of finding so many equidistant letter sequences, so accurate in their demographic profiling of notable rabbis—is very close to 0.

Versions of this argument for divine creation predate Fisher's formal development by a great while. The world is so richly structured and so perfectly ordered—how tremendously unlikely it would be for there to be a world like this one, under the null hypothesis that there's no primal designer who put the thing together!

The first person to have a go at making this argument mathematical was John Arbuthnot, royal physician, satirist, correspondent of Alexander Pope, and part-time mathematician. Arbuthnot studied the records of children born in London between 1629 and 1710, and found there a remarkable regularity: in every single one of those eighty-two years, more boys were born than girls. What are the odds, Arbuthnot asked, that such a coincidence could arise, under the null hypothesis that there was no God and all was random chance? Then the probability in any given year that London would welcome more boys than girls would be 1/2; and the p-value, the probability of the boys winning eighty-two times in a row, is

$$(1/2) \times (1/2) \times (1/2) \times. . . 82 \text{ times} . . . \times (1/2)$$

or a little worse than 1 in 4 septillion. In other words, more or less zero. Arbuthnot published his findings in a paper called "An Argument for Divine Providence, Taken from the Constant Regularity Observed in the Births of Both Sexes."

Arbuthnot's argument was widely praised and repeated by clerical worthies, but other mathematicians quickly pointed to flaws in his reasoning. Chief among them was the unreasonable specificity of his null hypothesis. Arbuthnot's data certainly puts the boot to the hypothesis that the sex of children is determined at random, with each child having an equal chance of being born male or female. But why should the chance be equal? Nicholas Bernoulli proposed a different null hypothesis: that the sex of a child is determined by chance, with an 18/35 chance of being a boy and 17/35 of being a girl. Bernoulli's null hypothesis is just as atheistic as Arbuthnot's, and it fits the data perfectly. If you flip a coin 82 times and get 82 heads, you ought to be thinking, "Something is biased about this coin," not "God loves heads."*

Though Arbuthnot's argument wasn't widely accepted, its spirit carried on. Arbuthnot is intellectual father not only to the Bible coders but to the "creation scientists," who argue, even today, that mathematics demands there must be a god, on the grounds that a godless world would be highly unlikely to look like the one we have.†

But significance testing is not restricted to theological apologetics. In some sense, Darwin, the creation scientists' shaggy godless devil, made arguments of substantially the same form on behalf of his own work:

It can hardly be supposed that a false theory would explain, in so satisfactory a manner as does the theory of natural selection, the several large classes of facts above specified. It has recently been objected that this is an unsafe method of arguing; but it is a method used in judging of the common events of life, and has often been used by the greatest natural philosophers.

In other words: if natural selection were false, think how unlikely it would be to encounter a biological world so thoroughly consistent with its predictions!

The contribution of R. A. Fisher was to make significance testing into a formal endeavor, a system by which the significance, or not, of an experimental result was a matter of objective fact. In the Fisherian form, the null hypothesis significance test has been a standard method for assessing the results of scientific research for nearly a century. A standard textbook calls the method "the backbone of psychological research." It's the standard by which we separate experiments into successes and failures. Every time you encounter the results of a medical, psychological, or economic research study, you're very likely reading about something that was vetted by a signficance test.

---

* Arbuthnot saw the propensity for the slight excess of boy children as itself an argument in favor of Providence: someone, or Someone, had to have set the knob just right to make extra infant boys to cancel out the extra adult men killed in wars and accidents.
† We'll assess this argument in more detail in chapter 9.

But the unease Darwin noted about this "unsafe method of arguing" has never really receded. For almost as long as the method has been standard, there have been people who branded it a colossal mistake. Back in 1966, the psychologist David Bakan wrote about the "crisis of psychology," which in his view was a "crisis in statistical theory":

> The test of significance does not provide the information concerning psychological phenomena characteristically attributed to it . . . a great deal of mischief has been associated with its use. . . . To say it "out loud" is, as it were, to assume the role of the child who pointed out that the emperor was really outfitted only in his underwear.

And here we stand, almost fifty years later, with the emperor still in office and still cavorting in the same birthday suit, despite the ever larger and more clamorous group of children broadcasting the news about his state of undress.

THE INSIGNIFICANCE OF SIGNIFICANCE

What's wrong with significance? To start with, there's the word itself. Mathematics has a funny relationship with the English language. Mathematical research articles, sometimes to the surprise of outsiders, are not predominantly composed of numerals and symbols; math is made of words. But the objects we refer to are often entities uncontemplated by the editors at Merriam-Webster. New things require new vocabulary. There are two ways to go. You can cut new words from fresh cloth, as we do when we speak of cohomology, syzygies, monodromy, and so on; this has the effect of making our work look forbidding and unapproachable. More commonly, we adapt existing words for our own purposes, based on some perceived resemblance between the mathematical object to be described and a thing in the so-called real world. So a "group," to a mathematician, is indeed a group of things, but a very special *kind* of group, like the group of whole numbers or the group of symmetries of a geometric figure; we mean by it not just an arbitrary collection of things, like OPEC or ABBA, but rather a collection of things with the property that any pair of them can be combined into a third, as a pair of numbers can be added, or a pair of symmetries can be carried out one after the other.* So too for schemes, bundles, rings, and stacks, mathematical objects which stand in only the most tenuous relation to the ordinary things referred to by those words. Sometimes the language we choose has a pastoral flavor: modern algebraic geometry, for instance, is largely concerned with fields, sheaves, kernels, and stalks. Other times it's more aggressive—it is not at all unusual to speak of an operator killing something, or, for a little more va-voom, annihilating it. I once had an uneasy moment with a colleague in an airport when he made the remark, unexceptional in a mathematical context, that it might be necessary to blow up the plane at one point.

So: significance. In common language it means something like "important" or "meaningful." But the significance test that scientists use doesn't measure importance. When we're testing the effect of a new drug, the null hypothesis is that there is no effect at all; so to reject the null hypothesis is merely to make a judgment that the effect of the drug is not zero. But the effect could still be very small—so small that the drug isn't effective in any sense that an ordinary non-mathematical Anglophone would call significant.

The lexical double booking of "significance" has consequences beyond making scientific papers hard to read. On October 18, 1995, the UK Committee on Safety of Medicines (CSM) issued a "Dear Doctor" letter to nearly 200,000 doctors and public health workers around Great Britain, with an alarming warning about certain brands of "third-generation" oral contraceptives. "New evidence has become available," the letter read, "indicating that the chance of a thrombosis occurring in a vein is increased around two-fold for some types of pill compared with others." A venous thrombosis is no joke; it means a clot is impeding the flow of the blood through the vein. If the clot breaks free, the bloodstream can carry it all the way to your lung, where, under its new identity as a pulmonary embolism, it can kill you.

The Dear Doctor letter was quick to assure readers that oral contra-

---

* And the actual mathematical definition of "group" has still more to it than that—but, sadly, this is another beautiful story we'll have to leave half told.

ception was safe for most women, and no one should stop taking the pill without medical advice. But details like that are easy to lose when the top-line message is "Pills kill." The AP story that ran October 19 led with "The government warned Thursday that a new type of birth control pill used by 1.5 million British women may cause blood clots. . . . It considered withdrawing the pills but decided not to, partly because some women cannot tolerate any other kind of pills."

The public, understandably, freaked out. One general practitioner found that 12% of pill users among her patients stopped taking their contraceptives as soon as they heard the government report. Presumably, many women switched to other versions of the pill not implicated in thrombosis, but any interruption makes the pill less effective. And less-effective birth control means more pregnancies. (What—you thought I was going to say there was a wave of abstinence?) After several successive years of decline, the conception rate in the United Kingdom jumped several percentage points the next year. There were 26,000 more babies conceived in 1996 in England and Wales than there had been one year previously. Since so many of the extra pregnancies were unplanned, that led to a lot more termination, too: 13,600 more abortions than in 1995.

This might seem a small price to pay to avoid a blood clot careening through your circulatory system, wreaking potentially lethal havoc. Think about all the women who were spared from death by embolism by the CSM's warning!

But how many women, exactly, is that? We can't know for sure. But one scientist, a supporter of the CSM decision to issue the warning, said the total number of embolism deaths prevented was "possibly one." The added risk posed by third-generation birth control pills, while significant in Fisher's statistical sense, was not so significant in the sense of public health.

The way the story was framed only magnified the confusion. The CSM reported a *risk ratio*: third-generation pills doubled women's risk of thrombosis. That sounds pretty bad, until you remember that thrombosis is really, really rare. Among women of childbearing age using first- and second-generation oral contraceptives, 1 in 7,000 could expect to suffer a thrombosis; users of the new pill indeed had twice as much risk, 2 in 7,000. But that's still a very small risk, because of this certified math fact: *twice a tiny number is a tiny number.* How good or bad it is to double something depends on how big that something is! Playing ZYMURGY on a double word score on the Scrabble board is a triumph; hitting the same square with NOSE is a waste of a move.

Risk ratios are much easier for the brain to grasp than tiny splinters of probability like 1 in 7,000. But risk ratios applied to small probabilities can easily mislead you. A study by sociologists at CUNY found that infants cared for at in-home day cares or by nannies had a fatality rate seven times that of kids in day-care centers. But before you fire your au pair, consider for a minute that American infants hardly ever die these days, and when they do it's almost never because a caregiver shook them to death. The annual rate of fatal accidents in home-based care was 1.6 per 100,000 babies: a lot higher, indeed, than the rate of 0.23 per 100,000 in day-care centers.* But both numbers are more or less zero. In the CUNY study, only a dozen or so babies a year died in accidents in family day cares, a tiny fraction of the 1,110 U.S. infants who died in accidents overall in 2010 (mostly by strangulation in bedclothes) or the 2,063 who died of sudden infant death syndrome. All things being equal, the results of the CUNY study provide a reason to prefer a day-care center to care in a family home; but all other things are usually *not* equal, and some inequalities matter more than others. What if the scrubbed and city-certified day-care center is twice as far from your house as the slightly questionable family-run in-home day care? Car accidents killed 79 infants in the U.S. in 2010; if your baby ends up spending 20% more time on the road per year thanks to the longer commute, you may have wiped out whatever safety advantage you gained by choosing the fancier day care.

A significance test is a scientific instrument, and like any other instrument, it has a certain degree of precision. If you make the test more sensitive—by increasing the size of the studied population, for example—you enable yourself to see ever-smaller effects. That's the power of the method, but also its danger. The truth is, the null hypothesis, if we take it literally, is probably just about always false. When you drop a powerful drug into a patient's bloodstream, it's hard to believe the intervention has

---

* The paper doesn't address the interesting question of what the corresponding rates are for children in the care of their own parents.

exactly zero effect on the probability that the patient will develop esoph-ageal cancer, or thrombosis, or bad breath. Every part of the body speaks to every other, in a complex feedback loop of influence and control. Everything you do either gives you cancer or prevents it. In principle, if you carry out a powerful enough study, you can find out which it is. But those effects are usually so minuscule that they can be safely ignored. Just because we can detect them doesn't always mean they matter.

If only we could go back in time to the dawn of statistical nomencla-ture and declare that a result passing Fisher's test with a p-value of less than 0.05 was "statistically noticeable" or "statistically detectable" in-stead of "statistically significant"! That would be truer to the meaning of the method, which merely counsels us about the existence of an effect but is silent about its size or importance. But it's too late for that. We have the language we have.*

THE MYTH OF THE MYTH OF THE HOT HAND

We know B. F. Skinner as a psychologist, in many ways *the* modern psy-chologist, the man who stared down the Freudians and led a competing psychology, behaviorism, concerned only with what was visible and what could be measured, requiring no hypotheses about unconscious or, for that matter, conscious motivations. For Skinner, a theory of mind just *was* a theory of behavior, and the interesting projects for psychologists thus did not concern thoughts or feelings at all, but rather the manipula-tion of behavior by means of reinforcement.

Less well known is Skinner's history as a frustrated novelist. Skinner was an English major at Hamilton College and spent much of his time with Percy Saunders, a chemistry professor and aesthete whose house was a kind of literary salon. Skinner read Ezra Pound, and listened to Schubert, and wrote adolescently heated poems ("At night, he stops, breathless / Murmuring to his earthly consort / 'Love exhausts me!'") for

---

* Not everyone has the language we have, of course. Chinese statisticians use 显著 *(xianzhu)* for significance in the statistical sense, which is closer to "notable"—but my Chinese-speaking friends tell me that the word carries a connotation of importance, as the English "significance" does. In Russian, the statistical term for significance is значимый, but the more typical way to express the English-language sense of "significant" would be значительный.

the college literary magazine. He did not take a single psychology course. After college, Skinner attended the Bread Loaf writer's conference, where he wrote "a one-act play about a quack who changed people's personalities with endocrines" and succeeded in pressing several of his short stories on Robert Frost. Frost wrote Skinner a very satisfactory letter praising his stories and counseling: "All that makes a writer is the ability to write strongly and directly from some unaccountable and almost invincible personal prejudice. . . . I take it that everybody has the prejudice and spends some time feeling for it to speak and write from. But most people end as they begin by acting out the prejudices of other people."

Thus encouraged, Skinner moved into his parents' attic in Scranton in the summer of 1926 and set out to write. But Skinner found it was not so easy to find his own personal prejudice, or, having found it, to put it in literary form. His time in Scranton came to nothing; he managed a couple of stories and a sonnet about labor leader John Mitchell, but spent his time mostly building model ships and tuning in to distant signals from Pittsburgh and New York on the radio, then a brand-new procrastination device.

"A violent reaction against all things literary was setting in," he later wrote of this period. "I had failed as a writer because I had nothing important to say, but I could not accept that explanation. It was literature which must be at fault." Or, more bluntly: "Literature must be demolished."

Skinner was a regular reader of the literary magazine *The Dial*; in its pages, he encountered the philosophical writings of Bertrand Russell, and via Russell was brought to John Watson, the first great advocate of the behaviorist outlook that would soon become almost synonymous with Skinner's name. Watson held that scientists were in the business of observing the results of experiments, and only that; there was no room for hypotheses about consciousness or souls. "No one has ever touched a soul or seen one in a test-tube," he famously wrote, by way of dismissing the notion. These uncompromising words must have thrilled Skinner, as he moved to Harvard as a graduate student in psychology, making ready to banish the vague, unruly self from the scientific study of behavior.

Skinner had been much struck by an experience of spontaneous verbal production he'd experienced in his lab; a machine in the background was making a repetitive, rhythmic sound, and Skinner found himself talking along with it, following the beat, silently repeating the phrase "You'll never get out, you'll never get out, you'll never get out." What seemed like speech, or even, in a small way, like poetry, was actually the result of a kind of autonomous verbal process, requiring nothing like a conscious author.* This provided just the idea Skinner needed to settle his score with literature. What if language, even the language of the great poets, was just another behavior, trained by exposure to stimuli, and manipulable in the lab?

In college, Skinner had written imitations of Shakespeare's sonnets; he retrospectively described this experience, in thoroughly behaviorist fashion, as "the strange excitement of emitting whole lines ready-made, properly scanned and rhymed." Now, as a young psychology professor in Minnesota, he recast Shakespeare himself as more emitter than writer. This approach was not as crazy then as it seems now; the dominant form of literary criticism at the time, "close reading," bore the mark of Watson's philosophy just as Skinner did, displaying a very behaviorist preference for the words on the page over the unobservable intentions of the author.

Shakespeare is famous as a master of the alliterative line, in which several words in close succession start with the same sound ("Full fathom five thy father lies . . ."). For Skinner, this argument by example was no kind of science. Did Shakespeare alliterate? If he did, then math could prove it so. "Proof that there is a process responsible for alliterative patterning," he wrote, "can be obtained only through a statistical analysis of all the arrangements of initial consonants in a reasonably large sample." And what form of statistical analysis? None other than a form of Fisher's p-value test. Here, the null hypothesis is that Shakespeare paid no heed to the initial sounds of words at all, so that the first letter of one word of poetry has no effect on other words in the same line. The protocol was much like that of a clinical trial, but with one big difference: the biomedical researcher testing a drug hopes with all his heart to see the null hypothesis refuted, and the effectiveness of the medicine demonstrated.

---

* It's said that David Byrne wrote the lyrics to "Burning Down the House" in a very similar way, barking nonsense syllables in rhythm with the instrumental track, then going back and writing down the words that the nonsense reminded him of.

For Skinner, aiming to knock literary criticism off its plinth, the null hypothesis was the attractive one.

Under the null hypothesis, the frequency with which initial sounds appeared multiple times in the same line would be unchanged if the words were put in a sack, shaken up, and laid out again in random order. And this is just what Skinner found in his sample of a hundred sonnets. Shakespeare failed the significance test. Skinner writes:

"In spite of the seeming richness of alliteration in the sonnets, there is no significant evidence of a process of alliteration in the behavior of the poet to which any serious attention should be given. So far as this aspect of poetry is concerned, Shakespeare might as well have drawn his words out of a hat."

"Seeming richness"—what chutzpah! It captures perfectly the spirit of the psychology that Skinner wanted to create. Where Freud had claimed to see what had previously been hidden, repressed, or obscured, Skinner wanted to do the opposite—to deny the existence of what seemed in plain view.

But Skinner was wrong; he hadn't proved that Shakespeare didn't alliterate. A significance test is an instrument, like a telescope. And some instruments are more powerful than others. If you look at Mars with a research-grade telescope, you'll see moons; if you look with binoculars, you won't. But the moons are still there! And Shakespeare's alliteration is still there. As documented by literary historians, it was a standard device of the time, known to and consciously deployed by nearly everyone writing in English.

What Skinner had proved is that Shakespeare's alliteration did not produce a surplus of repeated sounds so great as to show up on his test. But why would it? The use of alliteration in poetry is both positive and negative; in certain places you alliterate to create an effect, and in other places you intentionally avoid it, lest you create an effect you don't want. It may be that the overall tendency is to increase the number of alliterative lines, but even if so, the increase should be small. Stuff your sonnets with one or two extra alliterations each and you become one of the stone-footed poets mocked by Shakespeare's fellow Elizabethan George Gascoigne: "Many writers indulge in repeticion of sundrie wordes all beginning with one letter, the whiche (beyng modestly used) lendeth good grace to a verse; but they do so hunt a letter to death, that they make it *Crambe*, and *Crambe bis positum mors est*."

The Latin phrase means "Cabbage served twice is death." Shakespeare's writing is rich in effect, but always restrained. He would never pack in so much cabbage that Skinner's crude test could smell it.

A statistical study that's not refined enough to detect a phenomenon of the expected size is called *underpowered*—the equivalent of looking at the planets with binoculars. Moons or no moons, you get the same result, so you might as well not have bothered. You don't send binoculars to do a telescope's job. The problem of low power is the flip side to the problem of the British birth control scare. A high-powered study, like the birth control trial, may lead you to burst a vein about a small effect that isn't actually important. An underpowered one may lead you to wrongly dismiss a small effect that your method was simply too weak to see.

Consider Spike Albrecht. The freshman guard for Michigan's men's basketball team, standing at just five foot eleven and a bench player most of the season, wasn't expected to play a big role when the Wolverines faced Louisville in the 2013 NCAA final. But Albrecht made five straight shots, four of them three-pointers, in a ten-minute span in the first half, leading Michigan to a ten-point lead over the heavily favored Cardinals. He had what basketball fans call "the hot hand"—the apparent inability to miss a shot, no matter how great the distance or how fierce the defense.

Except there's supposed to be no such thing. In 1985, in one of the most famous contemporary papers in cognitive psychology, Thomas Gilovich, Robert Vallone, and Amos Tversky (hereafter GVT) did to basketball fans what B. F. Skinner had done to lovers of the Bard. They obtained records of every shot taken by the 1980–81 Philadelphia 76ers in their forty-eight home games and analyzed them statistically. If players tended toward hot streaks and cold streaks, you might expect a player to be more likely to hit a shot following a basket than a shot following a miss. And when GVT surveyed NBA fans, they found this theory had broad support; nine out of ten fans agreed that a player is more likely to sink a shot when he's just hit two or three baskets in a row.

But nothing of the kind was going on in Philadelphia. Julius Erving, the great Dr. J, was a 52% shooter overall. After three straight baskets, a situation that you'd think might indicate Erving was hot, his percentage went down to 48%. And after three straight misses, his field goal percentage didn't drop, but rather stayed right at 52%. For other players, like Darryl "Chocolate Thunder" Dawkins, the effect was even more extreme. After a hit, his overall 62% shooting percentage dipped to 57%; after a miss, it shot up to 73%, exactly the opposite of the fan predictions. (One possible explanation: a missed shot suggests Dawkins was facing effective defenders on the perimeter, inducing him to drive to the basket for one of his trademark backboard-shattering dunks, which he gave names like "In Your Face Disgrace" and "Turbo Sexophonic Delight.")

Does this mean there's no such thing as the hot hand? Not just yet. The hot hand, after all, isn't generally thought of as a universal tendency for hits to follow hits and misses to follow misses. It's an evanescent thing, a brief possession by a superior basketball being that inhabits a player's body for a short glorious interval on the court, giving no warning of its arrival or departure. Spike Albrecht is Ray Allen for ten minutes, mercilessly raining down threes, then he's Spike Albrecht again. Can a statistical test see this? In principle, why not? GVT devised a clever way to check for these short intervals of unstoppability. They broke up each player's season into sequences of four shots each; so if Dr. J's sequence of hits and misses looked like

HMHHHMHMMHHHHMMH

the sequences would be

HMHH, HMHM, MHHH, HMMH . . .

GVT then counted how many of the sequences were "good" (3 or 4 hits), "moderate" (2 hits), or "bad" (0 or 1 hits) for each of the nine players in the study. And then, good Fisherians, they considered the results of the null hypothesis—that there's no such thing as the hot hand.

There are sixteen possible sequences of four shots: the first shot can be either H or M, and for each of these options there are two possibilities for the second shot, giving us four options in all for the first two shots (here they are: HH, HM, MH, MM) and for each of *these* four there are two possibilities for the third shot, giving eight possible three-shot sequences, and doubling once more to account for the last shot in the sequence we get 16. Here they all are, divided into the good ones, the moderate ones, and the bad ones:

Good: HHHH, MHHH, HMHH, HHMH, HHHM
Moderate: HHMM, HMHM, HMMH, MHHM, MHMH, MMHH
Bad: HMMM, MHMM, MMHM, MMMH, MMMM

For a 50% shooter like Dr. J, all 16 possible sequences should then be equally likely, because each shot is equally likely to be an H or an M. So you'd expect about 5/16, or 31.25%, of Dr. J's four-shot sequences to be good, with 37.5% moderate and 31.25% bad.

But if Dr. J sometimes experienced the hot hand, you might expect a higher proportion of good sequences, contributed by those games where he just can't seem to miss. The more prone to hot and cold streaks you are, the more you're going to see HHHH and MMMM, and the less you're going to see HMHM.

The significance test asks us to address the following question: if the null hypothesis were correct and there were no hot hand, would we be unlikely to see the results that were actually observed? And the answer turns out to be no. The proportion of good, bad, and moderate sequences in the actual data is just about what chance would predict, any deviation falling well short of the statistically significant.

"If the present results are surprising," GVT write, "it is because of the robustness with which the erroneous belief in the 'hot hand' is held by experienced and knowledgeable observers." And indeed, while their result was quickly taken up as conventional wisdom by psychologists and economists, it has been slow to gain traction in the basketball world. This didn't faze Tversky, who relished a good fight, whatever the outcome. "I've been in a thousand arguments over this topic," he said. "I've won them all, and I've convinced no one."

But GVT, like Skinner before them, have answered only half the question: namely, what if the null hypothesis is true, and there is no hot hand? Then, as they demonstrate, the results would look very much like the ones observed in the real data.

But what if the null hypothesis is wrong? The hot hand, if it exists, is brief, and the effect, in strictly numerical terms, is small. The worst shooter in the league hits 40% of his shots and the best hits 60%; that's a big difference in basketball terms, but not so big statistically. What would the shot sequences look like if the hot hand were real?

Computer scientists Kevin Korb and Michael Stillwell worked out exactly that in a 2003 paper. They generated simulations with a hot hand built in: the simulated player's shooting percentage leaped up all the way to 90% for two ten-shot "hot" intervals over the course of the trial. In more than three-quarters of those simulations, the significance test used by GVT reported that there was no reason to reject the null hypothesis—*even though the null hypothesis was completely false*. The GVT design was underpowered, destined to report the nonexistence of the hot hand even if the hot hand was real.

If you don't like simulations, consider reality. Not all teams are equal when it comes to preventing shots; in the 2012–13 season, the stingy Indiana Pacers allowed opponents to make only 42% of their shots, while 47.6% of shots fell in against the Cleveland Cavaliers. So players really do have "hot spells" of a rather predictable kind: namely, they're more likely to hit a shot when they're playing the Cavs. But this mild heat—maybe we should call it "the warm hand"—is something the tests used by Gilovich, Vallone, and Tversky aren't sensitive enough to feel.

The right question isn't "Do basketball players sometimes temporarily get better or worse at making shots?"—the kind of yes/no question a significance test addresses. The right question is "How *much* does their ability vary with time, and to what extent can observers detect in real time whether a player is hot?" Here, the answer is surely "not as much as people think, and hardly at all." A recent study found that players who make the first of two free throws become slightly more likely to make the next one, but there's no convincing evidence supporting the hot hand in real-time game play, unless you count the subjective impressions of the players and coaches. The short life of the hot hand, which makes it so hard to disprove, makes it just as hard to reliably detect. Gilovich, Vallone, and Tversky are absolutely correct in their central contention that human beings are quick to perceive patterns where they don't exist and to overestimate their strength where they do. Any regular hoops watcher will routinely see one player or another sink five shots in a row. Most of the time, surely, this is due to some combination of indifferent defense, wise shot selection, or, most likely of all, plain good luck, not a sudden burst of basketball transcendence. Which means there's no reason to expect a guy who's just hit five in a row to be particularly likely to make the next one. Analyzing the performance of investment advisors presents the same problem. Whether there is such a thing as skill in investing or whether differences in performance between different funds are wholly due to luck has been a vexed, murky, unsettled question for years. But if there are investors with a temporary or permanent hot hand, they're rare, so rare that they make little to no dent in the kind of statistics contemplated by GVT. A fund that's beaten the market five years running is vastly more likely to have been lucky than good. Past performance is no guarantee of future returns. If Michigan fans were counting on Spike Albrecht to carry the team all the way to a championship, they were badly disappointed; Albrecht missed every shot he took in the second half, and the Wolverines ended up losing by 6.

A 2009 study by John Huizinga and Sandy Weil suggests that it might be a good idea for players to disbelieve in the hot hand, even if it really exists! In a much larger data set than GVT's, they found a similar effect; after making a basket, players were less likely to succeed on their next shot. But Huizinga and Weil had records of not only shot success but shot location. And *that* data showed a striking potential explanation; players who had just made a shot were more likely to take a more difficult shot on their next attempt. Yigal Attali, in 2013, found even more intriguing results along these lines. A player who made a layup was no more likely to shoot from distance than a player who just missed a layup. Layups are easy and shouldn't give the player a strong sense of being hot. But a player is much more likely to try a long shot after a three-point basket than after a three-point miss. In other words, the hot hand might "cancel itself out"—players, believing themselves to be hot, get overconfident and take shots they shouldn't.

The nature of the analogous phenomenon in stock investment is left as an exercise for the reader.

EIGHT

REDUCTIO AD UNLIKELY

The stickiest philosophical point in a significance test comes right at the beginning, before we run any of the sophisticated algorithms developed by Fisher and honed by his successors. It's right there at the beginning of step 2:

"Suppose the null hypothesis is true."

But what we're trying to prove, in most cases, is that the null hypothesis *isn't* true. The drug works, Shakespeare alliterates, the Torah knows the future. It seems very logically fishy to assume exactly what we're aiming to disprove, as if we're in danger of making a circular argument.

On this point, you can rest easy. Assuming the truth of something we quietly believe to be false is a time-honored method of argument that goes all the way back to Aristotle; it is the proof by contradiction, or reductio ad absurdum. The reductio is a kind of mathematical judo, in which we first affirm what we wish eventually to deny, with the plan of throwing it over our shoulder and defeating it by means of its own force. If a hypothesis implies a falsehood,* then the hypothesis itself must be false. So the plan goes like this:

---

* Some people will insist on the distinction that the argument is only a reductio if the consequence of the hypothesis is self-contradictory, while if the consequence is merely false the argument is a modus tollens.

- Suppose the hypothesis H is true.
- It follows from H that a certain fact F cannot be the case.
- But F is the case.
- Therefore, H is false.

Say someone exclaims to you that two hundred children were killed by gunfire in the District of Columbia in 2012. That's a hypothesis. But it might be somewhat hard to check (by which I mean that I typed "number of children killed by guns in DC in 2012" into the Google search bar and did not immediately learn the answer). On the other hand, if we assume the hypothesis is correct, then there cannot have been any fewer than two hundred homicides in total in DC in 2012. But there *were* fewer; in fact, there were only eighty-eight. So the exclaimer's hypothesis must have been wrong. There's no circularity here; we've "assumed" the false hypothesis in a kind of tentative, exploratory way, setting up the counterfactual mental world in which H is so and then watching it collapse under pressure from reality.

Put this way, the reductio sounds almost trivial, and in a sense, it is; but maybe it's more accurate to say it's a mental tool we've grown so used to handling that we forget how powerful it is. In fact, it's a simple reductio that drives the Pythagoreans' proof of the irrationality of the square root of 2; the one so awesomely paradigm-busting they had to kill its author; a proof so simple, refined, and compact that I can write it out whole in a page.

Suppose

H: the square root of 2 is a rational number

that is, $\sqrt{2}$ is a fraction m/n where m and n are whole numbers. We might as well write this fraction in *lowest terms*, which means that if there is a common factor between the numerator and denominator, we divide it out of both, leaving the fraction unchanged: no reason to write 10/14 instead of the simpler 5/7. So let's rephrase our hypothesis:

H: the square root of 2 is equal to m/n, where m and n are whole numbers with no factor in common.

In fact, this means we can be sure it's not the case that m and n are both even; for to say both numbers are even is exactly to say both have 2 as a factor. In that case, as in the case of 10/14, we could divide both numerator and denominator by 2 without changing the fraction, which is to say it was not in lowest terms after all. So

F: both m and n are even

is false.

Now since $\sqrt{2} = m/n$, then by squaring both sides we see that $2 = m^2 / n^2$ or, equivalently, that $2n^2 = m^2$. So m^2 is an even number, which means that m itself is even. A number is even just when it can be written as twice another whole number; so we can, and do, write m as 2k for some whole number k. Which means that $2n^2 = (2k)^2 = 4k^2$. Dividing both sides by 2, we find that $n^2 = 2k^2$.

What's the point of all this algebra? Simply to show that n^2 is twice k^2, and therefore an even number. But if n^2 is even, so must n be, just like m is. But that means that F is true! By assuming H we have arrived at a false-hood, even an absurdity; that F is false and true at once. So H must have been wrong. The square root of 2 is *not* a rational number. By assuming it was, we proved that it wasn't. It's a weird trick indeed, but it works.

You can think of the null hypothesis significance test as a sort of fuzzy version of the reductio:

- Suppose the null hypothesis H is true.
- It follows from H that a certain outcome O is very improbable (say, less than Fisher's 0.05 threshold).
- But O was actually observed.
- Therefore, H is very improbable.

Not a reductio ad absurdum, in other words, but a reductio ad un-likely.

A classical example comes from the eighteenth-century astronomer and clergyman John Michell, among the first to take a statistical ap-proach to the study of the heavenly bodies. The cluster of dim stars in one corner of the constellation Taurus has been observed by just about every civilization. The Navajo call them Dilyehe, "the sparkling figure"; the Maori call them Matariki, "the eyes of god." To the ancient Romans they were a bunch of grapes and in Japanese they're Subaru (in case you ever wondered where the car company's six-star logo came from). We call them the Pleiades.

All these centuries of observation and mythmaking couldn't answer the fundamental scientific question about the Pleiades: is the cluster actually a cluster? Or are the six stars separated by unfathomable distances, but arrayed by chance in almost the exact same direction from Earth? Points of light, placed at random in our frame of vision, look something like this:

You see some clumps, right? That's to be expected: there will inevitably be some groups of stars that wind up almost on top of one another, simply by happenstance. How can we be sure that's not what's going on with the Pleiades? It's the same phenomenon Gilovich, Vallone, and

Tversky pointed out: a perfectly consistent point guard, who enjoys no hot streaks and suffers no slumps, will nonetheless sometimes nail five shots in a row.

In fact, if there were no big visible clusters of stars, as in this picture:

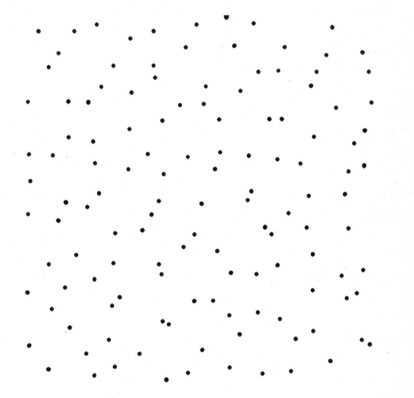

that itself would be evidence that some nonrandom process was at work. The second picture might look "more random" to the naked eye, but it is not; it testifies that the points have a built-in disinclination to crowd.

So the mere appearance of an apparent cluster shouldn't convince us that the stars in question are actually clumped together in space. On the other hand, a group of stars in the sky might be so tightly packed as to demand that one doubt it could have happened by chance. Michell showed that, were visible stars randomly strewn around in space, the chance that six would array themselves so neatly as to present a Pleiades-like cluster to our eyes was small indeed; about 1 in 500,000, by his computation. But there they are above us, the tightly packed bunch of grapes. Only a fool, Michell concluded, could believe it had happened by chance.

Fisher wrote approvingly of Michell's work, making explicit the analogy he saw there between Michell's argument and the classical reductio:

"The force with which such a conclusion is supported is logically that of a simple disjunction: Either an exceptionally rare chance has occurred, or the theory of random distribution is not true."

The argument is compelling, and its conclusion correct; the Pleiades are indeed no optical coincidence, but a real cluster—of several hundred adolescent stars, not just the six visible to the eye. The fact that we see many very tight clusters of stars like the Pleiades, much tighter than would be likely to exist by chance, is good evidence that the stars are not placed randomly, but rather are clumped by some real physical phenomenon out there in the void.

But here's the bad news: the reductio ad unlikely, unlike its Aristotelian ancestor, is not logically sound in general. It leads us into its own absurdities. Joseph Berkson, the longtime head of the medical statistics division at the Mayo Clinic, who cultivated (and loudly broadcast) a vigorous skepticism about methodology he thought shaky, offered a famous example demonstrating the pitfalls of the method. Suppose you have a group of fifty experimental subjects, who you hypothesize (H) are human beings. You observe (O) that one of them is an albino. Now, albinism is extremely rare, affecting no more than one in twenty thousand people. So given that H is correct, the chance you'd find an albino among your fifty subjects is quite small, less than 1 in 400,* or 0.0025. So the p-value, the probability of observing O given H, is much lower than .05.

We are inexorably led to conclude, with a high degree of statistical confidence, that H is incorrect: the subjects in the sample are not human beings.

It's tempting to think of "very improbable" as meaning "essentially impossible," and, from there, to utter the word "essentially" more and

* As a good rule of thumb, you can figure that each of the fifty subjects contributes a 1/20,000 chance of finding an albino in the sample, yielding 1/400; this isn't exactly right, but is usually close enough in cases like this one, where the result is very close to 0.

more quietly in our mind's voice until we stop paying attention to it.* But impossible and improbable are not the same—not even close. Impossible things never happen. But improbable things happen a lot. That means we're on quivery logical footing when we try to make inferences from an improbable observation, as reductio ad unlikely asks us to. That time in North Carolina when the lottery combo 4, 21, 23, 34, 39 came up twice in a week raised a lot of questions; was something wrong with the game? But each combination of numbers is exactly as likely to come up as any other. For the numbers to show 4, 21, 23, 34, 39 on Tuesday and 16, 17, 18, 22, 39 on Thursday is precisely as improbable as what actually took place—there's just one chance in 300 billion or so of getting those two draws on those two days. In fact, *any* particular outcome of the Tuesday and Thursday lottery draws is a one in 300 billion shot. If you're committed to the view that a highly improbable outcome should lead you to question the fairness of the game, you're going to be the person shooting off an angry e-mail to the lottery commissioner every Thursday of your life, no matter which numbered balls drop out of the cage.

Don't be that person.

PRIME CLUSTERS AND THE STRUCTURE OF STRUCTURELESSNESS

Michell's critical insight, that clusters of stars might appear to our eye even if stars were randomly distributed around our field of vision, doesn't apply only to the celestial sphere. This phenomenon was the hinge for the pilot episode of the math/cop drama *Numb3rs*.† A series of grisly attacks, marked by pins on the wall map at HQ, showed no clusters; ergo, a single cunning serial killer intentionally leaving space between victims, not an unconnected burst of psychos, was at work. It was somewhat contrived as a police story, but mathematically it was perfectly correct.

---

* Indeed, it's a general principle of rhetoric that when someone says "X is essentially Y," they generally mean "X is not Y, but it would be simpler for me if X were Y, so it'd be great if you could just go ahead and pretend X is Y, sound good?"

† Disclosure: I used to read *Numb3rs* scripts in advance to check their mathematical accuracy and provide comments. Only one line I suggested ever made it on the air: "trying to find a projection of affine three-space onto the sphere subject to some open constraints."

The appearance of clusters in random data offers insight even in situations where there is no real randomness at all, like the behavior of prime numbers. In 2013, Yitang "Tom" Zhang, a popular math lecturer at the University of New Hampshire, stunned the world of pure mathematics when he announced that he had proven the "bounded gaps" conjecture about the distribution of primes. Zhang had been a star student at Beijing University, but had never thrived after moving to the United States for his PhD in the 1980s. He hadn't published a paper since 2001. At one point, he left academic math entirely to sell sandwiches at Subway, until a fellow former student from Beijing tracked him down and helped him get an untenured lectureship at UNH. To all outward appearances, he was washed up. So it came as a great surprise when he released a paper proving a theorem some of the biggest names in number theory had tried, and failed, to conquer.

But the fact that the conjecture is *true* came as no surprise at all. Mathematicians have a reputation of being no-B.S. hard cases who don't believe a thing until it's locked down and proved. That's not quite true. All of us believed the bounded gaps conjecture before Zhang's big reveal, and we all believe the closely related twin primes conjecture, even though it remains unproven. Why?

Let's start with what the two conjectures say. The prime numbers are those numbers greater than 1 that aren't multiples of any number smaller than themselves and greater than 1; so 7 is a prime, but 9 is not, because it's divisible by 3. The first few primes are 2, 3, 5, 7, 11, and 13.

Every positive number can be expressed in just one way as a product of prime numbers. For instance, 60 is made up of two 2s, one 3, and one 5, because $60 = 2 \times 2 \times 3 \times 5$. (This is why we don't take 1 to be a prime, though some mathematicians have done so in the past; it breaks the uniqueness, because if 1 counts as prime, 60 could be written as $2 \times 2 \times 3 \times 5$ and $1 \times 2 \times 2 \times 3 \times 5$ and $1 \times 1 \times 2 \times 2 \times 3 \times 5 \ldots$) What about prime numbers themselves? They're fine; a prime number, like 13, is the product of a *single* prime, 13 itself. And what about 1? We've excluded it from our list of primes, so how can it be a product of primes, each one of which is larger than 1? Simple: 1 is the product of *no* primes.

At this point I'm sometimes asked, "Why is the product of no primes 1, and not 0?" Here's one slightly convoluted explanation: If you take the product of some set of primes, like 2 and 3, but then divide away the very primes you multiplied, you ought to be left with the product of nothing at all; and 6 divided by 6 is 1, not 0. (The *sum* of no numbers, on the other hand, is indeed 0.)

The primes are the atoms of number theory, the basic indivisible entities of which all numbers are made. As such, they've been the object of intense study ever since number theory started. One of the first theorems ever proved in number theory is that of Euclid, which tells us that the primes are infinite in number; we will never run out, no matter how far along the number line we let our minds range.

But mathematicians are greedy types, not inclined to be satisfied with a mere assertion of infinitude. After all, there's infinite and then there's *infinite*. There are infinitely many powers of 2, but they're very rare. Among the first one thousand numbers, there are only ten of them:

1, 2, 4, 8, 16, 32, 64, 128, 256, and 512.

There are infinitely many even numbers, too, but they're much more common: exactly 500 out of the first 1,000 numbers. In fact, it's pretty apparent that out of the first N numbers, just about (1/2)N will be even.

Primes, it turns out, are intermediate—more common than the powers of 2 but rarer than even numbers. Among the first N numbers, about N/log N are prime; this is the Prime Number Theorem, proven at the end of the nineteenth century by the number theorists Jacques Hadamard and Charles-Jean de la Vallée Poussin.

A NOTE ON THE LOGARITHM, AND THE FLOGARITHM

It has come to my attention that hardly anybody knows what the logarithm is. Let me take a step toward fixing this. The logarithm of a positive number N, called *log N*, is the number of digits it has.

Wait, really? That's it?

No. That's not *really* it. We can call the number of digits the "fake logarithm," or *flogarithm*. It's close enough to the real thing to give the general idea of what the logarithm means in a context like this one. The flogarithm (whence also the logarithm) is a very slowly growing function indeed: the flogarithm of a thousand is 4, the flogarithm of a million, a thousand times greater, is 7, and the flogarithm of a billion is still only 10.*

NOW BACK TO PRIME CLUSTERS

The Prime Number Theorem says that, among the first N integers, a proportion of about 1/log N of them are prime. In particular, prime numbers get less and less common as the numbers get bigger, though the decrease is very slow; a random number with twenty digits is half as likely to be prime as a random number with ten digits.

Naturally, one imagines that the more common a certain type of number, the smaller the gaps between instances of that type of number. If you're looking at an even number, you never have to travel farther than two numbers forward to encounter the next even; in fact, the gaps between the even numbers are always exactly of size 2. For the powers of 2, it's a different story. The gaps between successive powers of 2 grow exponentially, getting bigger and bigger with no retreats as you traverse the sequence; once you get past 16, for instance, you will never again see two powers of 2 separated by a gap of size 15 or less.

Those two problems are easy, but the question of gaps between consecutive primes is harder. It's so hard that, even after Zhang's breakthrough, it remains a mystery in many respects.

And yet we think we know what to expect, thanks to a remarkably fruitful point of view: we think of primes as *random numbers.* The reason the fruitfulness of this viewpoint is so remarkable is that the viewpoint is so very, very false. Primes are not random! Nothing about them is arbitrary or subject to chance. Quite the opposite: we take them as immutable features of the universe, and carve them on the golden records we shoot out into interstellar space to prove to the ETs that we're no dopes.

---

* Down here at the bottom of the page I can safely reveal the real definition of log N; it is that number x such that $e^x = N$. Here e is Euler's number, whose value is about 2.71828 . . . I say "e" and not "10" because the logarithm we mean to talk about is the *natural logarithm,* not the *common* or *base-10 logarithm.* The natural logarithm is the one you always use if you're a mathematician or if you have e fingers.

The primes are not random, but it turns out that in many ways *they act as if they were.* For example, when you divide a random whole number by 3, the remainder is either 0, 1, or 2, and each case arises equally often. When you divide a big prime number by 3, the quotient can't come out even; otherwise, the so-called prime would be divisible by 3, which would mean it wasn't really a prime at all. But an old theorem of Dirichlet tells us that remainder 1 shows up about equally as often as remainder 2, just as is the case for random numbers. So as far as "remainder when divided by 3" goes, prime numbers, apart from not being multiples of 3, look random.

What about the gaps between consecutive primes? You might think that, because prime numbers get rarer and rarer as numbers get bigger, that they also get farther and farther apart. On average, that's indeed the case. But what Zhang proved is that there are infinitely many pairs of primes that differ by at most 70 million. In other words, that the gap between one prime and the next is bounded by 70 million infinitely often—thus, the "bounded gaps" conjecture.

Why 70 million? Just because that's what Zhang was able to prove. In fact, the release of his paper set off an explosion of activity, with mathematicians from around the world working together in a "Polymath," a sort of frenzied online math kibbutz, to narrow the gap still more using variations on Zhang's method. By July 2013, the collective had shown that there were infinitely many gaps of size at most 5,414. In November, a just-fledged PhD in Montreal, James Maynard, knocked the bound down to 600, and Polymath scrambled into action to combine his insights with those of the hive. By the time you read this, the bound will no doubt be smaller still.

On first glance, the bounded gaps might seem a miraculous phenomenon. If the primes are tending to be farther and farther apart, what's causing there to be so many pairs that are close together? Is it some kind of prime gravity?

Nothing of the kind. If you strew numbers at random, it's very likely that some pairs will, by chance, land very close together, just as points dropped randomly in a plane form visible clusters.

It's not hard to compute that, if prime numbers behaved like random numbers, you'd see precisely the behavior that Zhang demonstrated.

Even more: you'd expect to see infinitely many pairs of primes that are separated by only 2, like 3-5 and 11-13. These are the so-called twin primes, whose infinitude remains conjectural.

(A short computation follows. If you're not on board, avert your eyes and rejoin the text where it says "And a lot of twin primes . . .")

Remember: among the first N numbers, the Prime Number Theorem tells us that about $N/\log N$ of them are primes. If these were distributed randomly, each number n would have a $1/\log N$ chance of being prime. The chance that n and $n + 2$ are both prime should thus be about $(1/\log N) \times (1/\log N) = (1/\log N)^2$. So how many pairs of primes separated by 2 should we expect to see? There are about N pairs $(n, n + 2)$ in the range of interest, and each one has a $(1/\log N)^2$ chance of being a twin prime, so one should expect to find about $N/(\log N)^2$ twin primes in the interval.

There are some deviations from pure randomness whose small effects number theorists know how to handle. The main point is that n being prime and $n + 1$ being prime are not independent events; n being prime makes it somewhat *more likely* that $n + 2$ is prime, which means our use of the product $(1/\log N) \times (1/\log N)$ isn't quite right. (One issue: if n is prime and bigger than 2, it's odd, which means $n + 2$ is odd as well, which makes $n + 2$ more likely to be prime.) G. H. Hardy, of the "unnecessary perplexities," together with his lifelong collaborator J. E. Littlewood, worked out a more refined prediction taking these dependencies into account, and predicting that the number of twin primes should in fact be about 32% greater than $N/(\log N)^2$. This better approximation gives a prediction that the number of twin primes less than a quadrillion should be about 1.1 trillion, a pretty good match for the actual figure of 1,177,209,242,304. That's a lot of twin primes.

And a lot of twin primes is exactly what number theorists expect to find, no matter how big the numbers get—not because we think there's a deep, miraculous structure hidden in the primes, but *precisely because we don't think so*. We expect the primes to be tossed around at random like dirt. If the twin primes conjecture were false, *that* would be a miracle, requiring that some hitherto unknown force was pushing the primes apart.

Not to pull back the curtain too much, but a lot of famous conjectures in number theory work this way. The Goldbach conjecture, that every even number greater than 2 is the sum of two primes, is another one that would have to be true if primes behaved like random numbers. So is the conjecture that the primes contain arithmetic progressions of any desired length, whose resolution by Ben Green and Terry Tao in 2004 helped win Tao a Fields Medal.

The most famous of all is the conjecture made by Pierre de Fermat in 1637, which asserted that the equation

$$A^n + B^n = C^n$$

has no solutions with A, B, C, and n positive whole numbers with n greater than 2. (When n is *equal* to 2, there are lots of solutions, like $3^2 + 4^2 = 5^2$.)

Everybody strongly believed the Fermat conjecture was true, just as we believe the twin primes conjecture now; but no one knew how to prove it[*] until the breakthrough of Princeton mathematician Andrew Wiles in the 1990s. We believed it because perfect nth powers are very rare, and the chance of finding two numbers that summed to a third in a *random* set of such extreme scarcity is next to nil. Even more: most people believe that there are no solutions to the *generalized Fermat equation*

$$A^p + B^q = C^r$$

when the exponents p, q, and r are big enough. A banker in Dallas named Andrew Beal will give you a million dollars if you can prove that the equation has no solutions for which p, q, and r are all greater than 3 and A, B, and C share no prime factor.[†] I fully believe that the statement is true, because it would be true if perfect powers were random; but I think we'll have to understand something truly new about numbers before we can make our way to a proof. I spent a couple of years, along with a bunch of collaborators, proving that the generalized Fermat equation

---

[*] Fermat wrote a note in a book claiming he had a proof, but that it was too long to fit in the margin; no one nowadays believes this.

[†] This condition may seem a bit out of the blue, but it turns out there's a cheap way to generate lots of "uninteresting" solutions if you allow common factors among A, B, and C.

has no solution with p = 4, q = 2, and r bigger than 4. Just for that one case, we had to develop some novel techniques, and it's clear they won't be enough to cover the full million-dollar problem.

Despite the apparent simplicity of the bounded gaps conjecture, Zhang's proof requires some of the deepest theorems of modern mathematics.[*] Building on the work of many predecessors, Zhang is able to prove that the prime numbers look random in the first way we mentioned, concerning the remainders obtained after division by many different integers. From there,[†] he can show that the prime numbers look random in a totally different sense, having to do with the sizes of the gaps between them. Random is random!

Zhang's success, along with related work of other contemporary big shots like Ben Green and Terry Tao, points to a prospect even more exciting than any individual result about primes: that we might, in the end, be on our way to developing a richer theory of randomness. Say, a way of specifying precisely what we mean when we say that numbers act as if randomly scattered with no governing structure, despite arising from completely deterministic processes. How wonderfully paradoxical: what helps us break down the final mysteries about prime numbers may be new mathematical ideas that structure the concept of structurelessness itself.

---

[*] Most notably, Pierre Deligne's results relating averages of number-theoretic functions with the geometry of high-dimensional spaces.

[†] Following a path laid out by Goldston, Pintz, and Yıldırım, the last people to make any progress on prime gaps.

NINE

THE INTERNATIONAL JOURNAL OF HARUSPICY

H ere's a parable I learned from the statistician Cosma Shalizi. Imagine yourself a haruspex; that is, your profession is to make predictions about future events by sacrificing sheep and then examining the features of their entrails, especially their livers. You do not, of course, consider your predictions to be reliable merely because you follow the practices commanded by the Etruscan deities. That would be ridiculous. You require evidence. And so you and your colleagues submit all your work to the peer-reviewed *International Journal of Haruspicy*, which demands without exception that all published results clear the bar of statistical signficance.

Haruspicy, especially rigorous evidence-based haruspicy, is not an easy gig. For one thing, you spend a lot of your time spattered with blood and bile. For another, a lot of your experiments don't work. You try to use sheep guts to predict the price of Apple stock, and you fail; you try to model Democratic vote share among Hispanics, and you fail; you try to estimate global oil supply, and you fail again. The gods are very picky and it's not always clear precisely which arrangement of the internal organs and which precise incantations will reliably unlock the future. Sometimes different haruspices run the same experiment and it works for one but not the other—who knows why? It's frustrating. Some days you feel like chucking it all and going to law school.

But it's all worth it for those moments of discovery, where everything works, and you find that the texture and protrusions of the liver really do predict the severity of the following year's flu season, and, with a silent thank-you to the gods, you publish.

You might find this happens about one time in twenty.

That's what I'd expect, anyway. Because I, unlike you, don't believe in haruspicy. I think the sheep's guts don't know anything about the flu data, and when they match up, it's just luck. In other words, in every matter concerning divination from entrails, I'm a proponent of the null hypothesis. So in my world, it's pretty unlikely that any given haruspectic experiment will succeed.

How unlikely? The standard threshold for statistical significance, and thus for publication in *IJoH*, is fixed by convention to be a p-value of .05, or 1 in 20. Remember the definition of the p-value; this says precisely that *if* the null hypothesis is true for some particular experiment, then the chance that that experiment will nonetheless return a statistically significant result is only 1 in 20. If the null hypothesis is always true—that is, if haruspicy is undiluted hocus-pocus—then only one in twenty experiments will be publishable.

And yet there are hundreds of haruspices, and thousands of ripped-open sheep, and even one in twenty divinations provides plenty of material to fill each issue of the journal with novel results, demonstrating the efficacy of the methods and the wisdom of the gods. A protocol that worked in one case and gets published usually fails when another haruspex tries it; but experiments without statistically significant results don't get published, so no one ever finds out about the failure to replicate. And even if word starts getting around, there are always small differences the experts can point to that explain why the follow-up study didn't succeed; after all, we *know* the protocol works, because we tested it and it had a statistically significant effect!

Modern medicine and social science are not haruspicy. But a steadily louder drum circle of dissident scientists has been pounding out an uncomfortable message in recent years: there's probably a lot more entrail reading in the sciences than we'd like to admit.

The loudest drummer is John Ioannidis, a Greek high school math star turned biomedical researcher whose 2005 paper "Why Most Published Research Findings Are False" touched off a fierce bout of self-criticism (and a second wave of self-defense) in the clinical sciences. Some papers plead for attention with a title more dramatic than the claims made in the body, but not this one. Ioannidis takes seriously the idea that entire specialties of medical research are "null fields," like haruspicy, in which there are simply no actual effects to be found. "It can be proven," he writes, "that most claimed research findings are false."

"Proven" is a little more than this mathematician is willing to swallow, but Ioannidis certainly makes a strong case that his radical claim is not implausible. The story goes like this. In medicine, most interventions we try won't work and most associations we test for are going to be absent. Think about tests of genetic association with diseases: there are lots of genes on the genome, and most of them don't give you cancer or depression or make you fat or have any recognizable direct effect at all. Ioannidis asks us to consider the case of genetic influence on schizophrenia. Such an influence is almost certain, given what we know about the heritability of the disorder. But where is it on the genome? Researchers might cast their net wide—it's the Big Data era, after all—looking at a hundred thousand genes (more precisely: genetic polymorphisms) to see which ones are associated with schizophrenia. Ioannidis suggests that around ten of these actually have some clinically relevant effect.

And the other 99,990? They've got nothing to do with schizophrenia. But one in twenty of them, or just about five thousand, are going to pass the p-value test of statistical significance. In other words, among the "OMG I found the schizophrenia gene" results that might get published, there are *five hundred times as many* bogus ones as real ones.

And that's assuming that all the genes that really do have an effect on schizophrenia pass the test! As we saw with Shakespeare and basketball, it's very possible for a real effect to be rejected as statistically insignificant if the study isn't high powered enough to find it. If the studies are underpowered, the genes that truly do make a difference might pass the significance test only half the time; but that means that of the genes certified by p-value to cause schizophrenia, only *five* really do so, as against the five thousand pretenders that passed the test by luck alone.

A good way to keep track of the relevant quantities is by drawing circles in a box:

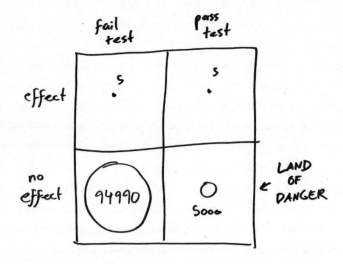

The size of each circle represents the number of genes in each category. On the left half of the box we have the negatives, the genes that don't pass the significance test, and on the right half we have the positives. The two top squares represent the tiny population of genes that actually do affect schizophrenia, so the genes in the top right are the true positives (genes that matter, and the test says they matter) while the top left represents the false negatives (genes that matter, but the test says they don't). In the bottom row, you have the genes that don't matter; the true negatives are the big circle on the bottom left, the false positives the circle on the bottom right.

You can see from the picture that the significance test isn't the problem. It's doing exactly the job it's built to do. The genes that don't affect schizophrenia very rarely pass the test, while the genes we're really interested in pass half the time. But the nonactive genes are so massively preponderant that the circle of false positives, while small relative to the true negatives, is much larger than the circle of true positives.

DOCTOR, IT HURTS WHEN I P

And it gets worse. A low-powered study is only going to be able to see a pretty big effect. But sometimes you know that the effect, if it exists, is small. In other words, a study that accurately measures the effect of a gene is likely to be rejected as statistically insignificant, while any result that passes the p < .05 test is either a false positive or a true positive that massively overstates the gene's effect. Low power is a special danger in fields where small studies are common and effect sizes are typically modest. A recent paper in *Psychological Science*, a premier psychological journal, found that married women were significantly more likely to support Mitt Romney, the Republican presidential candidate, when they were in the fertile portion of their ovulatory cycle: of those women queried during their peak fertility period, 40.4% expressed support for Romney, while only 23.4% of the married women polled at infertile times were pulling the lever for Mitt.* The sample is small, just 228 women, but the difference is big, big enough that the result passes the p-value test with a score of .03.

Which is just the problem—the difference is *too* big. Is it really plausible that, among married women who dig Mitt Romney, nearly *half* spend a large part of each month supporting Barack Obama? Wouldn't anyone notice?

If there's really a political swing to the right once ovulation kicks in, it seems likely to be substantially smaller. But the relatively small size of the study means a more realistic assessment of the strength of the effect would have been rejected, paradoxically, by the p-value filter. In other words, we can be quite confident that the large effect reported in the study is mostly or entirely just noise in the signal.

But noise is just as likely to push you in the *opposite* direction from the real effect as it is to tell the truth. So we're left in the dark by a result that offers plenty of statistical significance but very little confidence.

---

* I was disappointed to find that this study has not yet spawned any conspiracy videos claiming that Obama's support of birth control coverage was aimed at suppressing women's biological drive to vote GOP during ovulation. Get on the stick, conspiracy video producers!

Scientists call this problem "the winner's curse," and it's one reason that impressive and loudly touted experimental results often melt into disappointing sludge when the experiments are repeated. In a representative case, a team of scientists led by psychologist Christopher Chabris[*] studied thirteen single-nucleotide polymorphisms (SNPs) in the genome that had been observed in previous studies to have statistically significant correlations with IQ scores. We know that the ability to do well on IQ-type tests is somewhat heritable, so it's not unreasonable to look for genetic markers. But when Chabris's team tested those SNPs against IQ measures in large data sets, like the ten-thousand-person Wisconsin Longitudinal Study, every single one of these associations vanished into insignificance; if they're real at all, they're almost certainly too small for even a big trial to detect. Genomicists nowadays believe that heritability of IQ scores is probably not concentrated in a few smarty-pants genes, but rather accumulates from numerous genetic features, each one having a tiny effect. Which means that if you go hunting for large effects of individual polymorphisms, you'll succeed—at the same 1-in-20 rate as do the entrail readers.

Even Ioannidis doesn't really think that only one in a thousand published papers is correct. Most scientific studies don't consist of blundering around the genome at random; they test hypotheses that the researchers have some preexisting reason to think might be true, so the bottom row of the box is not quite so enormously dominant over the top. But the crisis of replicability is real. In a 2012 study, scientists at the California biotech company Amgen set out to replicate some of the most famous experimental results in the biology of cancer, fifty-three studies in all. In their independent trials, they were able to reproduce only six.

How can this have happened? It's not because genomicists and cancer researchers are dopes. In part, the replicability crisis is simply a reflection of the fact that science is hard and that most ideas we have are wrong—even most of those ideas that survive a first round of prodding.

But there are practices in the world of science that make the crisis

---

[*] Chabris is perhaps most famous for his immensely popular YouTube video demonstrating the cognitive principle of selective attention: viewers are asked to watch a group of students passing a basketball back and forth, and usually fail to notice an actor in a gorilla suit wandering in and out of the shot.

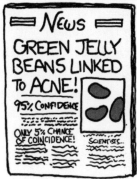

worse, and those can be changed. For one thing, we're doing publishing wrong. Consider the profound xkcd cartoon on page 151. Suppose you tested twenty genetic markers to see whether they were associated with some disorder of interest, and you found just one result that achieved p < .05 significance. Being a mathematical sophisticate, you'd recognize that one success in twenty is exactly what you'd expect if none of the markers had any effect, and you'd scoff at the misguided headline, just as the cartoonist intends you to.

All the more so if you tested the same gene, or the green jelly bean, twenty times and got a statistically significant effect just once.

But what if the green jelly bean were tested twenty times by twenty different research groups in twenty different labs? Nineteen of the labs find no statistically significant effect. They don't write up their results— who's going to publish the bombshell "green jelly beans irrelevant to your complexion" paper? The scientists in the twentieth lab, the lucky ones, find a statistically significant effect, because they got lucky— but *they don't know* they got lucky. For all they can tell, their green-jellybeans-cause-acne theory has been tested only once, and it passed.

If you decide what color jelly beans to eat based just on the papers that get published, you're making the same mistake the army made when they counted the bullet holes on the planes that came back from Germany. As Abraham Wald pointed out, if you want an honest view of what's going on, you also have to consider the planes that *didn't* come back.

This is the so-called file drawer problem—a scientific field has a drastically distorted view of the evidence for a hypothesis when public dissemination is cut off by a statistical significance threshold. But we've already given the problem another name. It's the Baltimore stockbroker. The lucky scientist excitedly preparing a press release about dermatological correlates of Green Dye #16 is just like the naive investor mailing off his life savings to the crooked broker. The investor, like the scientist, gets to see the one rendition of the experiment that went well by chance, but is blind to the much larger group of experiments that failed.

There's one big difference, though. In science, there's no shady con man and no innocent victim. When the scientific community file-drawers its failed experiments, it plays both parts at once. *They're running the con on themselves.*

And all this is assuming that the scientists in question are playing fair. But that doesn't always happen. Remember the wiggle-room problem that ensnared the Bible coders? Scientists, subject to the intense pressure to publish lest they perish, are not immune to the same wiggly temptations. If you run your analysis and get a p-value of .06, you're supposed to conclude that your results are statistically insignificant. But it takes a lot of mental strength to stuff years of work in the file drawer. After all, don't the numbers for that one experimental subject look a little screwy? Probably an outlier, maybe try deleting that line of the spreadsheet. Did we control for age? Did we control for the weather outside? Did we control for age *and* the weather outside? Give yourself license to tweak and shade the statistical tests you carry out on your results, and you can often get that .06 down to a .04. Uri Simonsohn, a professor at Penn who's a leader in the study of replicability, calls these practices "p-hacking." Hacking the p isn't usually as crude as I've made it out to be, and it's seldom malicious. The p-hackers truly believe in their hypotheses, just as the Bible coders do, and when you're a believer, it's easy to come up with reasons that the analysis that gives a publishable p-value is the one you should have done in the first place.

But everybody knows it's not really right. When they don't think anyone's listening, scientists call this practice "torturing the data until it confesses." And the reliability of the results are about what you'd expect from confessions extracted by force.

Assessing the scale of the p-hacking problem is not so easy—you can't examine the papers that are hidden in the file drawer or were simply never written, just as you can't examine the downed planes in Germany to see where they were hit. But you can, like Abraham Wald, make some inferences about data you can't measure directly.

Think again about the International Journal of Haruspicy. What would you see if you looked at every paper ever published there and recorded the p-values you found? Remember, in this case the null hypothesis is always true, because haruspicy doesn't work; so 5% of experiments will record a p-value of .05 or below; 4% will score .04 or below; 3% will score .03 or below, and so on. Another way to say this is that the number of experiments yielding a p-value between .04 and .05 should be about the *same* as the number scoring between .03 and .04, between .02

and .03, and so on. If you plotted all the p-values reported in all the papers you'd see a flat graph like this:

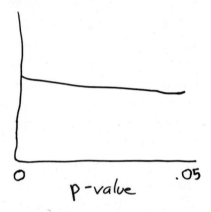

Now what if you looked at a real journal? Hopefully, a lot of the phenomena you're hunting for are actually real, which makes it more likely that your experiments will get a good (which means low) p-value score. So the graph of the p-values ought to slope downward:

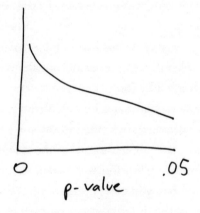

Except that's not exactly what happens in real life. In fields ranging from political science to economics to psychology to sociology, statistical detectives have found a noticeable *upward* slope as the p-value approaches the .05 threshold:

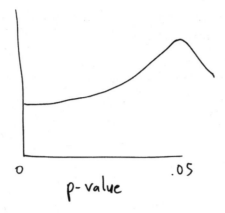

That slope is the shape of p-hacking. It tells you that a lot of experimental results that belong over on the unpublishable side of the p = .05 boundary have been cajoled, prodded, tweaked, or just plain tortured until, at last, they end up just on the happy side of the line. That's good for the scientists who need publications, but it's bad for science.

What if an author refuses to torture the data, or the torture fails to deliver the desired result, and the p-value stays stuck just above the all-important .05? There are workarounds. Scientists will twist themselves into elaborate verbal knots trying to justify reporting a result that doesn't make it to statistical significance: they say the result is "almost statistically significant," or "leaning toward significance," or "well-nigh significant," or "at the brink of significance," or even, tantalizingly, that it "hovers on the brink of significance."* It's easy to make fun of the anguished researchers who resort to such phrases, but we should be hating on the game, not the players—it's not *their* fault that publication is conditioned on an all-or-nothing threshold. To live or die by the .05 is to make a basic category error, treating a continuous variable (how much evidence do we have that the drug works, the gene predicts IQ, fertile women like Republicans?) as if it were a binary one (true or false? yes or no?). Scientists *should* be allowed to report statistically insignificant data.

In some settings, they may even be compelled to. In a 2010 opinion,

---

* All these examples are drawn from the immense collection at the blog of health psychologist Matthew Hankins, a connoisseur of nonsignificant results.

the U.S. Supreme Court ruled unanimously that Matrixx, the maker of the cold remedy Zicam, was required to reveal that some users of its product had suffered anosmia, a loss of the sense of smell. The court's opinion, written by Sonia Sotomayor, held that even though the reports of anosmia didn't pass the signficance test, they still contributed to the "total mix" of information investors in a company can reasonably expect to have available. A result with a weak p-value may provide only a little evidence, but a little is better than none; a result with a strong p-value might provide more evidence, but as we've seen, it's far from a certification that the claimed effect is real.

There is nothing special, after all, about the value .05. It's purely arbitrary, a convention chosen by Fisher. There's value in convention; a single threshold, agreed on by all, ensures that we know what we're talking about when we say the word "significant." I once read a paper by Robert Rector and Kirk Johnson of the conservative Heritage Foundation complaining that a rival team of scientists had falsely claimed that abstinence pledges made no difference in teen rates of sexually transmitted diseases. In fact, the teens in the study who'd pledged to wait for their wedding night *did* have a slightly lower rate of STDs than the rest of the sample, but the difference wasn't statistically significant. The Heritagists had a point; the evidence that pledges worked was weak, but not entirely absent.

On the other hand, Rector and Johnson write in another paper, concerning a statistically insignificant relationship between race and poverty that they wish to dismiss, "If a variable is not statistically significant, it means that the variable has no statistically discernable difference between the coefficient value and zero, so there is no effect." What's good for the abstinent goose is good for the racially charged gander! The value of convention is that it enforces some discipline on researchers, guarding them from the temptation to let their own preferences determine which results count and which don't.

But a conventional boundary, obeyed long enough, can be easily mistaken for an actual thing in the world. Imagine if we talked about the state of the economy this way! Economists have a formal definition of a "recession," which depends on arbitrary thresholds just as "statistical signficance" does. One doesn't say, "I don't care about the unemployment rate, or housing starts, or the aggregate burden of student loans, or the federal deficit; if it's not a recession, we're not going to talk about it." One would be nuts to say so. The critics—and there are more of them, and they are louder, each year—say that a great deal of scientific practice is nuts in just this way.

THE DETECTIVE, NOT THE JUDGE

It's clear that it's wrong to use "p < .05" as a synonym for "true" and "p > .05" to mean "false." Reductio ad unlikely, intuitively appealing as it is, just doesn't work as a principle for inferring the scientific truth underlying the data.

But what's the alternative? If you've ever run an experiment, you know scientific truth doesn't pop out of the clouds blowing a flaming trumpet at you. Data is messy, and inference is hard.

One simple and popular strategy is to report *confidence intervals* in addition to p-values. This involves a slight widening of conceptual scope, asking us to consider not only the null hypothesis but a whole range of alternatives. Perhaps you operate an online store that sells artisanal pinking shears. Being a modern person (except insofar as you make artisanal pinking shears) you set up an A/B test, where half your users see the current version of your website (A) and half see a revamped version (B) with an animated pair of shears that does a little song and dance on top of the "Buy Now" button. And you find that purchases go up 10% with option B. Great! Now, if you're a sophisticated type, you might be worried about whether this increase was merely a matter of random fluctuation—so you compute a p-value, finding that the chance of getting a result this good if the redesign weren't actually working (i.e., if the null hypothesis were correct) is a mere 0.03.*

But why stop there? If I'm going to pay a college kid to superimpose dancing cutlery on all my pages, I want to know not only whether it works, but how well. Is the effect I saw consistent with the hypothesis

---

* All the numbers in this example are made up, partially because the actual computation of confidence intervals is more complicated than I'm revealing in this small space.

that the redesign, in the long term, is really only improving my sales by 5%? Under that hypothesis, you might find that the probability of observing 10% growth is much more likely, say 0.2. In other words, the hypothesis that the redesign is 5% better is *not* ruled out by the reductio ad unlikely. On the other hand, you might optimistically wonder whether you got *unlucky*, and the redesign was actually making your shears 25% more appealing. You compute another p-value and get 0.01, unlikely enough to induce you to throw out that hypothesis.

The confidence interval is the range of hypotheses that the reductio doesn't demand that you trash, the ones that are reasonably consistent with the outcome you actually observed. In this case, the confidence interval might be the range from +3% to +17%. The fact that zero, the null hypothesis, is *not* included in the confidence interval is just to say that the results are statistically significant in the sense we described earlier in the chapter.

But the confidence interval tells you a lot more. An interval of [+3%, +17%] licenses you to be confident that the effect is positive, but not that it's particularly large. An interval of [+9%, +11%], on the other hand, suggests much more strongly that the effect is not only positive but sizable.

The confidence interval is also informative in cases where you don't get a statistically significant result—that is, where the confidence interval contains zero. If the confidence interval is [−0.5%, 0.5%], then the reason you didn't get statistical significance is because you have good evidence the intervention doesn't do anything. If the confidence interval is [−20%, 20%], the reason you didn't get statistical significance is because you have no idea whether the intervention has an effect, or in which direction it goes. Those two outcomes look the same from the viewpoint of statistical significance, but have quite different implications for what you should do next.

The development of the confidence interval is generally ascribed to Jerzy Neyman, another giant of early statistics. Neyman was a Pole who, like Abraham Wald, started as a pure mathematician in Eastern Europe before taking up the then-new practice of mathematical statistics and moving to the West. In the late 1920s, Neyman began collaborating with Egon Pearson, who had inherited from his father Karl both an academic position in London and a bitter academic feud with R. A. Fisher. Fisher was a difficult type, always ready for a fight, about whom his own daughter said, "He grew up without developing a sensitivity to the ordinary humanity of his fellows." In Neyman and Pearson he found opponents sharp enough to battle him for decades.

Their scientific differences are perhaps most starkly displayed in Neyman and Pearson's approach to the problem of inference.* How to determine the truth from the evidence? Their startling response is to unask the question. For Neyman and Pearson, the purpose of statistics isn't to tell us what to believe, but to tell us what to *do*. Statistics is about making decisions, not answering questions. A significance test is no more or less than a rule, which tells the people in charge whether to approve a drug, undertake a proposed economic reform, or tart up a website.

It sounds crazy at first to deny that the goal of science is to find out what's true, but the Neyman-Pearson philosophy is not so far from reasoning we use in other spheres. What's the purpose of a criminal trial? We might naively say it's to find out whether the defendant actually committed the crime they're on trial for. But that's obviously wrong. There are rules of evidence, which forbid the jury from hearing testimony obtained improperly, even if it might help them accurately determine the defendant's innocence or guilt. The purpose of a court is not truth, but justice. We have rules, the rules must be obeyed, and when we say that a defendant is "guilty" we mean, if we are careful about our words, not that he committed the crime he's accused of, but that he was convicted fair and square according to those rules. Whatever rules we choose, we're going to let some criminals go free and imprison some of the blameless. The less you do of the first, the more you're likely to do of the second. So we try to design the rules in whatever way society thinks we best handle that fundamental trade-off.

For Neyman and Pearson, science is like the court. When a drug fails a significance test, we don't say, "We are quite certain the drug didn't work," but merely "The drug wasn't shown to work." And then dismiss

---

* Oversimplification watch: Fisher, Neyman, and Pearson all lived and wrote for a long time, and their ideas and stances shifted over the decades; the rough sketch I draw of the philosophical gap between them ignores many important strands in each person's thinking. In particular, the view that the primary concern of statistics is making decisions is more closely associated with Neyman than with Pearson.

it, just as we would a defendant whose presence at the crime scene couldn't be established within reasonable doubt, even if every man and woman in the courthouse thinks he's guilty as sin.

Fisher wanted none of this—for him, Neyman and Pearson stunk of pure mathematics, insisting on an austere rationalism at the expense of anything resembling scientific practice. Most judges wouldn't have the stomach to let an obviously innocent defendant meet the hangman, even when the rules in the book require it. And most practicing scientists have no interest in following a rigid sequence of instructions, denying themselves the self-polluting satisfaction of forming an opinion about which hypotheses are actually true. In a 1951 letter to W. E. Hick, Fisher wrote:

> I am a little sorry that you have been worrying yourself at all with that unnecessary portentous approach to tests of significance represented by the Neyman and Pearson critical regions, etc. In fact, I and my pupils through the world would never think of using them. If I am asked to give an explicit reason for this I should say they approach the problem entirely from the wrong end, i.e. not from the point of view of a research worker, with a basis of well grounded knowledge on which a very fluctuating population of conjectures and incoherent observations is continually under examination. What he needs is a confident answer to the question "Ought I to take notice of that?" This question can, of course, and for refinement of thought should, be framed as "Is this particular hypothesis overthrown, and if so at what level of significance, by this particular body of observations?" It can be put in this form unequivocally only because the genuine experimenter already has the answers to all the questions that the followers of Neyman and Pearson attempt, I think vainly, to answer by merely mathematical considerations.

But Fisher certainly understood that clearing the significance bar wasn't the same thing as finding the truth. He envisions a richer, more iterated approach, writing in 1926: "A scientific fact should be regarded as experimentally established only if a properly designed experiment rarely fails to give this level of significance."

Not "succeeds once in giving," but "rarely fails to give." A statistically significant finding gives you a clue, suggesting a promising place to focus your research energy. *The significance test is the detective, not the judge.* You know how when you read an article about a breakthrough finding that this thing causes that thing, or that thing prevents the other thing, and at the end there's always a banal sort of quote from a senior scientist not involved in the study intoning some very minor variant of "The finding is quite interesting, and suggests that more research in this direction is needed"? And how you don't really even read that part because you think of it as an obligatory warning without content?

Here's the thing—the reason scientists always say that is because it's important and it's true! The provocative and oh-so-statistically-significant finding isn't the conclusion of the scientific process, but the bare beginning. If a result is novel and important, other scientists in other laboratories ought to test and retest the phenomenon and its variants, trying to figure out whether the result was a one-time fluke or whether it truly meets the Fisherian standard of "rarely fails." That's what scientists call *replication;* if an effect can't be replicated, despite repeated trials, science backs apologetically away. The replication process is supposed to be science's immune system, swarming over newly introduced objects and killing the ones that don't belong.

That's the ideal, at any rate. In practice, science is a bit immunosuppressed. Some experiments, of course, are hard to repeat. If your study measures a four-year-old's ability to delay gratification and then relates these measurements with life outcomes thirty years later, you can't just pop out a replication.

But even studies that *could* be replicated often aren't. Every journal wants to publish a breakthrough finding, but who wants to publish the paper that does the same experiment a year later and gets the same result? Even worse, what happens to papers that carry out the same experiment and *don't* find a significant result? For the system to work, those experiments need to be made public. Too, often they end up in the file drawer instead.

But the culture is changing. Reformers with loud voices like Ioannides and Simonsohn, who speak both to the scientific community and to the broader public, have generated a new sense of urgency about the danger of descent into large-scale haruspicy. In 2013, the Association for Psychological Science announced that they would start publishing a new genre of article, called Registered Replication Reports. These reports, aimed at reproducing the effects reported in widely cited studies, are treated differently from usual papers in a crucial way: the proposed experiment is accepted for publication *before* the study is carried out. If the outcomes support the initial finding, great news, but if not, they're published anyway, so the whole community can know the full state of the evidence. Another consortium, the Many Labs project, revisits high-profile findings in psychology and attempts to replicate them in large multinational samples. In November 2013, psychologists were cheered when the first suite of Many Labs results came back, finding that 10 of the 13 studies addressed were successfully replicated.

In the end, of course, judgments must be made, and lines drawn. What, after all, does Fisher really mean by the "rarely" in "rarely fails"? If we assign an arbitrary numerical threshold ("an effect is real if it reaches statistical significance in more than 90% of experiments") we may find ourselves in trouble again.

Fisher, at any rate, didn't believe in a hard and fast rule that tells us what to do. He was a distruster of pure mathematical formalism. In 1956, near the end of this life, he wrote that "in fact no scientific worker has a fixed level of significance at which from year to year, and in all circumstances, he rejects hypotheses; he rather gives his mind to each particular case in the light of his evidence and his ideas."

In the next chapter we will see one way in which "the light of the evidence" might be made more specific.

TEN

ARE YOU THERE, GOD?
IT'S ME, BAYESIAN
INFERENCE

The age of big data is frightening to a lot of people, and it's frightening in part because of the implicit promise that algorithms, sufficiently supplied with data, are better at inference than we are. Superhuman powers are scary: beings that can change their shape are scary, beings that rise from the dead are scary, and beings that can make inferences that we cannot are scary. It was scary when a statistical model deployed by the Guest Marketing Analytics team at Target correctly inferred based on purchasing data that one of its customers—sorry, *guests*—a teenaged girl in Minnesota, was pregnant, based on an arcane formula involving elevated rates of buying unscented lotion, mineral supplements, and cotton balls. Target started sending her coupons for baby gear, much to the consternation of her father, who, with his puny human inferential power, was still in the dark. Spooky to contemplate, living in a world where Google and Facebook and your phone, and, geez, even *Target*, know more about you than your parents do.

But it's possible we ought to spend less time worrying about eerily superpowered algorithms and more time worrying about crappy ones.

For one thing, crappy might be as good as it gets. Yes, the algorithms that drive the businesses of Silicon Valley get more sophisticated every year, and the data fed to them more voluminous and nutritious. There's a vision of the future in which Google *knows* you; where by aggregating millions of micro-observations ("How long did he hesitate before clicking on *this*. . . . how long did his Google Glass linger on *that*. . . . ") the central storehouse can predict your preferences, your desires, your actions, especially vis-à-vis what products you might want, or might be persuaded to want.

It might be that way! But it also might not. There are lots of mathematical problems where supplying more data improves the accuracy of the result in a fairly predictable way. If you want to predict the course of an asteroid, you need to measure its velocity and its position, as well as the gravitational effects of the objects in its astronomical neighborhood. The more measurements you can make of the asteroid and the more precise those measurements are, the better you're going to do at pinning down its track.

But some problems are more like predicting the weather. That's another situation where having plenty of fine-grained data, and the computational power to plow through it quickly, can really help. In 1950, it took the early computer ENIAC twenty-four hours to simulate twenty-four hours of weather, and that was an astounding feat of space-age computation. In 2008, the computation was reproduced on a Nokia 6300 mobile phone in less than a second. Forecasts aren't just faster now; they're longer-range and more accurate, too. In 2010, a typical five-day forecast was as accurate as a three-day forecast had been in 1986.

It's tempting to imagine that predictions will just get better and better as our ability to gather data gets more and more powerful; won't we eventually have the whole atmosphere simulated to a high precision in a server farm somewhere under The Weather Channel's headquarters? Then, if you wanted to know next month's weather, you could just let the simulation run a little bit ahead.

It's not going to be that way. Energy in the atmosphere burbles up very quickly from the tiniest scales to the most global, with the effect that even a minuscule change at one place and time can lead to a vastly different outcome only a few days down the road. Weather is, in the technical sense of the word, *chaotic*. In fact, it was in the numerical study of weather that Edward Lorenz discovered the mathematical notion of chaos in the first place. He wrote, "One meteorologist remarked that if the theory were correct, one flap of a sea gull's wing would be enough to alter the course of the weather forever. The controversy has not yet been settled, but the most recent evidence seems to favor the sea gulls."

There's a hard limit to how far in advance we can predict the weather, no matter how much data we collect. Lorenz thought it was about two weeks, and so far the concentrated efforts of the world's meteorologists have given us no cause to doubt that boundary.

Is human behavior more like an asteroid or more like the weather? It surely depends on what aspect of human behavior you're talking about. In at least one respect, human behavior ought to be even harder to predict than the weather. We have a very good mathematical model for weather, which allows us at least to get better at short-range predictions when given access to more data, even if the inherent chaos of the system inevitably wins out. For human action we have no such model and may never have one. That makes the prediction problem massively harder.

In 2006, the online entertainment company Netflix launched a $1 million competition to see if anyone in the world could write an algorithm that did a better job than Netflix's own at recommending movies to customers. The finish line didn't seem very far from the start; the winner would be the first program to do 10% better at recommending movies than Netflix did.

Contestants were given a huge file of anonymized ratings—about a million ratings in all, covering 17,700 movies and almost half a million Netflix users. The challenge was to predict how users would rate movies they *hadn't* seen. There's data—lots of data. And it's directly relevant to the behavior you're trying to predict. And yet this problem is really, really hard. It ended up taking three years before anyone crossed the 10% improvement barrier, and it was only done when several teams banded together and hybridized their almost-good-enough algorithms into something just strong enough to collapse across the finish line. Netflix never even used the winning algorithm in its business; by the time the contest was over, Netflix was already transitioning from sending DVDs in the mail to streaming movies online, which makes dud recommendations less of a big deal. And if you've ever used Netflix (or Amazon, or Facebook, or any other site that aims to recommend you products based on the data it's gathered about you), you know that the recommendations remain pretty comically bad. They might get a lot better as even more streams of data get integrated into your profile. But they certainly might not.

Which, from the point of view of the companies doing the gathering, is not so bad. It would be great for Target if they knew with absolute certainty whether or not you were pregnant, just from following the tracks of your loyalty card. They don't. But it would also be great if they could be 10% more accurate in their guesses about your gravidity than they are now. Same for Google. They don't have to know exactly what product you want; they just have to have a better idea than competing ad channels do. Businesses generally operate on thin margins. Predicting your behavior 10% more accurately isn't actually all that spooky for you, but it can mean a lot of money for them. I asked Jim Bennett, the vice president for recommendations at Netflix at the time of the competition, why they'd offered such a big prize. He told me I should have been asking why the prize was so small. A 10% improvement in their recommendations, small as that seems, would recoup the million in less time than it takes to make another *Fast and Furious* movie.

DOES FACEBOOK KNOW YOU'RE A TERRORIST?

So if corporations with access to big data are still pretty limited in what they "know" about you, what's to worry about?

Try worrying about this. Suppose a team at Facebook decides to develop a method for guessing which of its users are likely to be involved in terrorism against the United States. Mathematically, it's not so different from the problem of figuring out whether a Netflix user is likely to enjoy *Ocean's Thirteen*. Facebook generally knows its users' real names and locations, so it can use public records to generate a list of Facebook profiles belonging to people who have already been convicted of terroristic crimes or support of terrorist groups. Then the math starts. Do the terrorists tend to make more status updates per day than the general population, or fewer, or on this metric do they look basically the same? Are there words that appear more frequently in their updates? Bands or teams or products they're unusually prone or disinclined to like? Putting all this stuff together, you can assign to each user a score,* which represents your best estimate for the *probability* that the user has ties, or *will* have ties, to terrorist groups. It's more or less the same thing Target does when they cross-reference your lotion and vitamin purchases to estimate how likely it is that you're pregnant.

There's one important difference: pregnancy is very common, while terrorism is very rare. In almost all cases, the estimated probability that a given user would be a terrorist would be very small. So the result of the project wouldn't be a *Minority Report*–style precrime center, where Facebook's panoptic algorithm knows you're going to do some crime before you do. Think of something much more modest: say, a list of a hundred thousand users about whom Facebook can say, with some degree of confidence, "People drawn from this group are about twice as likely as the typical Facebook user to be terrorists or terrorism supporters."

What would you do if you found out a guy on your block was on that list? Would you call the FBI?

Before you take that step, draw another box.

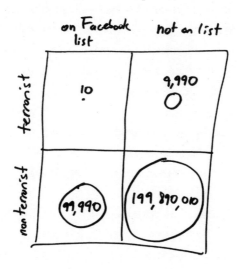

The contents of the box are the 200 million or so Facebook users in the United States. The line between the upper and lower halves separates future terrorists, on the top, from the innocent below. Any terrorist

---

* The basic method here is called *logistic regression*, if you're looking for further reading.

cells in the United States are surely pretty small—let's say, to be as paranoid as possible, that there are ten thousand people who the feds really ought to have their eye on. That's one in twenty thousand of the total user base.

The division between left and right is the one Facebook makes; on the left-hand side are the hundred thousand people Facebook reckons as having an elevated chance of terrorist involvement. We'll take Facebook at their word, that their algorithm is so good that the people who bear its mark are fully twice as likely as the average user to be terrorists. So among this group, one in ten thousand, or ten people, will turn out to be terrorists, while 99,990 will not.

If ten out of the 10,000 future terrorists are in the upper left, that leaves 9,990 for the upper right. By the same reasoning: there are 199,990,000 nonoffenders in Facebook's user base, 99,990 of whom were flagged by the algorithm and sit in the lower left box; that leaves 199,890,010 people in the lower right. If you add up all four quadrants, you get 200,000,000—that is, everybody.

Somewhere in the four-part box is your neighbor down the block.

But where? What you know is that he's in the left half of the box, because Facebook has identified him as a person of interest.

And the thing to notice is that almost nobody in the left half of the box is a terrorist. In fact, there's a 99.99% chance that your neighbor is innocent.

In a way, this is the birth control scare revisited. Being on the Facebook list doubles a person's chance of being a terrorist, which sounds terrible. But that chance starts out very small, so when you double it, it's *still* small.

But there's another way to look at it, which highlights even more clearly just how confusing and treacherous reasoning about uncertainty can be. Ask yourself this—if a person is in fact not a future terrorist, what's the chance that they'll show up, unjustly, on Facebook's list?

In the box, that means: if you're in the bottom row, what's the chance that you're on the left-hand side?

That's easy enough to compute; there are 199,990,000 people in the bottom half of the box, and of those, a mere 99,990 are on the left-hand side. So the chance that an innocent person will be marked as a potential terrorist by Facebook's algorithm is

99,990/199,990,000

or about 0.05%.

That's right—an innocent person has only a 1 in 2,000 chance of being wrongly identified as a terrorist by Facebook!

Now how do you feel about your neighbor?

The reasoning that governs p-values gives us clear guidance. The null hypothesis is that your neighbor is not a terrorist. Under that hypothesis—that is, presuming his innocence—the chance of him showing up on the Facebook red list is a mere 0.05%, well below the 1-in-20 threshold of statistical significance. In other words, under the rules that govern the majority of contemporary science, you'd be justified in rejecting the null hypothesis and declaring your neighbor a terrorist.

Except there's a 99.99% chance he's not a terrorist.

On the one hand, there's hardly any chance that an innocent person will be flagged by the algorithm. At the same time, the people the algorithm points to are almost all innocent. It seems like a paradox, but it's not. It's just how things are. And if you take a deep breath and keep your eye on the box, you can't go wrong.

Here's the crux. There are really two questions you can ask. They sound kind of the same, but they're not.

> Question 1: What's the chance that a person gets put on Face-book's list, given that they're not a terrorist?
>
> Question 2: What's the chance that a person's not a terrorist, given that they're on Facebook's list?

One way you can tell these two questions are different is that they have different answers. Really different answers. We've already seen that the answer to the first question is about 1 in 2,000, while the answer to the second is 99.99%. And it's the answer to the second question that you really want.

The quantities these questions contemplate are called *conditional probabilities*; "the probability that X is the case, given that Y is." And what we're wrestling with here is that the probability of X, given Y, is not the same as the probability of Y, given X.

If that sounds familiar, it should; it's exactly the problem we faced with the reductio ad unlikely. The p-value is the answer to the question

"The chance that the observed experimental result would
occur, given that the null hypothesis is correct."

But what we *want* to know is the other conditional probability:

"The chance that the null hypothesis is correct, given that
we observed a certain experimental result."

The danger arises precisely when we confuse the second quantity for the first. And this confusion is everywhere, not just in scientific studies. When the district attorney leans into the jury box and announces, "There is only a one in five million, I repeat, a ONE IN FIVE MILLLLLLLION CHANCE that an INNOCENT MAN would match the DNA sample found at the scene," he is answering question 1: How likely would an innocent person be to look guilty? But the jury's job is to answer question 2: How likely is this guilty-looking defendant to be innocent? That's a question the DA can't help them with.*

The example of Facebook and the terrorists makes it clear why you should worry about bad algorithms as much as good ones. Maybe more. It's creepy and bad when you're pregnant and Target knows you're pregnant. But it's even creepier and worse if you're not a terrorist and Facebook thinks you are.

You might well think that Facebook would never cook up a list of

---

* In this context, the confusion between question 1 and question 2 is usually called the *prosecutor's fallacy*. The book *Math on Trial*, by Coralie Colmez and Leila Schneps, treats several real-life cases of this kind in detail.

potential terrorists (or tax cheats, or pedophiles) or make the list public if they did. Why would they? Where's the money in it? Maybe that's right. But the NSA collects data on people in America, too, whether they're on Facebook or not. Unless you think they're recording the metadata of all our phone calls just so they can give cell phone companies good advice about where to build more signal towers, there's something like the red list going on. Big Data isn't magic, and it doesn't tell the feds who's a terrorist and who's not. But it doesn't have to be magic to generate long lists of people who are in some ways red-flagged, elevated-risk, "people of interest." Most of the people on those lists will have nothing to do with terrorism. How confident are you that you're not one of them?

RADIO PSYCHICS AND THE RULE OF BAYES

Where does the apparent paradox of the terrorist red list come from? Why does the mechanism of the p-value, which seems so reasonable, work so very badly in this setting? Here's the key. The p-value takes into account what proportion of people Facebook flags (about 1 in 2000) but it totally ignores the proportion of people who are terrorists. When you're trying to decide whether your neighbor is a secret terrorist, you have critical prior information, which is that most people aren't terrorists! You ignore that fact at your peril. Just as R.A. Fisher said, you have to evaluate each hypothesis in the "light of the evidence" of what you already know about it.

But how do you do that?

This brings us to the story of the radio psychics.

In 1937, telepathy was the rage. Psychologist J. B. Rhine's book *New Frontiers of the Mind*, which presented extraordinary claims about Rhine's ESP experiments at Duke in a soothingly sober and quantitative tone, was a best seller and a Book-of-the-Month Club selection, and psychic powers were a hot topic of cocktail conversation across the country. Upton Sinclair, the best-selling author of *The Jungle*, released in 1930 a whole book, *Mental Radio*, about his experiments in psychic communication with his wife, Mary; the subject was mainstream enough that

Albert Einstein contributed a preface to the German edition, stopping short of endorsing telepathy, but writing that Sinclair's book "deserves the most earnest consideration" from psychologists.

Naturally, the mass media wanted in on the craze. On September 5, 1937, the Zenith Radio Corporation, in collaboration with Rhine, launched an ambitious experiment of the kind only the new communication technology they commanded made possible. Five times, the host spun a roulette wheel, with a panel of self-styled telepaths looking on. With each spin, the ball landed either in the black or in the red, and the psychics concentrated with all their might on the appropriate color, transmitting that signal across the country over their own broadcast channel. The station's listeners were implored to use their own psychic powers to pick up the mental transmission and to mail the radio station the sequence of five colors they'd received. More than forty thousand listeners responded to the first request, and even for later programs, after the novelty was gone, Zenith was getting thousands of responses a week. It was a test of psychic powers on a scale Rhine could never have carried out subject by subject in his office at Duke, a kind of proto–Big Data event.

The results of the experiment were not, in the end, favorable to telepathy. But the accumulated data of the responses turned out to be useful for psychologists in a totally different way. The listeners were trying to reproduce sequences of blacks and reds (hereafter Bs and Rs) generated by five spins of the roulette wheel. There are 32 possible sequences:

BBBBB	BBRBB	BRBBB	BRRBB
BBBBR	BBRBR	BRBBR	BRRBR
BBBRB	BBRRB	BRBRB	BRRRB
BBBRR	BBRRR	BRBRR	BRRRR
RBBBB	RBRBB	RRBBB	RRRBB
RBBBR	RBRBR	RRBBR	RRRBR
RBBRB	RBRRB	RRBRB	RRRRB
RBBRR	RBRRR	RRBRR	RRRRR

all of which are equally likely to come up, since each spin is equally likely to land red or black. And since the listeners weren't actually receiving any psychic emanations, you might expect that their responses, too, would be drawn equally from the thirty-two choices.

But no. In fact, the cards the listeners mailed in were highly nonuniform. Sequences like BBRBR and BRRBR were offered much more frequently than chance would predict, while sequences like RBRBR are less frequent than they ought to be, and RRRRR almost never showed up.

This probably doesn't surprise you. RRRRR somehow doesn't *feel* like a random sequence the way BBRBR does, even though the two are equally likely to occur when we spin the wheel. What's going on? What do we really mean when we say that one sequence of letters is "less random" than another?

Here's another example. Quick, think of a number from 1 to 20.

Did you pick 17?

Okay, that trick doesn't always work—but if you ask people to pick a number between 1 and 20, 17 is the most common choice. And if you ask people for a number between 0 and 9, they most frequently pick 7. Numbers ending in 0 and 5, by contrast, are chosen much more rarely than chance would lead you to expect—they just *seem* less random to people. This leads to an irony. Just as the radio psychic contestants tried to match random sequences of Rs and Bs and produced notably nonrandom results, so people who choose random numbers tend to make choices that visibly deviate from randomness.

In 2009, Iran held a presidential election, which incumbent Mahmoud Ahmadinejad won by a large margin. There were widespread accusations that the vote had been fixed. But how could you hope to test the legitimacy of the vote count in a country whose government allowed for almost no independent oversight?

Two graduate students at Columbia, Bernd Beber and Alexandra Scacco, had the clever idea to use the numbers themselves as evidence of fraud, effectively compelling the official vote count to testify against itself. They looked at the official total amassed by the four main candidates in each of Iran's twenty-nine provinces, a total of 116 numbers. If these were true vote counts, there should be no reason for the last digits of those numbers to be anything but random. They should be distributed just about evenly among the digits 0, 1, 2, 3, 4, 5, 6, 7, 8, and 9, each one appearing 10% of the time.

That's not how the Iranian vote counts looked. There were too many 7s, almost twice as many as their fair share; not like digits derived from a random process, but very much like digits written down by humans trying to make them *look* random. This, by itself, isn't proof that the election was fixed, but it's evidence in that direction.[*]

Human beings are always inferring, always using observations to refine our judgments about the various competing theories that jostle around inside our mental representation of the world. We are very confident, almost unshakably confident, about some of our theories ("The sun will rise tomorrow," "When you drop things, they fall") and less sure about others ("If I exercise today, I'll sleep well tonight," "There's no such thing as telepathy"). We have theories about big things and little things, things we encounter every day and things we've run into only once. As we encounter evidence for and against those theories, our confidence in them bobs up and down.

Our standard theory about roulette wheels is that they're fairly balanced, and that the ball is equally likely to land on red or black. But there are competing theories—say, that the wheel is biased in favor of one color or the other.[†] Let's simplify matters and suppose there are just three theories available to you:

> RED: The wheel is biased to make the ball land on red 60% of the time.
> FAIR: The wheel is fair, so the ball lands on red half the time and on black half the time.
> BLACK: The wheel is biased to make the ball land on black 60% of the time.

How much credence do you assign to these three theories? You prob-

* Complicating factors: Beber and Scacco found that numbers ending in 0 were slightly rarer than would be expected by chance, but not nearly as rare as in human-produced digits; what's more, in another data set of apparently fraudulent election data from Nigeria, there were lots of extra numbers ending in 0. Like most forms of detective work, this is far from an exact science.
† Admittedly, this is not a very compelling theory about conventional roulette wheels, where the slots alternate in color. But for a roulette wheel you can't see, you might theorize that it actually has more red slots than black.

ably tend to think roulette wheels are fair, unless you have reason to believe otherwise. Maybe you think there's a 90% chance that FAIR is the right theory, and only a 5% chance for each of BLACK and RED. We can draw a box for this, just like we did for the Facebook list:

BLACK	FAIR	RED
0.05	0.9	0.05

The box records what we call in probability lingo the *a priori* probabilities that the different theories are correct; the *prior*, for short. Different people might have different priors; a hardcore cynic might assign a 1/3 probability to each theory, while someone with a really firm pre-existing belief in the rectitude of roulette-wheel makers might assign only a 1% probability to each of RED and BLACK.

But those priors aren't fixed in place. If we're presented with evidence favoring one theory over another—say, the ball landing red five times in a row—our levels of belief in the different theories can change. How might that work in this case? The best way to figure it out is to compute more conditional probabilities and draw a bigger box.

How likely is it that we'll spin the wheel five times and get RRRRR? The answer depends on which theory is true. Under the FAIR theory, each spin has a 1/2 chance of landing on the red, so the probability of seeing RRRRR is

$$(1/2) \times (1/2) \times (1/2) \times (1/2) \times (1/2) = 1/32 = 3.125\%$$

In other words, RRRRR is exactly as likely as any of the other 31 possibilities.

But if BLACK is true, there's only an 40%, or 0.4 chance of getting red on each spin, so the chance of RRRRR is

$(0.4) \times (0.4) \times (0.4) \times (0.4) \times (0.4) = 1.024\%$

And if RED is true, so that each spin has a 60% chance of landing red, the chance of RRRRR is

$(0.6) \times (0.6) \times (0.6) \times (0.6) \times (0.6) = 7.76\%.$

Now we're going to expand the box from three parts to six.

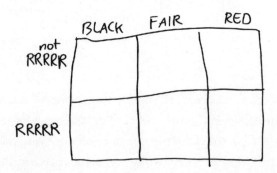

The columns still correspond to the three theories, BLACK, FAIR, and RED. But now we've split each column into two boxes, one corresponding to the outcome of getting RRRRR and the other to the outcome of *not* getting RRRRR. We've already done all the math we need to figure out what numbers go in the boxes. For instance, the a priori probability that FAIR is the correct theory is 0.9. And 3.125% of this probability, 0.9×0.03125 or about 0.0281, goes in the box where FAIR is correct and the balls fall RRRRR. The other 0.8719 goes in the "FAIR correct, not RRRRR" box, so that the FAIR column still adds up to 0.9 in all.

The a priori probability of being in the RED column is 0.05. So the chance that RED is true and the balls fall RRRRR is 7.76% of 5%, or 0.0039. That leaves 0.0461 to sit in the "RED true, not RRRRR" box.

The BLACK theory also has an a priori probability of 0.05. But that theory doesn't jibe nearly as well with seeing RRRRR. The chance that BLACK is true and the balls fall RRRRR is just 1.024% of 5%, or .0005.

Here's the box, filled in:

	BLACK	FAIR	RED
not RRRRR	.0495	.872	.0461
RRRRR	.0005	.028	.0039

(Notice that the numbers in all six boxes together add up to 1; that's as it must be, because the six boxes represent all possible situations.)

What happens to our theories if we spin the wheel and we *do* get RRRRR? That ought to be good news for RED and bad news for BLACK. And that's just what we see. Getting five reds in a row means we're in the bottom row of the six-part box, where there's 0.0005 attached to BLACK, 0.028 attached to FAIR, and 0.0039 attached to RED. In other words, given that we saw RRRRR, our new judgment is that FAIR is about seven times as likely as RED, and RED is about eight times as likely as BLACK.

If you want to translate those proportions into probabilities, you just need to remember that the total probability of *all* the possibilities has to be 1. The sum of the numbers in the bottom row is about 0.0325, so to make those numbers sum to one without changing their proportions to one another, we can just divide each number by 0.0325. This leaves you with

1.5% chance that BLACK is correct
86.5% chance that FAIR is correct
12% chance that RED is correct.

The extent to which you believe in RED has more than doubled, while your belief in BLACK has been almost totally wiped out. As is appropriate! You see five reds in a row, why shouldn't you start to suspect a little more seriously than before that the game is rigged?

That "dividing everything by 0.0325" step might seem a bit of an ad hoc trick. But it's really the correct thing to do. In case your intuition doesn't swallow it right away, here's another picture some people like better. Imagine there are ten thousand roulette wheels. And there are ten thousand rooms, each with a different roulette wheel, each roulette wheel with a person playing it. One of those people, following one of those wheels, is you. But you don't know which wheel you've got! So your state of unknowledge about the wheel's true nature can be modeled by supposing that, of the original ten thousand, five hundred were biased toward the black, five hundred were biased toward the red, and nine thousand were fair.

The computation we just did above tells you to expect about 281 of the FAIR wheels, about 39 of the RED wheels, and only 5 of the BLACK wheels to come up RRRRR. So if you do get RRRRR, you still don't know which of the ten thousand rooms you're in, but you've narrowed it down a hell of a lot; you're in one of the 325 rooms where the ball landed on red five times in a row. And of those rooms, 281 of them (about 86.5%) have FAIR wheels, 39 (12%) have RED wheels, and only 5 (1.5%) have BLACK wheels.

The more balls that fall red, the more favorably you're going to look on that RED theory (and the less credence you'll give to BLACK). If you saw ten reds in a row instead of five, the same computation would raise your estimation of the chance of RED to 25%.

What we've done is to compute how our degrees of belief in the various theories ought to change once we see five reds in a row—what are known as the *posterior* probabilities. Just as the prior describes your beliefs before you see the evidence, the posterior describes your beliefs afterward. What we're doing here is called *Bayesian inference*, because the passage from prior to posterior rests on an old formula in probability called *Bayes's Theorem*. That theorem is a short algebraic expression and I could write it down for you right here and now. But I'm going to try not doing that. Because sometimes a formula, if you train yourself to apply it mechanically without thinking about the situation in front of you, can obscure what's really going on. And everything you need to know about what's going on here can already be seen in the box.*

---

* Of course, if we were doing this for real, we'd have to consider more than three theories. We'd also want to include the theory that the wheel is weighted to come up 55% red, or 65%, or 100%, or 93.756%, and so on and so on. There are infinitely many potential theories, not just three, and

———

The posterior is affected by the evidence you encounter, but also by your prior. The cynic, who started out with a prior that assigned probability 1/3 to each of BLACK, FAIR, and RED, would respond to five reds in a row with a posterior judgment that RED had a 65% chance of being correct. The trusting soul who starts out assigning only 1% probability to the RED will still only give it a 2.5% chance of being right, even after seeing five reds in a row.

In the Bayesian framework, how much you believe something *after* you see the evidence depends not just on what the evidence shows, but on how much you believed it to begin with.

That may seem troubling. Isn't science supposed to be objective? You'd like to say that your beliefs are based on evidence alone, not on some prior preconceptions you walked in the door with. But let's face it—no one actually forms their beliefs this way. If an experiment provided statistically significant evidence that a new tweak of an existing drug slowed the growth of certain kinds of cancer, you'd probably be pretty confident the new drug was actually effective. But if you got the exact same results by putting patients inside a plastic replica of Stonehenge, would you grudgingly accept that the ancient formations were actually focusing vibrational earth energy on the body and stunning the tumors? You would not, because that's nutty. You'd think Stonehenge probably got lucky. You have different priors about those two theories, and as a result you interpret the evidence differently, despite it being numerically the same.

It's just the same with Facebook's terrorist-finding algorithm and the next-door neighbor. The neighbor's presence on the list really does offer some evidence that he's a potential terrorist. But your prior for that hypothesis ought to be very small, because most people aren't terrorists. So, despite the evidence, your posterior probability remains small as well, and you don't—or at least shouldn't—worry.

Relying purely on null hypothesis significance testing is a deeply non-Bayesian thing to do—strictly speaking, it asks us to treat the cancer

when scientists carry out Bayesian computations in real life, they need to grapple with infinities and infinitesimals, to compute integrals instead of simple sums, and so on. But these complications are merely technical; in essence the process is no deeper than the one we carried out.

drug and the plastic Stonehenge with exactly the same respect. Is that a blow to Fisher's view of statistics? On the contrary. When Fisher says that "no scientific worker has a fixed level of significance at which from year to year, and in all circumstances, he rejects hypotheses; he rather gives his mind to each particular case in the light of his evidence and his ideas," he is saying exactly that scientific inference can't, or at least shouldn't, be carried out purely mechanically; our preexisting ideas and beliefs must always be allowed to play a part.

Not that Fisher was a Bayesian statistician. That phrase, nowadays, refers to a cluster of practices and ideologies in statistics, once unfashionable but now rather mainstream, which includes a general sympathy toward arguments based on Bayes's Theorem, but is not simply a matter of taking both previous beliefs and new evidence into account. Bayesianism tends to be most popular in genres of inference, like teaching machines to learn from large-scale human input, that are poorly suited to the yes-or-no questions Fisher's approach was set up to adjudicate. In fact, Bayesian statisticians often don't think about the null hypothesis at all; rather than asking "Does this new drug have any effect?" they might be more interested in a best guess for a predictive model governing the drug's effects in various doses on various populations. And when they do talk about hypotheses, they're relatively at ease with talking about the probability that a hypothesis—say, that the new drug works better than the existing one—is true. Fisher was not. In his view, the language of probability was appropriately used only in a context where some actual chance process is taking place.

At this point, we've arrived at the shore of a great sea of philosophical difficulty, into which we'll dip one or two toes, max.

First of all: when we call Bayes's Theorem a theorem it suggests we are discussing incontrovertible truths, certified by mathematical proof. That's both true and not. It comes down to the difficult question of what we mean when we say "probability." When we say that there's a 5% chance that RED is true, we *might* mean that there actually is some vast global population of roulette wheels, of which exactly one in twenty is biased to fall red 3/5 of the time, and that any given roulette wheel we encounter is randomly picked from the roulette wheel multitude. If that's what we mean, then Bayes's Theorem is a plain fact, akin to the Law of

Large Numbers we saw in the last chapter; it says that, in the long run, under the conditions we set up in the example, 12% of the roulette wheels that come up RRRRR are going to be of the red-favoring kind.

But this isn't actually what we're talking about. When we say that there's a 5% chance that RED is true, we are making a statement not about the global distribution of biased roulette wheels (how could we know?) but rather about our own mental state. Five percent is the *degree to which we believe* that a roulette wheel we encounter is weighted toward the red.

This is the point at which Fisher totally got off the bus, by the way. He wrote an unsparing pan of John Maynard Keynes's *Treatise on Probability*, in which probability "measures the 'degree of rational belief' to which a proposition is entitled in the light of given evidence." Fisher's opinion of this viewpoint is well summarized by his closing lines: "If the views of the last section of Mr. Keynes's book were accepted as authoritative by mathematical students in this country, they would be turned away, some in disgust, and most in ignorance, from one of the most promising branches of applied mathematics."

For those who are willing to adopt the view of probability as degree of belief, Bayes's Theorem can be seen not as a mere mathematical equation but as a form of numerically flavored advice. It gives us a rule, which we may choose to follow or not, for how we should update our beliefs about things in the light of new observations. In this new, more general form, it is naturally the subject of much fiercer disputation. There are hard-core Bayesians who think that *all* our beliefs should be formed by strict Bayesian computations, or at least as strict as our limited cognition can make them; others think of Bayes's rule as more of a loose qualitative guideline.

The Bayesian outlook is already enough to explain why RBRRB looks random while RRRRR doesn't, even though both are equally improbable. When we see RRRRR, it strengthens a theory—the theory that the wheel is rigged to land red—to which we've already assigned some prior probability. But what about RBRRB? You could imagine someone walking around with an unusually open-minded stance concerning roulette wheels, which assigns some modest probability to the theory that the roulette wheel was fitted with a hidden Rube Goldberg apparatus designed to produce the outcome red, black, red, red, black. Why not? And such a person, observing RBRRB, would find this theory very much bolstered.

But this is not how real people react to the spins of a roulette wheel coming up red, black, red, red, black. We don't allow ourselves to consider every cockamamie theory we can logically devise. Our priors are not *flat*, but *spiky*. We assign a lot of mental weight to a few theories, while others, like the RBRRB theory, get assigned a probability almost indistinguishable from zero. How do we choose our favored theories? We tend to like simpler theories better than more complicated ones, theories that rest on analogies to things we already know about better than theories that posit totally novel phenomena. That may seem like an unfair prejudice, but without some prejudices we would run the risk of walking around in a constant state of astoundedness. Richard Feynman famously captured this state of mind:

> You know, the most amazing thing happened to me tonight. I was coming here, on the way to the lecture, and I came in through the parking lot. And you won't believe what happened. I saw a car with the license plate ARW 357. Can you imagine? Of all the millions of license plates in the state, what was the chance that I would see that particular one tonight? Amazing!

If you've ever used America's most popular sort-of-illegal psychotropic substance, you know what it feels like to have too-flat priors. Every single stimulus that greets you, no matter how ordinary, seems *intensely meaningful*. Each experience grabs hold of your attention and demands that you take notice. It's a very interesting mental state to be in. But it's not conducive to making good inferences.

The Bayesian point of view explains why Feynman wasn't actually amazed; it's because he assigns a very low prior probability to the hypothesis that a cosmic force intended him to see the license plate ARW 357 that night. It explains why five reds in a row feels "less random" than RBRRB to us; it's because the former activates a theory, RED, to which we assign some non-negligible prior probability, and the latter doesn't. And a number ending in 0 feels less random than a number ending in 7, because the former supports the theory that the number we're seeing is not a precise count, but an estimate.

This framework also helps unwind some of the conundrums we've already encountered. Why are we surprised and a little suspicious when the lottery comes up 4, 21, 23, 34, 39 twice in a row, but not when it comes up 4, 21, 23, 34, 39 one day and 16, 17, 18, 22, 39 the next day, even though both events are equally improbable? Implicitly, you have some kind of theory in the back of your mind, a theory that lottery games are for some reason unusually likely to spit out the same numbers twice in close succession; maybe because you think lottery games are rigged by the proprietors, maybe because you think a cosmic synchronicity-loving force has a thumb on the scale, doesn't matter. You might not believe in this theory very strongly; maybe in your heart you think there's a one-in-a-hundred-thousand chance that there really is such a bias in favor of repeated numbers. But that's much more than the prior you assign the theory that there's a weird conspiracy in favor of the 4, 21, 23, 34, 39–16, 17, 18, 22, 39 combo. *That* theory is crazy, and you are not stoned, so you pay it no mind.

If you do happen to find yourself partially believing a crazy theory, don't worry—probably the evidence you encounter will be inconsistent with it, driving down your degree of belief in the craziness until your beliefs come into line with everyone else's. Unless, that is, the crazy theory is designed to survive this winnowing process. That's how conspiracy theories work.

Suppose you learn from a trusted friend that the Boston Marathon bombing was an inside job carried out by the federal government in order to, I don't know, garner support for NSA wiretapping. Call that theory T. At first, because you trust your friend, maybe you assign that theory a reasonably high probability, say 0.1. But then you encounter other information: police located the suspected perpetrators, the surviving suspect confessed, etc. Each of these pieces of information is pretty unlikely, given T, and each one knocks down your degree of belief in T until you hardly credit it at all.

That's why your friend isn't going to give you theory T; he's going to add to it theory U, which is that the government and the news media are in on the conspiracy together, with the newspapers and cable networks feeding false information to support the story that the attack was carried out by Islamic radicals. The combined theory, T + U, should start out with a smaller prior probability; it is by definition harder to believe than T, because it asks you to swallow both T and another theory at the same time. But as the evidence flows in, which would tend to kill T alone,* the combined theory T + U remains untouched. Dzhokar Tsarnaev convicted? Well, sure, that's *exactly* what you'd expect from a federal court— the Justice Department is totally in on it! The theory U acts as a kind of Bayesian coating to T, keeping new evidence from getting to it and dissolving it. This is a property most successful crackpot theories have in common; they're encased in just enough protective stuff that they're equally consistent with many possible observations, making them hard to dislodge. They're like the multi-drug-resistant *E. coli* of the information ecosystem. In a weird way you have to admire them.

THE CAT IN THE HAT, THE CLEANEST MAN IN SCHOOL, AND THE CREATION OF THE UNIVERSE

When I was in college, I had a friend with entrepreneurial habits who had the idea of making a little extra money at the beginning of the school year by selling T-shirts to first-year students. At that time you could buy a large lot of T-shirts from the screen-printing shop for about four dollars each, while the going rate on campus was ten bucks. It was the early 1990s, and it was fashionable to go to parties wearing a hat modeled after the one worn by the Cat in the Hat.† So my friend got together eight hundred dollars and printed up two hundred shirts with a picture of the Cat in the Hat drinking a mug of beer. These shirts sold fast.

My friend was entrepreneurial, but not *that* entrepreneurial. In fact, he was kind of lazy. And once he'd sold eighty shirts, making back his initial investment, he started to lose his desire to hang out on the quad all day making sales. So the box of shirts went under his bed.

A week later, laundry day came around. My friend, as I mentioned,

---

* More precisely, it tends to kill T + not-U.
† No, seriously, this was actually fashionable.

was lazy. He really didn't feel like washing his clothes. And then he remembered that he had a box of clean, brand-new beer-swigging-Cat-in-the-Hat T-shirts under his bed. So that solved the problem of laundry day.

As it turned out, it also solved the problem of the day after laundry day.

And so on.

So here was the irony. Everyone around thought my friend was the dirtiest man in school, because he wore the same T-shirt every single day. But in fact, he was the *cleanest* man in school, dressed every day in a new-from-the-store, never-worn shirt!

The lesson about inference: you have to be careful about the universe of theories you consider. Just as there may be more than one solution to a quadratic equation, there may be multiple theories that give rise to the same observation, and if we don't consider them all, our inferences may lead us badly astray.

This brings us back to the Creator of the Universe.

The most famous argument in favor of a God-made world is the so-called argument by design, which, in its simplest form, simply says, holy cow, just *look around you*—everything is so complex and amazing, and you think it just glommed together that way by dumb luck and physical law?

Or, phrased more formally, by the liberal theologian William Paley, in his 1802 book *Natural Theology; or, Evidences of the Existence and Attributes of the Deity, Collected from the Appearances of Nature*:

> In crossing a heath, suppose I pitched my foot against a *stone*, and were asked how the stone came to be there: I might possibly answer that, for any thing I knew to the contrary, it had lain there for ever; nor would it perhaps be very easy to shew the absurdity of this answer. But suppose I had found a *watch* upon the ground, and it should be inquired how the watch happened to be in that place; I should hardly think of the answer which I had before given,—that, for any thing I knew, the watch might have always been there. . . . The inference, we think, is inevitable, that the watch must have had a maker: that there must have existed, at some time, and at some place or other, an artificer or artificers who formed it for the purpose which

we find it actually to answer: who comprehended its construction, and designed its use.

If this is true of a watch, how much more so of a sparrow, or a human eye, or a human brain?

Paley's book was a tremendous success, going through fifteen editions in fifteen years. Darwin read it closely in college, later saying, "I do not think I hardly ever admired a book more than Paley's *Natural Theology*: I could almost formerly have said it by heart." And updated forms of Paley's argument form the backbone of the modern intelligent design movement.

It is, of course, a classic reductio ad unlikely:

- If there's no God, it would be unlikely for things as complex as human beings to have developed;
- Humans have developed;
- Therefore, it's unlikely there's no God.

This is much like the argument that the Bible coders used; if God didn't write the Torah, it's unlikely that the text on the scroll would so faithfully record the birthdays of the rebbes!

You may be sick of hearing me say it by now, but reductio ad unlikely doesn't always work. If we *really* mean to compute in numerical terms how confident we should be that God created the universe, we'd better draw another Bayes box.

The first difficulty is to understand the priors. This is a hard thing to get your head around. For the roulette wheels, we were asking: How likely do we think it is that the wheel is rigged, before we see any of the spins? Now we're asking: How likely would we think it was that there was a God, if we didn't know that the universe, the Earth, or we ourselves exist?

At this point, the usual move is to throw up one's hands and invoke the charmingly named *principle of indifference*—since there can be no principled way to pretend we don't know we exist, we just divvy up the prior probability evenly, 50% for GOD and 50% for NO GOD.

If NO GOD is true, then complex beings like humans must have arisen by pure chance, perhaps spurred along by natural selection. Designists then and now agree that this is phenomenally unlikely; let's make up numbers and say it was a one-in-a-billion-billion shot. So what goes in the bottom right box is one-billion-billionth of 50%, or one in two billion billion.

What if GOD is true? Well, there are lots of ways God could be; we don't know in advance that a God who made the universe would care to create human beings, or any thinking entities at all, but certainly any God worth the name would have the *ability* to whip up intelligent life. Perhaps if there's a God there's a one in a million chance God would make creatures like us.

So the box now looks like this:

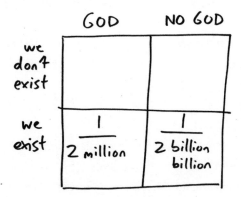

At this point we can examine the evidence, which is that we exist. So the truth lies somewhere in the bottom row. And in the bottom row, you can plainly see that there is a lot more probability—a trillion times more!—in the GOD box than in the NO GOD box.

This, in essence, is Paley's case, the "argument by design," as a modern Bayesian type would express it. There are many solid objections to the argument by design, and there are also two billion billion fighty books on the topic of "you should totally be a cool atheist like me" where you can read those arguments, so let me stick here to the one that's closest to the math at hand: the "cleanest man in school" objection.

You probably know what Sherlock Holmes had to say about inference, the most famous thing he ever said that wasn't "Elementary!":

"It is an old maxim of mine that when you have excluded the impossible, whatever remains, however improbable, must be the truth."

Doesn't that sound cool, reasonable, indisputable?

But it doesn't tell the whole story. What Sherlock Holmes *should* have said was:

"It is an old maxim of mine that when you have excluded the impossible, whatever remains, however improbable, must be the truth, unless the truth is a hypothesis it didn't occur to you to consider."

Less pithy, more correct. The people who inferred that my friend was the dirtiest man in school were considering only two hypotheses:

CLEAN: my friend was rotating through his shirts, washing them, then starting the rotation over, like a normal person
DIRTY: my friend was a filthy savage who wore dirty clothes.

You may start with some prior; based on my memory of college, assigning a probability of 10% to DIRTY is about right. But it doesn't really matter what your prior is: CLEAN is ruled out by the observation that my friend wears the same shirt every day. "When you have excluded the impossible . . ."

But hold up, Holmes—the true explanation, LAZY ENTREPRENEUR, was a hypothesis not on the list.

The argument by design suffers from much the same problem. If the only two hypotheses you admit are NO GOD and GOD, the rich structure of the living world might well be taken as evidence in favor of the latter against the former.

But there are other possibilities. What about GODS, where the world was put together in a hurry by a squabbling committee? Many distinguished civilizations have believed as much. And you can't deny that there are aspects of the natural world—I'm thinking pandas here—that seem more likely to have resulted from grudging bureaucratic compromise than from the mind of an all-knowing deity with total creative control. If we start by assigning the same prior probability to GOD and GODS—and why not, if we're going with the principle of indifference?—then Bayesian inference should lead us to believe in GODS much more than GOD.*

Why stop there? There's no end to the making of origin stories. Another theory with some adherents is SIMS, where we're not actually people at all, but simulations running on an ultracomputer built by other people.† That sounds bizarre, but plenty of people take the idea seriously (most famously, the Oxford philosopher Nick Bostrom), and on Bayesian grounds, it's hard to see why you shouldn't. People like to build simulations of real-world events; surely, if the human race doesn't extinguish itself, our power to simulate will only increase, and it doesn't seem crazy to imagine that those simulations might one day include conscious entities that believed themselves to be people.

If SIMS is true, and the universe is a simulation constructed by people in a realer world, then it's pretty likely there'd be people in the universe, because people are people's favorite things to simulate! I'd call it a near certainty (for the sake of the example, let's say an absolute certainty) that a simulated world created by technologically advanced humans would have (simulated) humans in it.

If we assign each of the four hypotheses we've met so far a prior probability of 1/4, the box looks something like this:

	GOD	NO GOD	GODS	SIMS
we don't exist				
we exist	$\frac{1}{4\ million}$	$\frac{1}{4\ billion\ billion}$	$\frac{1}{400,000}$	$\frac{1}{4}$

* Paley himself was surely aware of this issue; note how careful he is to say "artificer or artificers."
† People who, of course, might themselves actually be simulations engineered by a yet higher order of people!

Given that we actually do exist, so that the truth is in the bottom row, almost all the probability is sitting in SIMS. Yes, the existence of human life is evidence for the existence of God; but it's much *better* evidence that our world was programmed by people much smarter than us.

Advocates of "scientific creationism" hold that we should argue in the classroom for the existence of a world-designer, not because the Bible says so—that would be unconstitutionally naughty!—but on coolly reasonable grounds, founded on the astonishing unlikelihood of the existence of humanity under the NO GOD hypothesis.

But if we took this approach seriously, we would tell our tenth graders something like this: "Some have argued that it's highly unlikely for something as complex as the Earth's biosphere to have arisen purely by natural selection without any intervention from outside. By far the most likely such explanation is that we are actually not physical beings at all, but residents of a computer simulation being carried out by humans with unthinkably advanced technology, to what purpose we can't exactly know. It's also possible that we were created by a community of gods, something like those worshiped by the ancient Greeks. There are even some people who believe that one single God created the universe, but that hypothesis should be considered less strongly supported than the alternatives."

Think the school board would go for this?

I had better hasten to point out that I don't *actually* think this constitutes a good argument that we're all sims, any more than I think Paley's argument is a good one for the existence of the deity. Rather, I take the queasy feeling these arguments generate as an indication that we've reached the limits of quantitative reasoning. It's customary to express our uncertainty about something as a number. Sometimes it even makes sense to do so. When the meteorologist on the nightly news says, "There's a 20% chance of rain tomorrow," what he means is that, among some large population of past days with conditions similar to those currently obtaining, 20% of them were followed by rainy days. But what can we mean when we say, "There's a 20% chance that God created the universe?" It can't be that one in five universes was made by God and the rest popped up on their own. The truth is, I've never seen a method I find satisfying for assigning numbers to our uncertainty about ultimate questions of this kind. As much as I love numbers, I think people ought to stick to "I don't believe in God," or "I do believe in God," or just "I'm not sure." And as much as I love Bayesian inference, I think people are probably best off arriving at their faith, or discarding it, in a non-quantitative way. On this matter, math is silent.

If you don't buy it from me, take it from Blaise Pascal, the seventeenth-century mathematician and philosopher who wrote in his *Pensées*, " 'God is, or He is not.' But to which side shall we incline? Reason can decide nothing here."

This is not quite all Pascal had to say on the subject. We return to his thoughts in the next chapter. But first, the lottery.

PART III

. . . . .

Expectation

Includes: MIT kids game the Massachusetts State Lottery, how Voltaire got rich, the geometry of Florentine painting, transmissions that correct themselves, the difference between Greg Mankiw and Fran Lebowitz, "I'm sorry, was that bofoc or bofog?," parlor games of eighteenth-century France, where parallel lines meet, the other reason Daniel Ellsberg is famous, why you should be missing more planes

ELEVEN

WHAT TO EXPECT
WHEN YOU'RE
EXPECTING TO WIN
THE LOTTERY

Should you play the lottery?

It's generally considered canny to say no. The old saying tells us lotteries are a "tax on the stupid," providing government revenue at the expense of people misguided enough to buy tickets. And if you see the lottery as a tax, you can see why lotteries are so popular with state treasuries. How many other taxes will people line up at convenience stores to pay?

The attraction of lotteries is no novelty. The practice dates back to seventeenth-century Genoa, where it seems to have evolved by accident from the electoral system. Every six months, two of the city's *governatori* were drawn from the members of the Petty Council. Rather than hold an election, Genoa carried out the election by lot, drawing two slips from a pile containing the names of all 120 councilors. Before long, the city's gamblers began to place extravagant side bets on the election outcome. The bets became so popular that gamblers started to chafe at having to wait until Election Day for their enjoyable game of chance; and they quickly realized that if they wanted to bet on paper slips drawn from a pile, there was no need for an election at all. Numbers replaced names of politicians, and by 1700 Genoa was running a lottery that would look very familiar to modern Powerball players. Bettors tried to guess five randomly drawn numbers, with a bigger payoff the more numbers a player matched.

Lotteries quickly spread throughout Europe, and from there to North America. During the Revolutionary War, both the Continental Congress and the governments of the states established lotteries to fund the fight against the British. Harvard, back in the days before it enjoyed a nine-figure endowment, ran lotteries in 1794 and 1810 to fund two new college buildings. (They're still used as dorms for first-year students today.)

Not everyone applauded this development. Moralists thought, not wrongly, that lotteries amounted to gambling. Adam Smith, too, was a lottery naysayer. In *The Wealth of Nations*, he wrote:

> That the chance of gain is naturally overvalued, we may learn from the universal success of lotteries. The world neither ever saw, nor ever will see, a perfectly fair lottery, or one in which the whole gain compensated the whole loss; because the undertaker could make nothing by it. . . . In a lottery in which no prize exceeded twenty pounds, though in other respects it approached much nearer to a perfectly fair one than the common state lotteries, there would not be the same demand for tickets. In order to have a better chance for some of the great prizes, some people purchase several tickets; and others, small shares in a still greater number. There is not, however, a more certain proposition in mathematics, than that the more tickets you adventure upon, the more likely you are to be a loser. Adventure upon all the tickets in the lottery, and you lose for certain; and the greater the number of your tickets, the nearer you approach to this certainty.

The vigor of Smith's writing and his admirable insistence on quantitative considerations shouldn't blind you to the fact that his conclusion is not, strictly speaking, correct. Most lottery players would say buying two tickets instead of one doesn't make you more likely to be a loser, but twice as likely to be a winner. And that's right! In a lottery with a simple prize structure, it's easy to check for yourself. Suppose the lottery has 10 million combinations of numbers and just one is a winner. Tickets cost $1 and the jackpot is $6 million.

The person who buys every single ticket spends $10 million and gets a $6 million prize; in other words, just as Smith says, this strategy is a certain loser, to the tune of $4 million. The small-time operator who buys a single ticket is better off—at least she has a 1 in 10 million chance of coming out ahead!

But what if you buy two tickets? Then your chance of losing shrinks, though admittedly only from 9,999,999 in 10 million to 9,999,998 in 10 million. Keep buying tickets and your chance of being a loser keeps going down, until the point where you've purchased 6 million tickets. In that case, your chance of winning the jackpot, and thus breaking even, is a solid 60%, and there's only a 40% chance of you ending up a loser. Contrary to Smith's claim, you've made yourself less likely to lose money by buying more tickets.

Purchase one more ticket, though, and you're *sure* to lose money (though whether it's $1 or $4,000,001 depends on whether you hold the winning ticket).

It's hard to reconstruct Smith's reasoning here, but he may have been a victim of the all-curves-are-lines fallacy, reasoning that if buying all the tickets made you certain to lose money, then buying more tickets must make you more likely to lose money.

Buying 6 million tickets minimizes the chance of losing money, but that doesn't mean it's the right play; it matters how *much* money you lose. The one-ticket player suffers a near certainty of losing money; but she knows she won't lose a lot. The buyer of 6 million tickets, despite the lower chance of losing, is in a much more dangerous position. And probably you still feel that neither choice seems very wise. As Smith points out, if the lottery is a winning proposition for the state, it seems like it must be a bad idea for whoever takes the other side of the bet.

What Smith's argument against lotteries is missing is the notion of *expected value*, the mathematical formalism that captures the intuition Smith is trying to express. It works like this. Suppose we possess an item whose monetary worth is uncertain—like, say, a lottery ticket:

9,999,999/10,000,000 times: ticket is worth nothing

1/10,000,000 times: ticket is worth $6 million

Despite our uncertainty, we still might want to assign the ticket a definite value. Why? Well, what if a guy comes around offering to pay $1.20 for people's tickets? Is it wise to make the deal and pocket the 20-cent profit, or should I hold on to my ticket? That depends whether I've assigned the ticket a worth of more or less than $1.20.

Here's how you compute the expected value of a lottery ticket. For each possible outcome, you multiply the chance of that outcome by the ticket's value given that outcome. In this simplified case, there are only two outcomes: you lose, or you win. So you get

$$9{,}999{,}999/10{,}000{,}000 \times \$0 = \$0$$

$$1/10{,}000{,}000 \times \$6{,}000{,}000 = \$0.60$$

Then you add the results up:

$$\$0 + \$0.60 = \$0.60.$$

So the expected value of your ticket is 60 cents. If a lottophile comes to your door and offers $1.20 for your ticket, expected value says you ought to make the deal. In fact, expected value says you shouldn't have paid a dollar for it in the first place!

EXPECTED VALUE IS NOT THE VALUE YOU EXPECT

Expected value is another one of those mathematical notions saddled, like significance, with a name that doesn't quite capture its meaning. We certainly don't "expect" the lottery ticket to be worth 60 cents: on the contrary, it's either worth 10 million clams or zilch, nothing in between.

Similarly: suppose I make a $10 bet on a dog I think has a 10% chance of winning its race. If the dog wins, I get $100; if the dog loses, I get nothing. The expected value of the bet is then

$$(10\% \times \$100) + (90\% \times \$0) = \$10.$$

But this is not, of course, what I expect to happen. Winning $10 is, in fact, not even a possible outcome of my bet, let alone the expected one. A better name might be "average value"—for what the expected value of the bet really measures is what I'd expect to happen if I made *many* such bets on *many* such dogs. Let's say I laid down a thousand $10 bets like that. I'd probably win about a hundred of them (the Law of Large Numbers again!) and make $100 each time, totaling $10,000; so my thousand bets are returning, on average, $10 per bet. In the long run, you're likely to come out about even.

Expected value is a great way to figure out the right price of an object, like a gamble on a dog, whose true value isn't certain. If I pay $12 apiece for those tickets, I'm very likely to lose money in the long run; if I can get them for $8, on the other hand, I should probably buy as many as I can.* Hardly anybody plays the dogs anymore, but the machinery of expected value is the same whether you're pricing race tickets, stock options, lottery tickets, or life insurance.

THE MILLION ACT

The notion of expected value started to come into mathematical focus in the mid-1600s, and by the end of that century, the idea was understood well enough to be used by practical scientists like Edmond Halley, the Astronomer Royal of England.† Yep, the comet guy! But he was also one of the first scientists to study the correct pricing of insurance, which in the reign of William III was a matter of critical military importance. England had thrown itself enthusiastically into war on the continent, and war required capital. Parliament proposed to raise the necessary funds via the "Million Act" of 1692, which aimed to raise a million pounds by selling life annuities to the population. Signing up for an annuity meant

---

* A more refined analysis of "the right price" would also take into account my feelings about risk; we'll return to this issue in the next chapter.
† This job still exists! But it is now a largely honorary position, since the annual salary of one hundred pounds sterling has remained unchanged since Charles II established the post in 1675.

paying the Crown a lump sum, in exchange for a guaranteed lifetime
annual payout. This is a kind of life insurance in reverse; purchasers of
such an annuity are essentially betting that they won't die anytime soon.
As a measure of the rudimentary state of the actuarial science of the
time, the cost of the annuity was set without reference to the annuitant's
age!* A lifetime annuity for a grandfather, likely to require funding for at
most a decade, cost as much as one for a child.

Halley was scientist enough to understand the absurdity of the age-
independent pricing scheme. He determined to work out a more rational
accounting of the value of a lifetime annuity. The difficulty is that people don't arrive and depart on a rigid schedule, as comets do. But by
using birth and death statistics, Halley was able to estimate the *probability* of various life spans for each annuitant, and thereby to compute the
expected value of the annuity: "It is plain that the purchaser ought to
pay for only such a part of the value of the annuity, as he has chances
that he is living; and this ought to be computed yearly, and the sum of all
those yearly values being added together, will amount to the value of the
annuity for the life of the person proposed."

In other words: Grandpa, with his shorter expected life span, pays
less for an annuity than Junior.

"EET EES OBVIOUS."

Digression: when I tell people the story of Edmond Halley and the price
of annuities, I often get interrupted: "But it's *obvious* that you should
charge younger people more!"

It is not obvious. Rather, it is obvious if you already know it, as modern people do. But the fact that people who administered annuities failed
to make this observation, again and again, is proof that it's not *actually*
obvious. Mathematics is filled with ideas that seem obvious now—that
negative quantities can be added and subtracted, that you can usefully
represent points in a plane by pairs of numbers, that probabilities of un-

* Other states, as far back as third-century Rome, had understood that the proper price of an annuity needed to be greater when the purchaser is younger.

certain events can be mathematically described and manipulated—but are in fact not obvious at all. If they were, they would not have arrived so late in the history of human thought.

This reminds me of an old story from the Harvard math department, concerning one of the grand old Russian professors, whom we shall call O. Professor O is midway through an intricate algebraic derivation when a student in the back row raises his hand.

"Professor O, I didn't follow that last step. Why do those two operators commute?"

The professor raises his eyebrows and says, "Eet ees obvious."

But the student persists: "I'm sorry, Professor O, I really don't see it."

So Professor O goes back to the board and adds a few lines of explanation. "What we must do? Well, the two operators are both diagonalized by . . . well, it is not exactly diagonalized but . . . just a moment . . ." Professor O pauses for a little while, peering at what's on the board and scratching his chin. Then he retreats to his office. About ten minutes go by. The students are about to start leaving when Professor O returns, and again assumes his station in front of the chalkboard.

"Yes," he says, satisfied. "Eet ees obvious."

DON'T PLAY POWERBALL

The nationwide lottery game Powerball is currently playable in forty-two U.S. states, the District of Columbia, and the U.S. Virgin Islands. It's extremely popular, sometimes selling as many as 100 million tickets for a single drawing. Poor people play Powerball and people who are already rich play Powerball. My father, a former president of the American Statistical Association, plays Powerball, and since he usually gets me a ticket, I guess I've played, too.

Is this wise?

On December 6, 2013, as I write this, the jackpot stands at a handsome $100 million. And the jackpot isn't the only way to win. Like many lotteries, Powerball features many levels of prizes; the smaller, more frequent prizes help keep people feeling the game's worth playing.

With expected value, we can check those feelings against the mathematical facts. Here's how you compute the expected value of a $2 ticket. When you buy that ticket, you're buying a:

> 1/175,000,000 chance of an $100 million jackpot
> 1/5,000,000 chance of a $1 million prize
> 1/650,000 chance of a $10,000 prize
> 1/19,000 chance of a $100 prize
> 1/12,000 chance of a different $100 prize
> 1/700 chance of a $7 prize
> 1/360 chance of a different $7 prize
> 1/110 chance of a $4 prize
> 1/55 chance of a different $4 prize

(You can get all these details from Powerball's website, which also offers a surprisingly spunky Frequently Asked Questions page, filled with material like "**Q: Do powerball tickets expire?** A: Yes. The Universe is decaying and nothing lasts forever.")

So the expected amount you'll win is

100 million / 175 million + 1 million / 5 million + 10,000 / 650,000 + 100 / 19,000 + 100 / 12,000 + 7 / 700 + 7 / 360 + 4 / 110 + 4 / 55

which comes to just under 94 cents. In other words: according to expected value, the ticket isn't worth your two bucks.

That's not the end of the story, because not all lottery tickets are the same. When the jackpot is $100 million, as it is today, the expected value of a ticket is scandalously low. But each time the jackpot goes unclaimed, more money enters the prize pool. And the bigger the jackpot gets, the more people buy tickets, and the more people buy tickets, the more likely it is that one of those tickets is going to make somebody a multimillionaire. In August 2012, Donald Lawson, a Michigan railroad worker, took home a $337 million jackpot.

When the top prize is that big, the expected value of a ticket gets bigger too. Same computation as above, but substituting in the $337 million jackpot:

337 million / 175 million + 1 million / 5 million + 10,000 / 650,000 + 100 / 19,000 + 100 / 12,000 + 7 / 700 + 7 / 360 + 4 / 110 + 4 / 55

which is $2.29. Suddenly, playing the lottery doesn't seem like such a bad bet after all. How big does the jackpot have to be before the expected value of a ticket exceeds the two dollars it costs? Now you can finally go back to your eighth-grade math teacher and tell her you figured out what algebra is for. If we call the value of the jackpot J, the expected value of a ticket is

J / 175 million + 1 million / 5 million + 10,000 / 650,000 + 100 / 19,000 + 100 / 12,000 + 7 / 700 + 7 / 360 + 4 / 110 + 4 / 55

or, to make it a little simpler,

J / 175 million + 36.7 cents.

Now here comes the algebra. For the expected value to be more than the two dollars you spent, you need J / 175 million to be bigger than $1.63 or so. Multiplying both sides by 175 million, you find that the threshold value of the jackpot is a little over $285 million. That's not a once-in-a-lifetime occurrence; the pot got that big three times in 2012. So it sounds like the lottery might be a good idea after all—if you're careful to play only when the jackpot gets high enough.

But that's not the end of the story either. You are not the only person in America who knows algebra. And even people who don't know algebra instinctively understand that a lottery ticket is more enticing when the jackpot is $300 million than when it's $80 million—as usual, the mathematical approach is a formalized version of our natural mental reckonings, an extension of common sense by other means. A typical $80 million drawing might sell about 13 million tickets. But when Donald Lawson won $337 million, he was up against some 75 million other players.[*]

---

[*] Or so it seems to me. I wasn't able to get official statistics for ticket sales, but you can get pretty good estimates of the number of players from the data Powerball releases about the number of winners of the lower-tier prizes.

The more people who play, the more people win prizes. But there's only one jackpot. And if two people hit all six numbers, they have to share the big money.

How likely is it that you'll win the jackpot and not have to share it? Two things have to happen. First, you have to hit all six numbers; your chance of doing so is 1 in 175 million. But it is not enough to win—*everyone else must lose.*

The chance of any particular player missing out on the jackpot is pretty good—just about 174,999,999 in 175 million. But when 75 million other players are in the game, there starts to be a substantial chance one of those folks will hit the jackpot.

How substantial? We use a fact we've already encountered several times; that if we want to know the probability that thing one happens, and we know the probability that thing two happens, and if the two things are independent—the occurrence of one has no effect on the likelihood of the other—then the probability of thing one *and* thing two happening is the product of the two probabilities.

Too abstract? Let's do it with the lottery.

There's a 174,999,999 / 175,000,000 chance that I lose, and a 174,999,999 / 175,000,000 chance that my dad loses. So the probability that we *both* lose is

$$174,999,999 / 175,000,000 \times 174,999,999 / 175,000,000$$

or 99.9999994%. In other words, as I tell my dad every single time, we'd better not quit our jobs.

But what's the chance that *all 75 million* of your competitors lose? All I have to do is multiply 174,999,999 / 175,000,000 by itself 75 million times. That sounds like an incredibly brutal detention assignment. But you can make the problem a lot simpler by phrasing it as an exponential, which your computer can calculate for you instantaneously:

$$(174,999,999 / 175,000,000)^{75 \text{ million}} = 0.651 \ldots$$

So there's a 65% chance that none of your fellow players will win, which means there's a 35% chance at least one of them will. If that happens, your share of the $337 million prize drops to a puny $168 million. That cuts the expected value of the jackpot to

65% × $337 million + 35% × $168 million = $278 million

which is just below the threshold value of $285 million that makes the jackpot worth it. And that doesn't even take into account the possibility that *more* than two people will hit the jackpot, divvying up the big prize even further. The possibility of jackpot-splitting means the lottery ticket has an expected value less than what it costs you, even when the jackpot tops $300 million. If the jackpot were bigger still, the expected value might tip into the "worth it" zone—or it might not, if the big jackpot attracted an even higher level of ticket sales.* The biggest Powerball jackpot yet, $588 million, was won by two players, and the biggest lottery jackpot in U.S. history, a $688 million Mega Millions prize, was split three ways.

And we haven't even considered the taxes you'll pay on your winnings, or the fact that the prize is distributed to you in yearly chunks—if you want all the money up front, you get a substantially smaller payout. And remember, the lottery is a creature of the state, and the state knows a lot about you. In many states, back taxes or other outstanding financial obligations get paid off from lottery winnings before you see a dime. An acquaintance who works at a state lottery told me the story of a man who came to the lottery office with his girlfriend to cash in his $10,000 ticket and spend a wild weekend on the town. When he turned in his ticket, the lottery official on duty told the couple that all but a few hundred dollars of the prize was already committed to delinquent child support the man owed his ex-girlfriend.

This was the first the man's current girlfriend had heard of the man's child. The weekend did not go as planned.

* For readers who want to go even deeper into the decision-theoretic details of the lottery, "Finding Good Bets in the Lottery, and Why You Shouldn't Take Them," by Aaron Abrams and Skip Garibaldi (*The American Mathematical Monthly*, vol. 117, no. 1, January 2010, pp. 3–26) is a great resource. The title of the article serves as an executive summary of their conclusions.

So what's your best strategy for making money playing Powerball? Here's my mathematically certified three-point plan:

1. Don't play Powerball.
2. If you do play Powerball, don't play Powerball unless the jackpot is really big.
3. And if you buy tickets for a massive jackpot, try to reduce the odds you'll have to share your haul; pick numbers other players won't. Don't pick your birthday. Don't pick the numbers that won a previous draw. Don't pick numbers that form a nice pattern on the ticket. And for God's sake, don't pick numbers you find in a fortune cookie. (You know they don't put different numbers in every cookie, right?)

Powerball isn't the only lottery, but all lotteries have one thing in common; they're bad bets. A lottery, just as Adam Smith observed, is designed to return a certain proportion of ticket sales to the state; for that to work, the state has to take in more money in tickets than it gives out in prizes. Turning that on its head, lottery players, on average, are spending more money than they win. So the expected value of a lottery ticket *has* to be negative.

Except when it's not.

THE LOTTERY SCAM THAT WASN'T

On July 12, 2005, the Compliance Unit of the Massachusetts State Lottery received an unusual phone call from an employee at a Star Market in Cambridge, the northern suburb of Boston that houses both Harvard and MIT. A college student had come into the supermarket to buy tickets for the state's new Cash WinFall game. That wasn't strange. What was unusual was the size of the order; the student had presented fourteen thousand order slips, each one filled out by hand, for a total of $28,000 in lottery tickets.

No problem, the lottery told the store; if the slips are filled out properly, anybody can play as much as they want. Stores were required to get a waiver from the lottery office if they wanted to sell more than $5,000 in tickets per day, but those waivers were easily granted.

That was a good thing, because the Star wasn't the only Boston-area lottery agent doing a vigorous business that week. Twelve more stores contacted the lottery in advance of the July 14 drawing to ask for waivers. Three of those were concentrated in a heavily Asian-American neighborhood of Quincy, just south of Boston on the bay shore. Tens of thousands of Cash WinFall tickets were being sold to a small group of buyers at a handful of stores.

What was going on? The answer wasn't secret; it was in plain sight, right there in the rules for Cash WinFall. The new game, launched in the fall of 2004, was a replacement for Mass Millions, which had been phased out after going an entire year without paying out a jackpot. Players were getting discouraged, and sales were down. Massachusetts needed to shake up its lottery, and state officials hit on the idea of adapting WinFall, a game from Michigan. In Cash WinFall, the jackpot didn't pile higher and higher with each week it went unclaimed; instead, every time the pot went over $2 million, the money "rolled down" to enhance the lesser prizes that weren't so hard to win. The jackpot reset to its minimum $500,000 value for the following drawing. The lottery commission hoped the new game, which made it possible to take in serious winnings without hitting the jackpot, would seem like a good deal.

They did their job too well. In Cash WinFall, Massachusetts had inadvertently designed a game that actually *was* a good deal. And by the summer of 2005, a few enterprising players had figured that out.

On a normal day, here's how the prize distribution for Cash WinFall looked:

match all 6 numbers	1 in 9.3 million	variable jackpot
match 5 of 6	1 in 39,000	$4,000
match 4 of 6	1 in 800	$150
match 3 of 6	1 in 47	$5
match 2 of 6	1 in 6.8	free lottery ticket

If the jackpot is $1 million, the expected value of a two-dollar ticket is pretty poor:

($1 million / 9.3 million) + ($4,000 / 39,000)
+ ($150 / 800) + ($5 / 47) + ($2 / 6.8) = 79.8 cents.

That's a rate of return so pathetic it makes Powerball players look like canny investors. (And we've generously valued a free ticket at the $2 it would cost you instead of the substantially smaller expected value it brings you.)

But on a roll-down day, things look very different. On February 7, 2005, the jackpot stood near $3 million. Nobody won that jackpot—unsurprising, considering that only about 470,000 people played Cash WinFall that day, and matching all six numbers was about a 1-in-10 million long shot.

So all that money rolled down. The state's formula rolled $600,000 to the match-5 and match-3 prize pools and $1.4m into the match-4s. The probability of getting four out of six WinFall numbers right is about 1 in 800, so there must have been about six hundred match-4 winners that day out of the 470,000 players. That's a lot of winners, but $1.4 million dollars is a lot of money. Dividing it into six hundred pieces leaves more than $2,000 for each match-4 winner. In fact, you'd expect the payout for matching 4 out of 6 numbers that day to be around $2,385. That's a much more attractive proposition than the measly $150 you'd win on a normal day. A 1-in-800 chance of a $2,385 payoff has an expected value of

$2364 / 800 = $2.98

In other words, the match-4 prize *alone* makes the ticket worth its two-dollar price. Throw in the other prizes, and the story gets even sweeter.

Prize	Chance of winning	Expected number of winners	Roll-down allocation	Roll-down per prize
match 5 of 6	1 in 39,000	12	$600,000	$50,000
match 4 of 6	1 in 800	587	$1.4m	$2,385
match 3 of 6	1 in 47	10,000	$600,000	$60

So the average ticket could be expected to bring home cash winnings of

$$\$50,000 / 39,000 + \$2385 / 800 + \$60 / 47 = \$5.53.$$

An investment where you make three and a half bucks of profit on a $2 investment is not one to pass up.[*]

Of course, if one lucky person hits the jackpot, the game turns back into a pumpkin for everybody else. But Cash WinFall was never popular enough to make that outcome likely. Out of forty-five roll-down days during the lifetime of the game, only once did a player match all six numbers and stop the roll-down in its tracks.[†]

Let's be clear—this computation doesn't mean that a $2 bet is sure to win you money. On the contrary, when you buy a Cash WinFall ticket on a roll-down day, your ticket is most likely a loser, just as it is on any other day. The expected value is not the value you expect! But on roll-down day, the prizes, in the unlikely event that you do win, are bigger—a lot bigger. The magic of expected value is that the *average* payout of a hundred, or a thousand, or ten thousand tickets is very likely to be close to $5.53. Any given ticket is probably worthless, but if you've got a thousand tickets, it's essentially certain that you'll make your money back and then some.

Who buys a thousand lottery tickets at a time?

Kids at MIT, that's who.

The reason I can tell you the WinFall payoffs on February 7, 2005, down to the last dollar is because this figure is recorded in the exhaustive and, frankly, kind of thrilling account of the WinFall affair submitted to the state in July 2012 by Gregory W. Sullivan, the inspector general of the Commonwealth of Massachusetts. I think I'm safe in saying this is

---

[*] As it happened, only seven people matched five numbers that day, so each of those luckies shared a prize of over $80,000. But the scarcity of those winners seems to have been just bad luck, and not something you could fairly have anticipated when computing the expected value of a ticket in advance.

[†] Given the popularity of Cash WinFall, this is actually somewhat surprising; there was about a 10% chance per roll-down that somebody would win the jackpot, so it should have happened four or five times. That it happened only once was, as far as I can tell, plain bad luck—or, if you like, good luck for the people counting on those lesser roll-down prizes.

history's only state fiscal oversight document which inspires the reader to wonder: Does someone have the movie rights to this?

And the reason it's *this particular day* for which this data is recorded is that February 7 was the first roll-down day after James Harvey, an MIT senior working on an independent study project comparing the merits of various state lottery games, realized that Massachusetts had accidentally created an insanely profitable investment vehicle for anyone quantitatively savvy enough to notice it. Harvey got a group of friends together (at MIT, it's not hard to get a group of friends together who can all compute expected value) and bought a thousand tickets. Just as you might expect, one of those 1-in-800 shots came through, and Harvey's group took home one of those $2,000 prizes. They won a bunch of match-3s too; in all, they just about tripled their initial investment.

It won't surprise you to hear that Harvey and his co-investors didn't stop playing Cash WinFall. Or that he never did get around to finishing that independent study—at least not for course credit. In fact, his research project quickly developed into a thriving business. By summer, Harvey's confederates were buying tens of thousands of tickets at a time—it was a member of his group who placed the mammoth order at the Cambridge Star Market. They called their team Random Strategies, though their approach was anything but scattershot; the name referred to Random Hall, the MIT dorm where Harvey had originally cooked up his plan to make money on WinFall.

And the MIT students weren't alone. At least two more betting clubs formed up to take advantage of the WinFall windfall. Ying Zhang, a medical researcher in Boston with a PhD from Northeastern, formed the Doctor Zhang Lottery Club. It was the DZLC that accounted for the spike in sales in Quincy. Before long, the group was buying $300,000 worth of tickets for each roll-down. In 2006, Doctor Zhang quit doctoring to devote himself full-time to Cash WinFall.

Still another betting group was led by Gerald Selbee, a retiree in his seventies with a bachelor's degree in math. Selbee lived in Michigan, the original home of WinFall; his group of thirty-two bettors, mostly made up of his relatives, played WinFall there for about two years until the game shut down in 2005. When Selbee found out the gravy train was getting back on the tracks out East, his course was clear; in August 2005, he and his wife Marjorie drove to Deerfield, in the western part of Massachusetts, and placed their first bet—sixty thousand tickets. They took home a little over $50,000 in pure profit. Selbee, with the benefit of his experience playing the game in Michigan, added an extra profit-making venture to his Cash WinFall tickets. Stores in Massachusetts got a 5% commission on lottery ticket sales. Selbee cut deals directly with one store, offering hundreds of thousands of dollars worth of business at a time in exchange for going halfsies on the 5% commission. That move alone made Selbee's team thousands of dollars in extra profit every roll-down.

You don't need an MIT degree to see how the influx of high-volume players affected the game. Remember: the reason the roll-down payoffs were so swollen was that a lot of money was being split among just a few winners. By 2007, a million or more tickets were being sold for each roll-down drawing, most of them to the three high-volume syndicates. The days of the $2,300 prize for matching four out of six numbers were long gone; if a million and a half people bought tickets, and one person in eight hundred matched 4, then you'd typically see almost two thousand match-4 winners. So each share of the $1.4 million kitty was now more like $800.

It's pretty easy to figure out how much a big player stood to gain from Cash WinFall—the trick is to look at it from the point of view of the lottery itself. If it's roll-down day, the state has (at least!) $2 million of accumulated jackpot money it's got to get rid of. Let's say a million and a half people buy tickets for the roll-down. That's $3 million more in revenue, of which 40%, or $1.2 million, goes into the state's coffers, and the other $1.8 million gets plowed into the jackpot fund, all of which is to be disbursed to bettors before the day is through. So the state takes in $3 million that day and hands out $3.8 million:* $2 million from the money already in the jackpot fund and $1.8 million from that day's ticket receipts. On any given day, whatever the state makes, the players, on

---

* As long as we ignore all the prize money that doesn't come from the roll-down; but as we've seen, that money doesn't amount to much.

average, lose, and vice versa. So this day is a good day to play; ticket buyers, in the aggregate, took $800,000 from the state.

If players buy 3.5 million tickets, it's a different story; now the Lottery takes $2.8 million as its share and pays out the remaining $4.2 million. On top of the $2 million already in the kitty, that amounts to $6.2 million, less than the $7 million of revenue the state took in. In other words, despite the generosity of the roll-down, the lottery has gotten so popular that the state still ends up making money at the expense of the players.

This makes the state very, very happy.

The break-even point comes when the 40% share of the roll-down day revenue exactly matches the $2 million already in the pot (that is, the money contributed by the players who were unsophisticated or risk-loving enough to play WinFall without a roll-down). That's $5 million, or 2.5 million tickets. More sales than that, and WinFall is a bad bet. But any fewer—and over the life span of the WinFall game, it always *was* fewer—and WinFall offers players a way to make some money.

What we're really using here is a wonderful, while at the same time commonsensical, fact called *additivity of expected value.* Suppose I own a McDonald's franchise and a coffee shop, and the McDonald's has an expected annual profit of $100,000, while the coffee shop's expected net is $50,000. The money might go up and down from year to year, of course; the expected value means that, in the long run, the average amount of money the McDonald's makes will be about $100,000 a year, and the average amount from the coffee shop $50,000.

Additivity says that, on average, my total take from Big Macs and mochaccinos together is going to average out to $150,000, the sum of the expected profits from each of my two businesses.

In other words:

ADDITIVITY: *The expected value of the sum of two things is the sum of the expected value of the first thing with the expected value of the second thing.*

Mathematicians like to sum up that reasoning in a formula, just as we summed up the commutativity of addition ("*this* many rows of *that* many holes is the same thing as *that* many columns of *this* many holes) by the formula a × b = b × a. In this case, if X and Y are two numbers whose values we're uncertain about, and E(X) is short for "the expected value of X," then additivity just says

$$E(X+Y) = E(X) + E(Y).$$

Here's what this has to do with the lottery. The value of all the tickets in a given drawing is the amount of money handed out by the state. And that value isn't subject to uncertainty at all;* it's just the amount of roll-down money, $3.8 million in the first example above. The expected value of a sure $3.8 million is, well, just what you expect—$3.8 million.

In that example, there were 1 million players on roll-down day. Additivity tells you that the sum of the expected values of all 1.5 million lottery tickets is the expected value of the total value of all the tickets, or $3.8 million. But each ticket (at least before you know what the winning numbers are) is worth the same. So you're summing 1.5 million copies of the same number and getting $3.8 million; that number must be $2.53. Your expected profit on your $2 ticket is 53 cents, more than 25% of your wager, a handsome profit on what's supposed to be a sucker's bet.

The principle of additivity is so intuitively appealing that it's easy to think it's obvious. But, just like the pricing of life annuities, it's not obvious! To see that, substitute other notions in place of expected value and watch everything go haywire. Consider:

The most likely value of the sum of a bunch of things is the sum of the most likely values of each of the things.

That's totally wrong. Suppose I choose randomly which of my three children to give the family fortune to. The most likely value of each child's share is zero, because there's a two in three chance I'm disinheriting them. But the most likely value of the sum of those three allotments—in fact, its *only possible* value—is the amount of my whole estate.

---

* Still ignoring the money that doesn't come from the jackpot fund.

BUFFON'S NEEDLE, BUFFON'S NOODLE, BUFFON'S CIRCLE

We have to interrupt the story of the college nerds versus the lottery for a minute, because once we're talking about additivity of expected value I can't not tell you about one of the most beautiful proofs I know, which is based on the very same idea.

It starts with the game of *franc-carreau*, which, like the Genoese lottery, reminds you that people in olden times would gamble on just about anything. All you need for *franc-carreau* is a coin and a floor with square tiles. You throw the coin on the floor and make a bet: will it land wholly within one tile, or end up touching one of the cracks? (*"Franc-carreau"* translates roughly as "squarely within the square"—the coin used for this game was not a franc, which wasn't in circulation at the time, but the *ecu*.)

Georges-Louis LeClerc, Comte de Buffon, was a provincial aristocrat from Burgundy who developed academic ambitions early on. He went to law school, perhaps with the aim of following his father into the magistracy, but as soon as he finished his degree he threw aside legal matters in favor of science. By 1733, at the age of twenty-seven, he was ready to stand for membership in the Royal Academy of Sciences in Paris.

Buffon would later gain fame as a naturalist, writing a massive, forty-four-volume *Natural History* that laid out his proposal for a theory intended to explain the origin of life as universally and parsimoniously as Newton's theory had explained motion and force. But as a young man, influenced by a brief meeting and long exchange of letters with the Swiss mathematician Gabriel Cramer,* Buffon's interests lay in pure mathematics, and it was as a mathematician that he offered himself to the Royal Academy.

The paper Buffon presented was an ingenious juxtaposition of two mathematical fields that had been thought of as separate: geometry and probability. Its subject wasn't a grand question about the mechanics of the planets in their orbits or the economies of the great nations, but rather the humble game of *franc-carreau*. What was the probability, Buffon[†]

* Of Cramer's rule, for all the linear algebra fans in the house.

† Actually, it's not totally clear to me that he was actually "Buffon" at the time of his Academy presentation; his father, who had bought the title of comte de Buffon in the first place, had mis-

asked, that the franc would land entirely within a single tile? And how large should the floor tiles be to make the game a fair bet for both players?

Here's how Buffon did it. If the coin has radius r and the square tile has a side of length L, then the coin touches a crack exactly when its center lands inside a smaller square, whose side has length L – 2r:

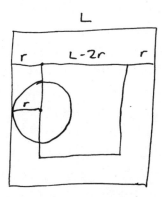

The smaller square has area $(L - 2r)^2$, while the bigger square has area L^2; so if you're betting on the coin landing "squarely in the square," your chance of winning is the fraction $(L - 2r)^2 / L^2$. For the game to be fair, this chance needs to be 1/2; which means that

$$(L - 2r)^2 / L^2 = 1/2$$

Buffon solved this equation (and so can you, if that's the kind of thing you're into), finding that *franc-carreau* was a fair game just when the side of the *carreau* was $4 + 2\sqrt{2}$ times the radius of the coin, a ratio of just under seven. This was conceptually interesting, in that the combination of probabilistic reasoning with geometric figures was novel; but it wasn't difficult, and Buffon knew it wouldn't be enough to get him into the academy. So he pressed forward:

"But if instead of throwing in the air a round piece, as an *ecu*, one would throw a piece of another shape, as a squared Spanish *pistole*, or a needle, a stick, etc., the problem demands a little more geometry."

This was an understatement; the problem of the needle is the one for

managed his business and had to sell the Buffon holdings, and meanwhile had remarried a twenty-two-year-old; Georges-Louis sued and apparently managed to divert his mother's childless uncle's fortune directly to himself, allowing him to buy back both land and title.

which Buffon's name is remembered in mathematical circles even today. Let me explain it more precisely than Buffon did:

Buffon's Needle Problem: Suppose you have a hardwood floor made of long, skinny slats, and you happen to have in your possession a needle exactly as long as the slats are wide. Throw the needle on the floor. What's the chance that the needle crosses one of the cracks separating the slats?

Here's why this problem is so touchy. When you throw the *ecu* on the floor, it doesn't matter which direction Louis XV's face ends up pointing. A circle looks the same from every angle; the chance that it crosses a crack doesn't depend on its orientation.

Buffon's needle is a different story. A needle oriented nearly parallel to the slats is very unlikely to cross a crack:

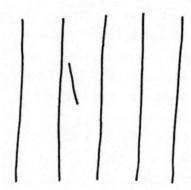

but if the needle lands crosswise to the slats, it's almost certain to do so:

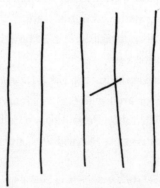

The *franc-carreau* is highly symmetric—in technical terms, we say it is *invariant* under rotation. In the needle problem, that symmetry has been broken. And that makes the problem much harder; we need to keep track of not just where the center of the needle falls, but also what direction it's pointing.

In the two extreme cases, the chance the needle crosses a crack is 0 (if the needle is parallel to the slat) or 1 (if the needle and the crack are perpendicular). So you might split the difference and guess that the needle touches a crack exactly half the time.

But that's wrong; in fact, the needle crosses a crack substantially more often than it lands wholly within a single slat. Buffon's needle problem has a beautifully unexpected answer: the probability is $2/\pi$, or about 64%. Why π, when there's no circle in sight? Buffon found his answer using a somewhat intricate argument involving the area under a curve called the cycloid. Computing this area requires a bit of calculus; nothing a modern-day sophomore math major couldn't handle, but not exactly enlightening.

But there's another solution, discovered by Joseph-Émile Barbier more than a century after Buffon's entry into the Royal Academy. No formal calculus is needed; in fact, you don't need computation of any kind. The argument, while a little involved, uses no more than arithmetic and basic geometric intuition. And the crucial point is, of all things, the additivity of expected value!

The first step is to rephrase Buffon's problem in terms of expected value. We can ask: What is the expected *number* of cracks the needle crosses? The number Buffon aimed to compute was the probability p that the thrown-down needle crosses a crack. Thus there is a probability of $1 - p$ that the needle doesn't cross any cracks. But if the needle crosses a crack, it crosses *exactly* one.[*] So expected number of crossings is obtained the same way we always compute expected value: by summing up each possible number of crossings, multiplied by the probability of observing that number. In this case the only possibilities are 0 (observed with probability $1 - p$) and 1 (observed with probability p) so we add up

---

[*] You might complain that since the needle is exactly as long as the slat is wide, it is possible that the needle touches two cracks. But this requires that the needle span the slat *exactly*; it is possible, but the probability that it happens is 0, and we can safely ignore it.

$$(1 - p) \times 0 = 0$$
and
$$p \times 1 = p$$

and get p. So the expected number of crossings is simply p, the same number Buffon computed. We seem to have made no progress. How can we figure out the mystery number?

When you're faced with a math problem you don't know how to do, you've got two basic options. You can make the problem easier, or you can make it harder.

Making it easier sounds better—you replace the problem with a simpler one, solve that, and then hope that the understanding gained by solving the easier problem gives you some insight about the actual problem you're trying to solve. This is what mathematicians do every time we model a complex real-world system by a smooth, pristine mathematical mechanism. Sometimes this approach is very successful; if you're tracking the path of a heavy projectile, you can do pretty well by ignoring air resistance and thinking of the moving body as subject only to a constant force of gravity. Other times, your simplification is so simple that it eliminates the interesting features of the problem, as in the old joke about the physicist tasked with optimizing dairy production: he begins, with great confidence, "Consider a spherical cow . . ."

In this spirit, one might try to get some ideas about Buffon's needle via the solution of the easier *franc-carreau* problem: "Consider a circular needle . . ." But it's not clear what useful information one can draw from a coin, whose rotational symmetry robs it of the very feature that makes the needle problem interesting.

Instead, we turn to the other strategy, which is the one Barbier used: *make the problem harder.* That doesn't sound promising. But when it works, it works like a charm.

Let's start small. What if we ask, more generally, about the expected number of crack crossings by a needle that's two slats wide? That sounds like a more complicated question, because now there are three possible outcomes instead of two. The needle could land entirely within one slat, it could cross one crack, or it could cross two. So to compute the expected number of crossings it seems we'd have to compute the probability of three separate events instead of just two.

But thanks to additivity, this harder problem is easier than you think. Draw a dot in the center of the long needle and label the two halves "1" and "2," like so:

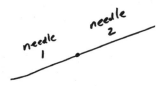

Then the expected number of crossings on the long needle is just the sum of the expected number of crossings by half-needle 1 and the expected number of crossings by half-needle 2. In algebraic terms, if X is the number of cracks crossed by half-needle 1, and Y the number of cracks crossed by half-needle 2, then the total number of cracks the long needle crosses is X + Y. But each of the two pieces is a needle of the length originally considered by Buffon; so each of those needles, on average, crosses the cracks p times; that is, E(X) and E(Y) are both equal to p. Thus the expected number of crossings of the whole needle, E(X+Y), is just E(X) + E(Y), which is p + p, which is 2p.

And the same reasoning applies to a needle of length three, or four, or a hundred times the width of a slat. If a needle has length N (where we now take the width of a slat to be our unit of measure) its expected number of crossings is Np.

This works for short needles as well as long ones. Suppose I throw a needle whose length is 1/2—that is, it's just half as long as a slat is wide. Since Buffon's length-1 needle can be split into two length-1/2 needles, his expected value p must be the twice the expected number of crossings on the length-1/2 needle. So the length-1/2 needle has (1/2)p expected crossings. In fact, the formula

Expected number of crossings of a length-N needle = Np

holds for *any* positive real number N, large or small.

(At this point, we've left rigorous proof behind—some technical argument is necessary to justify why the statement above is okay when N is some hideous irrational quantity like the square root of 2. But I promise you that the essential ideas of Barbier's proof are the ones I'm putting on the page.)

Now comes a new angle, so to speak—*bend the needle:*

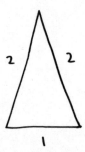

This needle is the longest yet, length 5 in all. But it's bent in two places, and I've closed the ends to form a triangle. The straight segments have length 1, 2, and 2; so the expected number of crossings on each segment is p, 2p, and 2p respectively. The number of crossings on the whole needle is the sum of the number of crossings on each segment. So additivity tells us that the expected number of crossings on the whole needle is

p + 2p + 2p = 5p.

In other words, the formula

Expected number of crossings of a length-N needle = Np

holds for bent needles, too.

Here's one such needle:

And another:

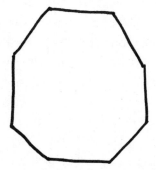

And another:

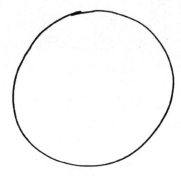

We've seen those pictures before. They're the same ones Archimedes and Eudoxus used two millennia ago, when they were developing the method of exhaustion. That last picture looks like a circle with diameter 1. But it's really a polygon made out of 65,536 tiny little needles. Your eye can't tell the difference—and neither can the floor. Which means the expected number of crossings of the diameter-1 circle is just about exactly the same as the expected number of crossings of the 65,536-gon. And by our bent-needle rule, that's Np, where N is the perimeter of the polygon. What's that perimeter? It must be almost exactly that of the circle; the circle has radius 1/2, so its circumference is π. So the expected number of times the circle crosses a crack is πp.

How's making the problem harder working out for you? Doesn't it seem like we're making the problem more and more abstract, and more and more general, without ever addressing the fundamental issue: what is p?

Well, guess what: we just computed it.

Because how many crossings does the circle have? All of a sudden, what looked like a hard problem becomes easy. The symmetry we lost when we went from coin to needle has now been restored by bending the needle into a circular hoop. And this simplifies matters tremendously. It doesn't matter where the circle falls—it crosses the lines in the floor exactly twice.

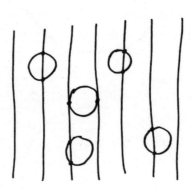

So the expected number of crossings is 2; and it is also πp; and so we have discovered that $p = 2 / \pi$, just as Buffon said. In fact, the argument above applies to any needle, however polygonal and curvy it might be; the expected number of crossings is Lp, where L is the length of the needle in slat-width units. Throw a mass of spaghetti on the tile floor and I can tell you exactly how many times to expect a strand to cross a line. This generalized version of the problem is called, by mathematical wags, *Buffon's noodle problem*.

THE SEA AND THE STONE

Barbier's proof reminds me of what the algebraic geometer Pierre Deligne wrote of his teacher, Alexander Grothendieck: "Nothing seems to happen, and yet at the end a highly nontrivial theorem is there."

Outsiders sometimes have an impression that mathematics consists of applying more and more powerful tools to dig deeper and deeper into the unknown, like tunnelers blasting through the rock with ever more powerful explosives. And that's one way to do it. But Grothendieck, who remade much of pure mathematics in his own image in the 1960s and '70s, had a different view: "The unknown thing to be known appeared to me as some stretch of earth or hard marl, resisting penetration . . . the sea advances insensibly in silence, nothing seems to happen, nothing moves, the water is so far off you hardly hear it . . . yet it finally surrounds the resistant substance."

The unknown is a stone in the sea, which obstructs our progress. We can try to pack dynamite in the crevices of rock, detonate it, and repeat until the rock breaks apart, as Buffon did with his complicated computations in calculus. Or you can take a more contemplative approach, allowing your level of understanding gradually and gently to rise, until after a time what appeared as an obstacle is overtopped by the calm water, and is gone.

Mathematics as currently practiced is a delicate interplay between monastic contemplation and blowing stuff up with dynamite.

AN ASIDE ON MATHEMATICIANS AND INSANITY

Barbier published his proof of Buffon's theorem in 1860, when he was just twenty-one, a promising student at the École Normale Supérieure in Paris. By 1865, troubled by a nervous condition, he'd left town with no forwarding address. No mathematician saw him again until an old teacher of his, Joseph Bertrand, located him in a mental asylum in 1880. As for Grothendieck, he too left academic mathematics, in the 1980s; he now lives in Salingeresque seclusion somewhere in the Pyrenees. No one really knows what math he's working on, if any. Some say he herds sheep.

These stories resonate with a popular myth about mathematics—that it drives you crazy, or is itself a species of craziness. David Foster Wallace, the most mathematical of modern novelists (he once took a break from fiction to write a whole book about transfinite set theory!) described the myth as the "Math Melodrama," and described its protagonist as "a kind of Prometheus-Icarus figure whose high-altitude genius is also hubris and Fatal Flaw." Movies like *A Beautiful Mind*, *Proof*, and π use math as a shorthand for obsession and flight from reality. And a best-selling murder mystery, Scott Turow's *Presumed Innocent,* turned on the twist that the hero's own wife, a mathematician, was actually the demented killer. (In this case, the myth comes with a chaser of off-kilter sexual politics, the book's strong implication being that the difficulty of stretching a *woman's* brain onto a mathematical frame is what sent the murderess over the edge.) You can find a more recent version of the myth in *The Curious Incident of the Dog in the Night-Time,* in which mathematical ability presents itself as just another color in the autism spectrum.

Wallace rejects this melodramatic picture of the mental life of mathematicians, and so do I. In real life, mathematicians are a pretty ordinary bunch, no madder than the average, and it's not actually very common for us to slink off into isolation to wage lonely battles in unforgiving abstract realms. Mathematics tends to strengthen the mind rather than strain it to its breaking point. If anything, I've found that in moments of emotional extremity there is nothing like a math problem to quiet the complaints the rest of the psyche serves up. Math, like meditation, puts you in direct contact with the universe, which is bigger than you, was here before you, and will be here after you. It might drive me crazy *not* to do it.

"TRYING TO MAKE IT ROLL"

Meanwhile, in Massachusetts:

The more people played Cash WinFall, the less profitable it was. Each big purchaser who entered the game split the prizes into more pieces. At one point, Gerald Selbee told me, Yuran Lu from Random Strategies suggested that they and the Selbee group agree to take turns playing the roll-downs, guaranteeing each group a higher profit margin. Selbee paraphrased Yuran's proposal as "You're a big player, I'm a big player, we can't control these other players who are fleas in our hair." By cooperating, Selbee and Lu could at least control each other. The plan made sense, but Selbee didn't bite. He was comfortable with exploiting a quirk in the game, since the rules of the game were public, just as available to any other player as they were to him. But colluding with other players—though it's not clear this would have violated any lottery rules—felt too much like cheating. So the cartels settled into an equilibrium, all three pouring money into every roll-down drawing. With the high-volume bettors buying 1.2 to 1.4 million tickets a drawing, Selbee estimated that lottery tickets on roll-down days had an expected value of just 15% more than their cost.

That's still a pretty nice profit. But Harvey and his confederates weren't satisfied. The life of a professional lottery winner isn't the cartoon of leisure you might imagine. For Harvey, running Random Strategies was a full-time job, and not a particularly fulfilling one. Before roll-down day, tens of thousands of lottery tickets had to be purchased and bubbled in by hand; on the day itself, Harvey had to manage the logistics of multiple team members scanning all those slips at the convenience stores that agreed to handle the team's megapurchases. And after the winning numbers were announced, there was still the long slog of sifting the winning tickets from the worthless losers. Not that you could throw the losing tickets in the trash; Harvey saved those in storage boxes, because when you win the lottery a lot, the IRS audits you a lot, and Harvey needed to be able to document his gambling activities. (Gerald Selbee still has twenty-some plastic laundry tubs full of losing lottery tickets, about $18 million worth, occupying the back of a pole barn on his property.) The winning tickets required some effort, too. Each member of the group had to fill out an individual W-2G tax form for each drawing, no matter how small their share. Does it sound like fun yet?

The inspector general estimated that Random Strategies made $3.5 million, before taxes, over the seven-year life of Cash WinFall. We don't know how much of that money went to James Harvey, but we do know he bought a new car.

It was a used 1999 Nissan Altima.

The good times, the early days of Cash WinFall, when you could double your money with ease, weren't so far in the past; surely Harvey and his team wanted to get back there. But how could they, with the

Selbee family and the Doctor Zhang Lottery Club buying up hundreds
of thousands of tickets for every roll-down drawing?

The only time the other high-volume bettors took a break was when
the jackpot wasn't large enough to trigger the roll-down. But Harvey,
too, sat those drawings out, for a good reason: without the roll-down
money, the lottery was a crappy bet.

On Friday, August 13, 2010, the lottery projected the jackpot for the
next Monday's drawing at $1.675 million, well short of the roll-down
threshold. The Zhang and Selbee cartels were quiet, waiting for the jackpot to creep up over the roll-down level. But Random Strategies made a
different play. Over the previous months, they'd quietly prepared hundreds of thousands of extra tickets, waiting for a day when the projected
jackpot was close to $2 million, but not quite there. This was the day.
And over the weekend, their members fanned across Greater Boston,
buying up more tickets than anyone had before; around 700,000 in all.
With the unexpected infusion of cash from Random Strategies, the jackpot on Monday, August 16, stood at $2.1 million. It was a roll-down,
payday for lottery players, and nobody except the MIT students knew
it was coming. Almost 90% of the tickets for the drawing were held
by Harvey's team. They were standing in front of the money spigot, all
alone. And when the drawing was over, Random Strategies had made
$700,000 on their $1.4 million investment, a cool 50% profit.

This trick wasn't going to work twice. Once the lottery realized what
had happened, they put an early-warning system in place to notify top
management if it looked like one of the teams was trying to push the
jackpot over the roll-down line single-handedly. When Random Strategies tried again in late December, the lottery was ready. On the morning
of December 24, three days before the drawing, the chief of staff of the
lottery got an e-mail from his team saying "Cash WinFall guys are trying
to make it roll again." If Harvey was betting on the lottery being off-duty
for the holiday, he wagered wrong; early Christmas morning, the lottery
updated its estimated jackpot to announce to the world that a roll-down
was coming. The other cartels, still smarting from their August snooker-
ing, canceled their Christmas vacations and bought hundreds of thousands of tickets, bringing profits back down to normal levels.

At any rate, the game was almost up. Sometime shortly afterward, a friend of *Boston Globe* reporter Andrea Estes noticed something funny in the "20-20 list" of winners that the lottery makes public: there were a lot of people in Michigan winning prizes, and they were all winning one particular game, Cash WinFall. Did Estes think there was anything to it? Once the *Globe* started asking questions, the whole picture quickly came clear. On July 31, 2011, the *Globe* ran a front-page story by Estes and Scott Allen explaining how the three betting clubs had been monopolizing the Cash WinFall prizes. In August, the lottery changed the rules of WinFall, capping at $5,000 the total ticket sales any individual retailer could disburse in a day, effectively blocking the cartels from making their high-volume purchases. But the damage was done. If the point of Cash WinFall was to seem like a better deal for ordinary players, the game was now pointless. The last WinFall drawing—fittingly, a roll-down—was held on January 23, 2012.

IF GAMBLING IS EXCITING, YOU'RE DOING IT WRONG

James Harvey wasn't the first person to take advantage of a poorly designed state lottery. Gerald Selbee's group made millions on Michigan's original WinFall game before the state got wise and shut it down in 2005. And the practice goes back much further. In the early eighteenth century, France financed government spending by selling bonds, but the interest rate they offered wasn't enticing enough to drive sales. To spice the pot, the government attached a lottery to the bond sales. Every bond gave its holder the right to buy a ticket for a lottery with a 500,000-livre prize, enough money to live on comfortably for decades. But Michel Le Peletier des Forts, the deputy finance minister who conceived the lottery plan, had botched the computations; the prizes to be disbursed substantially exceeded the money to be gained in ticket receipts. In other words, the lottery, like Cash WinFall on roll-down days, had a positive expected value for the players, and anyone who bought enough tickets was due for a big score.

One person who figured this out was the mathematician and explorer Charles-Marie de La Condamine; just as Harvey would do almost three centuries later, he gathered his friends into a ticket-buying cartel.

One of these was the young writer François-Marie Arouet, better known as Voltaire. While he may not have contributed to the mathematics of the scheme, Voltaire placed his stamp on it. Lottery players were to write a motto on their ticket, to be read aloud when a ticket won the jackpot; Voltaire, characteristically, saw this as a perfect opportunity to epigrammatize, writing cheeky slogans like "All men are equal!" and "Long live M. Peletier des Forts!" on his tickets for public consumption when the cartel won the prize.

Eventually, the state caught on and canceled the program, but not before La Condamine and Voltaire had taken the government for enough money to be rich men for the rest of their lives. What—you thought Voltaire made a living writing perfectly realized essays and sketches? Then, as now, that's no way to get rich.

Eighteenth-century France had no computers, no phones, no rapid means of coordinating information about who was buying lottery tickets and where: you can see why it took the government some months to catch on to Voltaire and Le Condarmine's scheme. What was Massachusetts's excuse? The *Globe* story came out *six years* after the lottery first noticed college students making bizarrely large purchases in supermarkets near MIT. How could they not have known what was going on?

That's simple: they did know what was going on.

They didn't even have to sleuth it out, because James Harvey had come to the lottery offices in Braintree in January 2005, before his betting cartel placed its first bet, before it even had a name. His plan seemed too good to be true, such a sure thing that there must be some regulatory barrier to carrying it out. He went to the lottery to see whether his high-volume betting scheme fell within the rules. We don't know exactly what conversation took place, but it seems to have amounted to "Sure, kid, knock yourself out." Harvey and company placed their first big bet just a few weeks later.

Gerald Selbee arrived not long after; he told me he met with lottery lawyers at Braintree in August 2005, to let them know his Michigan corporation would be buying lottery tickets in Massachusetts. The existence of high-volume betting was no secret to the state.

But why would Massachusetts allow Harvey, Doctor Zhang, and the

Selbees to cart off state money by the millions? What kind of casino lets the players beat the house, week after week, and takes no action?

To unravel this requires thinking a little more closely about how the lottery actually works. Out of every $2 lottery ticket sold, Massachusetts kept 80 cents. Some of that money was used to pay commissions to stores that sell tickets and to operate the lottery itself, and the rest was sent out to city and town governments across the state; almost $900 million in 2011, paying police officers, funding school programs, and generally spackling over the holes in municipal budgets.

The other $1.20 was plowed back into the prize pool, to be distributed among the players. But remember the computation we did at the very beginning? The expected value of a ticket, on a normal day, is just 80 cents, meaning the state is giving back, on average, 80 cents per ticket sold. What happens to the extra 40 cents? That's where the roll-down comes in. Giving out 80 cents per ticket isn't enough to exhaust the prize pool, so the jackpot grows bigger each week until it hits $2 million and rolls down. And that's when the lottery changes its nature; the floodgates are opened and the accumulated money pours out, into the hands of whoever's smart enough to be waiting.

It might look like the state's losing money that day, but that's taking a limited view. Those millions never belonged to Massachusetts; they were earmarked as prize money from the beginning. The state takes its 80 cents out of each ticket and gives back the rest. The more tickets sold, the more revenue comes in. The state doesn't care who wins. The state just cares how many people play.

So when the betting cartels cashed in the fat profits on their roll-down bets, they weren't taking money from the state. They were taking it from the other players, especially the ones who made the bad decision to play the lottery on days without a roll-down. The cartels weren't beating the house. *They were the house.*

Like the operators of a Las Vegas casino, the high-volume bettors weren't totally impervious to bad luck. Any roulette player can go on a hot streak and take the casino for a lot of money, and the same thing could have happened to the cartels if an ordinary bettor had hit all six numbers, diverting all the roll-down money to their own jackpot. But

Harvey and the others had done the math carefully enough to make this outcome rare enough to tolerate. Only once in the whole course of Cash WinFall did somebody win the jackpot on a roll-down day. If you make enough bets with the odds tilted in your favor, the sheer volume of your advantage dilutes any bad luck you might experience.

That makes playing the lottery less exciting, to be sure. But for Harvey and the other high-volume bettors, excitement wasn't the point. Their approach was governed by a simple maxim: *if gambling is exciting, you're doing it wrong.*

If the betting cartels were the house, then what was the state? The state was . . . the state. Just as Nevada charges the casinos on the Strip a percentage of their profits, in exchange for maintaining the infrastructure and regulation that allows their business to thrive, Massachusetts took its steady cut from the money the cartels were raking in. When Random Strategies bought 700,000 tickets to trigger the roll-down, the towns of Massachusetts got 40 cents out of each of those tickets, a $560,000 take. States don't like to gamble, good odds or no. States like to collect taxes. That, in essence, is what the Massachusetts State Lottery was doing. And not unsuccessfully, either. According to the inspector general's report, the lottery took in $120 million of revenue on Cash WinFall. When you walk away with a nine-figure haul, you probably didn't get scammed.

So who did get scammed? The obvious answer is "the other players." It was their cash, after all, that ended up rolling into the cartels' pockets. But Inspector General Sullivan concludes his report in a tone of voice that suggests no one got scammed at all:

> As long as the Lottery announced to the public an impending $2 million jackpot that would likely trigger a roll-down, an ordinary bettor buying a single ticket or any number of tickets was not disadvantaged by high-volume betting. In short, no one's odds of having a winning ticket were affected by high-volume betting. Small bettors enjoyed the same odds as high-volume bettors. When the jackpot hit the roll-down threshold, Cash WinFall became a good bet for everyone, not just the big-time bettors.

Sullivan is right that the presence of Harvey and the other cartels didn't affect the chance of another player's ticket being a winner. But he's making the same mistake Adam Smith did—the relevant question isn't just how likely you are to win, but how *much*, on average, you can expect to win or lose. The cartels' purchases of hundreds of thousands of tickets substantially increased the number of pieces into which each roll-down prize would be sliced, which makes each winning ticket less valuable. In that sense, the cartels were hurting the average player.

Analogy: if hardly anyone shows up for the church raffle, it's pretty likely that I'll win the casserole pot. When a hundred new people show up and buy raffle tickets, my chance of winning the casserole pot goes way down. That might make me unhappy. But is it unfair? What if I find out that those hundred people are actually all working for one mastermind, who really, really wants a casserole pot and has calculated that the cost of a hundred raffle tickets is about 10% less than the retail price? That's unsporting, somehow—but I can't really say I'd feel cheated. And of course the crowded raffle is a lot better than the empty raffle at making money for the church, which is, in the end, the point of the enterprise.

Still, even if the high-volume bettors aren't scammers, there's something discomfiting about the Cash WinFall story. By virtue of the game's quirky rules, the state ended up doing the equivalent of licensing James Harvey as the proprietor of a virtual casino, taking money month after month from less sophisticated players. But doesn't that mean the rules were bad? As William Galvin, the Massachusetts secretary of state, told the *Globe*: "It's a private lottery for skilled people. The question is why?"

If you go back to the numbers, a possible answer suggests itself. Remember, the point of switching to WinFall was to increase the lottery's popularity. And they succeeded—but maybe not as well as they'd planned. What if the buzz around Cash WinFall had gotten so strong that the lottery started selling 3.5 million tickets to ordinary Bay Staters each time roll-down day arrived? Remember, the more people who play, the bigger the state's 40% cut. As we computed before, if the state sells 3.5 million tickets, it comes out ahead even on the roll-down days. Under those circumstances, high-volume betting isn't profitable anymore: the loophole closes, the cartels dissolve, and everybody, except maybe the high-volume players themselves, winds up happy.

Selling that many tickets would have been a long shot, but lottery officials in Massachusetts might have thought that if they got lucky they could pull it off. In a way, the state liked to gamble after all.

TWELVE

MISS MORE PLANES!

George Stigler, the 1982 Nobelist in economics, used to say, "If you never miss the plane, you're spending too much time in airports." That's a counterintuitive slogan, especially if you've actually missed a flight recently. When I'm stuck in O'Hare, eating a cruddy $12 chicken Caesar wrap, I seldom find myself applauding my economic good sense. But as weird as Stigler's slogan sounds, an expected value computation shows it's completely correct—at least for people who fly a lot. To simplify matters, we can just consider three choices:

Option 1: arrive 2 hours before flight, miss flight 2% of the time
Option 2: arrive 1.5 hours before flight, miss flight 5% of the time
Option 3: arrive 1 hour before flight, miss flight 15% of the time

How much it costs you to miss a flight depends very strongly on context, of course; it's one thing to miss the shuttle to DC and hop on the next one, quite another to miss the last flight out when you're trying to get to a family wedding at ten the next morning. In the lottery, both the cost of the ticket and the size of the prize are denominated in dollars. It's much less clear how to weigh the cost of the time we might waste sitting in the terminal against the cost of missing the flight. Both are annoying, but there's no universally recognized currency of annoyingness.

Or at least there's no such currency on paper. But decisions must be made, and economists aspire to tell us how to make them, and so some version of the annnoyingness dollar must be constructed. The standard economic story is that human beings, when they're acting rationally, make decisions that maximize their *utility*. Everything in life has utility; good things, like dollars and cake, have positive utility, while bad things, like stubbed toes and missed planes, have negative utility. Some people even like to measure utility in standard units, called *utils*.* Let's say an hour of your time at home is worth one util; then arriving two hours before your flight costs you two utils, while arriving one hour before costs you only one. Missing a plane is clearly worse than wasting an hour of your time. If you think it's worth about six hours of your time, you can think of a missed plane as costing you six utils.

Having translated everything into utils, we can now compare the expected values of the three strategies.

Option 1	$-2 + 2\% \times (-6) = -2.12$ utils
Option 2	$-1.5 + 5\% \times (-6) = -1.8$ utils
Option 3	$-1 + 15\% \times (-6) = -1.9$ utils

Option 2 is the one that costs you the least utility on average, even though it comes with a nontrivial chance of missing your flight. Yes, getting stuck in the airport is painful and unpleasant—but is it so painful and unpleasant that it's worth spending an extra half hour at the terminal, time after time, in order to reduce the already small chance of missing your plane?

Maybe you say yes. Maybe you *hate* missing your plane, and missing a plane costs you twenty utils, not six. Then the computation above changes, and the conservative option 1 becomes the preferred choice, with an expected value of

$$-2 + 2\% \times (-20) = -2.4 \text{ utils.}$$

---

* Customarily pronounced "you-tills," but in my experience it's much more fun to say "yoodles."

But that doesn't mean Stigler is wrong; it just moves the tradeoff to a different place. You could reduce your chance of missing the plane even further by arriving three hours earlier; but doing so, even if it reduced your chance of missing the plane essentially to zero, would come with a guaranteed cost of 3 utils for the flight, making it a worse choice than option 1. If you graph the number of hours you leave yourself at the airport against your expected utility, you get a picture that looks like this:

It's the Laffer curve again! Showing up fifteen minutes before the plane leaves is going to slam you with a very high probability of missing the plane, with all the negative utility that implies. On the other hand, arriving many hours before also costs you many utils. The optimal course of action falls somewhere in between. Exactly *where* it falls depends on how you personally feel about the relative merits of missing planes and wasting time. But that optimal strategy always assigns you some positive probability of missing the flight—it might be small, but it's not zero. If you literally never miss a flight, you may be off to the left of the best strategy. Just as Stigler says, you should save your utils and miss more planes.

Of course, this kind of computation is necessarily subjective; your extra hour in the airport might not cost you as many utils as mine does. (I *really* hate those airport chicken Caesar wraps.) So you can't ask the theory to spit out an exact optimal time to arrive at the airport or an optimal number of planes to miss. The output is qualitative, not quantitative. I don't know what your ideal probability of missing a plane is; I just know it's not zero.

One warning: in practice, a probability that's close to zero can be hard to distinguish from a probability that actually is zero. If you're a global jet-setting economist, accepting a 1% risk of missing a plane might really mean missing a flight every year. For most people, such a low risk might well mean going your whole life without missing a plane—so if 1% is the right level of risk for you, always catching the plane doesn't mean you're doing anything wrong. Similarly, one doesn't take Stigler's argument to make a good case for "If you've never totaled your car, you drive too slow." What Stigler would say is that if you have *no risk at all* of totaling your car, you're driving too slow, which is trivially true: the only way to have no risk is to not drive at all!

Stigler-style argument is a handy tool for all sorts of optimization problems. Take government waste: you don't go a month without reading about a state worker who gamed the system to get an outsized pension, or a defense contractor who got away with absurdly inflated prices, or a city agency that has long outlived its function but persists at the public expense thanks to inertia and powerful patrons. Typical of the form is an item from the *Wall Street Journal*'s *Washington Wire* blog of June 24, 2013:

> The Social Security Administration's inspector general on Monday said the agency improperly paid $31 million in benefits to 1,546 Americans believed to be deceased.
>
> And potentially making matters worse for the agency, the inspector general said the Social Security Administration had death certificate information on each person filed in the government database, suggesting it should have known the Americans had died and halted payments.

Why do we allow this kind of thing to persist? The answer is simple— eliminating waste has a cost, just as getting to the airport early has a cost. Enforcement and vigilance are worthy goals, but eliminating *all* the waste, just like eliminating even the slightest chance of missing a plane, carries a cost that outweighs the benefit. As blogger (and former mathlete) Nicholas Beaudrot observed, that $31 million represents .004% of the benefits disbursed annually by the SSA. In other words, the agency is already *extremely good* at knowing who's alive and who's no more. Getting even better at that distinction, in order to eliminate those last few mistakes, might be expensive. If we're going to count utils, we shouldn't be asking, "Why are we wasting the taxpayer's money?," but "What's the right amount of the taxpayer's money to be wasting?" To paraphrase Stigler: if your government isn't wasteful, you're spending too much time fighting government waste.

ONE MORE THING ABOUT GOD, THEN I PROMISE WE'RE DONE

One of the first people to think clearly about expected value was Blaise Pascal; puzzled by some questions posed to him by the gambler Antoine Gombaud (self-styled the Chevalier de Méré), Pascal spent half of 1654 exchanging letters with Pierre de Fermat, trying to understand which bets, repeated over and over, would tend to be profitable in the long run, and which would lead to ruin. In modern terminology, he wished to understand which kinds of bets had positive expected value and which kinds were negative. The Pascal-Fermat correspondence is generally thought of as marking the beginning of probability theory.

On the evening of November 23, 1654, Pascal, already a pious man, experienced an intense mystical experience, which he documented in words as best he could:

> FIRE.
> God of Abraham, God of Isaac, God of Jacob
> Not of the philosophers and the scholars . . .
> I have cut myself off from him, shunned him, denied him,
>   crucified him.
> Let me never be cut off from him!
> He can only be kept by the ways taught in the Gospel.
> Sweet and total renunciation.
> Total submission to Jesus Christ and my director.
> Everlasting joy in return for one day's effort on earth.

Pascal sewed this page of notes into the lining of his coat and kept it there the rest of his life. After his "night of fire," Pascal largely with-

The *Mémorial*, Parchment Copy. Photograph © Bibliothèque Nationale de France, Paris.

drew from mathematics, devoting his intellectual effort to religious topics. By 1660, when his old friend Fermat wrote to propose a meeting, he
replied:

> For, to talk frankly with you about Geometry, is to me the very best
> intellectual exercise: but at the same time I recognize it to be so use-
> less that I can find little difference between a man who is nothing
> else but a geometrician and a clever craftsman . . . my studies have
> taken me so far from this way of thinking, that I can scarcely remem-
> ber that there is such a thing as geometry.

Pascal died two years later, at thirty-nine, leaving behind a collection
of notes and short essays meant for a book defending Christianity. These
were later collected as the *Pensées* ("Thoughts") which appeared eight
years after his death. It's a remarkable work, aphoristic, endlessly quot-
able, in many ways despairing, in many ways inscrutable. Much of it
comes in short, numbered bursts:

> 199. Let us imagine a number of men in chains, and all condemned
> to death, where some are killed each day in the sight of the others,
> and those who remain see their own fate in that of their fellows, and
> wait their turn, looking at each other sorrowfully and without hope.
> It is an image of the condition of men.
>   209. Art thou less a slave by being loved and favored by thy mas-
> ter? Thou art indeed well off, slave. Thy master favors thee; he will
> soon beat thee.

But what the *Pensées* are most famous for is thought 233, which Pascal titled *"Infinite-rien"* ("Infinity-nothing") but which is universally
known as "Pascal's wager."
As we've mentioned, Pascal held the question of God's existence to
be one that logic couldn't touch: " 'God is, or He is not.' But to which side
shall we incline? Reason can decide nothing here." But Pascal doesn't
stop there. What is the question of belief, he asks, if not a kind of gamble,
a game with the highest possible stakes, a game you have no choice but
to play? And the analysis of wagers, the distinction between the smart

play and the foolish one, was a subject Pascal understood better than almost anyone on earth. He had not quite left his mathematical work behind him after all.

How does Pascal compute the expected value of the game of faith?
The key is already present in his mystic revelation:

Everlasting joy in return for one day's effort on earth.

What is this, but a reckoning of the costs and benefits of adopting
faith? Even in the middle of ecstatic communion with his savior, Pascal
was still doing math! I love this about him.

To compute Pascal's expected value, we still need the probability that
God exists; say for a moment we are pretty fervent doubters and assign
this hypothesis a probability of only 5%. If we believe in God, and we
turn out to be right, then our reward is "everlasting joy," or, in the economists' terms, infinitely many utils.* If we believe in God and we turn out
to be *wrong*—an outcome we are 95% sure will be the case—then we pay
a price; maybe more than the "one day's effort" that Pascal suggests, since
we have to count not only the time spent in worship but the opportunity
cost of all the libertine pleasures we forwent in our quest for salvation.
Still, it's a certain fixed sum, let's say a hundred utils.

Then the expected value of belief is

$$(5\%) \times infinity + (95\%)(-100)$$

Now, 5% is a small number. But infinite joy is a lot of joy; 5% of it is
still infinite. So it swamps whatever finite cost imposed on us by adopting religion.

We've already discussed the perils of trying to assign a numerical
probability to a proposition like "God exists." It is not clear any such assignment makes sense. But Pascal doesn't make any such dodgy numerical
move. He doesn't need to. Because it doesn't matter whether that number
is 5% or something else. One percent of infinite bliss is still infinite bliss,

---

* Although I've heard at least one economist argue that since a certain amount of future happiness is worth less than the same amount of happiness now, the value of eternal joy in the bosom
of Abraham is actually finite.

and outweighs whatever finite costs attach to a life of piety. The same goes for 0.1% or 0.000001%. All that matters is that the probability God exists is *not zero*. Don't you have to concede that point? That the existence of the Deity is at least *possible*? If so, then the expected value computation seems unequivocal: it is worth it to believe. The expected value of that choice is not only positive, but infinitely positive.

Pascal's argument has serious flaws. The gravest is that it suffers from the Cat in the Hat problem we saw in chapter 10, failing to consider all possible hypotheses. In Pascal's setup, there are only two options: that the God of Christianity is real and will reward that particular sector of the faithful, or that God doesn't exist. But what if there's a God who damns Christians eternally? Such a God is surely possible too, and this possibility alone suffices to kill the argument: now, by adopting Christianity, we are wagering on a chance of infinite joy but also taking on the risk of infinite torment, with no principled way to weight the relative odds of the two options. We're back to our starting point, where reason can decide nothing.

Voltaire raised a different objection. You might have expected him to be sympathetic to Pascal's wager—as we've already seen, he had no objection to gambling. And he admired mathematics; his attitude toward Newton approached worship (he once called him "the god to whom I sacrifice") and he was romantically entangled for many years with the mathematician Émilie du Châtelet. But Pascal was not quite Voltaire's sort of thinker. The two men stood at odds across a gulf as much temperamental as philosophical. Voltaire's generally chipper outlook had no room for Pascal's dark, introspective, mystical emissions. Voltaire dubbed Pascal "the sublime misanthrope" and devoted a long essay to knocking down the gloomy *Pensées* piece by piece. His attitude toward Pascal is that of the popular smart kid toward the bitter and nonconforming nerd.

As for the wager, Voltaire said it was "a little indecent and puerile: the idea of a game, and of loss and gain, does not befit the gravity of the subject." More substantively: "The interest I have to believe a thing is no proof that such a thing exists." Voltaire himself, typically sunny, leans toward an informal argument by design: look at the world, look how amazing it is, God is real, QED!

Voltaire has missed the point. Pascal's wager is curiously modern, so much so that Voltaire has not caught up to it. Voltaire is right that, unlike Witztum and the Bible coders, or Arbuthnot, or the contemporary advocates of intelligent design, Pascal is not offering *evidence* for God's existence at all. He is indeed proposing a reason to believe, but the reason has to do with the utility of believing, not the justifiability of believing. In a way, he anticipates the austere stance of Neyman and Pearson we saw in chapter 9. Just like them, he was skeptical that the evidence we encounter will provide a reliable means of determining what is true. Nonetheless, we have no choice but to decide what to *do*. Pascal is not trying to convince you God exists; he is trying to convince you that it would be to your benefit to believe so, and thus that your best course of action is to hang out with Christians and obey the forms of piety, until, just by force of propinquity, you start to truly believe. Can I put Pascal's argument in modern terms better than David Foster Wallace did in *Infinite Jest*? I cannot.

> The desperate, newly sober White Flaggers are always encouraged to invoke and pay empty lip-service to slogans they don't yet understand or believe—e.g. "Easy Does It!" and "Turn It Over!" and "One Day at a Time!" It's called "Fake It Till You Make It," itself an oft-invoked slogan. Everyone on a Commitment who gets up publicly to speak starts out saying he's an alcoholic, says it whether he believes he is yet or not; then everybody up there says how Grateful he is to be sober today and how great it is to be Active and out on a Commitment with his Group, even if he's not grateful or pleased about it at all. You're encouraged to keep saying stuff like this until you start to believe it, just like if you ask somebody with serious sober time how long you'll have to keep schlepping to all these goddamn meetings he'll smile that infuriating smile and tell you just until you start to *want* to go to all these goddamn meetings.

ST. PETERSBURG AND ELLSBERG

Utils are useful when making decisions about items that don't have well-defined dollar values, like wasted time or unpleasant meals. But you also need to talk about utility when dealing with items that *do* have well-defined dollar values—like dollars.

This realization arrived very early in the development of probability theory. Like many important ideas, it entered the conversation in the form of a puzzle. Daniel Bernoulli famously described the conundrum in his 1738 paper "Exposition on a New Theory of the Measurement of Risk": "Peter tosses a coin and continues to do so until it should land 'heads' when it comes to the ground. He agrees to give Paul one ducat if he gets 'heads' on the very first throw, two ducats if he gets it on the second, four if on the third, eight if on the fourth, and so on, so that with each additional throw the number of ducats he must pay is doubled."

This is obviously a rather attractive scenario for Paul, a game he should be willing to ante up some entrance fee to play. But how much? The natural answer, given our experience with lotteries, is to compute the expected value of the amount of money Paul gets from Peter. There's a 50/50 chance that the first throw of the coin lands heads, in which case Paul gets one ducat. If the first throw is tails and the second is heads, an event which happens 1/4 of the time, Paul gets two ducats. To get four, the first three throws have to fall tails, tails, heads, which happens with probability 1/8. Carrying on and adding up, Paul's expected profit is

$$(1/2) \times 1 + (1/4) \times 2 + (1/8) \times 4 + (1/16) \times 8$$
$$+ (1/32) \times 16 + \ldots$$

or

$$1/2 + 1/2 + 1/2 + 1/2 + \ldots$$

That sum is not a number. It's *divergent*; the more terms you add, the bigger the sum gets, growing without bound past any finite threshold.* This seems to suggest that Paul should be willing to spend *any* number of ducats for the right to play this game.

That sounds nuts. And it is! But when the math tells us something

---

* Though remember from chapter 2 that divergent series aren't just the ones which shoot off to infinity; they also include those which fail to settle down in other ways, like Grandi's series $1 - 1 + 1 - 1 + \ldots$

that sounds nuts, mathematicians don't just shrug and walk away. We go hunting for the kink in the tracks where either the math or our intuition has gone off the rails. The condundrum, known as the St. Petersburg paradox, had been devised by Nicolas Bernoulli, Daniel's cousin, some thirty years before, and many of the probabilists of the time had puzzled over it without coming to any satistfying conclusion. The younger Bernoulli's beautiful untwisting of the paradox is a landmark result, and one that has formed the foundation of economic thinking about uncertain values ever since. The mistake, Bernoulli said, is to say that a ducat is a ducat is a ducat. A ducat in the hand of a rich man is not worth the same as a ducat in the hand of a peasant, as is plainly visible from the different levels of care with which the two men treat their cash. In particular, having two thousand ducats isn't twice as good as having one thousand; it is less than twice as good, because a thousand ducats is worth less to a person who already has a thousand ducats than it is to the person who has none. Twice as many ducats doesn't translate into twice as many utils; not all curves are lines, and the relation between money and utility is governed by one of those nonlinear curves.

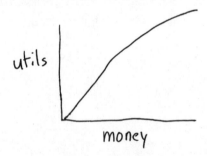

Bernoulli thought that utility grew like the logarithm, so that the kth prize of 2^k ducats was worth just k utils. Remember, we can think of the logarithm as more or less the number of digits: so in dollar terms, Bernoulli's theory is saying that rich people measure the value of their pile by the number of digits after the dollar sign—a billionaire is as much richer than a hundred-millionaire as the hundred-millionaire is richer than the ten-millionaire.

In Bernoulli's formulation, the expected utility of the St. Petersburg game is the sum

$$(1/2) \times 1 + (1/4) \times 2 + (1/8) \times 3 + (1/16) \times 4 + \ldots$$

This tames the paradox; this sum, it turns out, is no longer infinite, or even very large. In fact, there's a beautiful trick that allows us to compute it exactly:

$$\frac{1}{2} + \frac{1}{4} + \frac{1}{8} + \frac{1}{16} + \frac{1}{32} + \ldots = 1$$

$$\frac{1}{4} + \frac{1}{8} + \frac{1}{16} + \frac{1}{32} + \ldots = \frac{1}{2}$$

$$\frac{1}{8} + \frac{1}{16} + \frac{1}{32} + \ldots = \frac{1}{4}$$

$$\frac{1}{16} + \frac{1}{32} + \ldots = \frac{1}{8}$$

$$\frac{1}{32} + \ldots = \frac{1}{16}$$

$$\frac{1}{2} + \frac{2}{4} + \frac{3}{8} + \frac{4}{16} + \frac{5}{32} + \ldots = 2$$

The sum of the first row, $(1/2) + (1/4) + (1/8) + \ldots$, is 1; this is the very infinite series that Zeno encountered in chapter 2. The second row is the same as the first, but with every entry divided by 2; so its sum must be half the sum of the first row, or 1/2. By the same reasoning, the third row, which is just the second row with each term halved, must have half the sum of the second row; so 1/4. Now the sum of *all* the numbers in the triangle is $1 + 1/2 + 1/4 + 1/8 + \ldots$; just one more than Zeno's sum, which is to say, 2.

But what if we sum down the columns first instead of the rows? Just as with the holes in my parents' stereo set, it can't matter whether we start counting vertically or horizontally; the sum is what the sum is.* In the first column there is just a single 1/2; in the second, there are two

---

* Warning: great dangers await when using this kind of intuitive argument with infinite sums. It's okay in the case at hand, but wildly wrong for knottier infinite sums, especially those with both positive and negative terms.

copies of 1/4, making (1/4) × 2; in the third, three copies of 1/8, making (1/8) × 3, and so on. The series formed by the column sums is none other than the sum Bernoulli set up to study the St. Petersburg problem. And its sum is the sum of all the numbers in the infinite triangle, which is to say: 2. So the amount Paul should pay is the number of ducats his personal utility curve tells him 2 utils is worth.[*]

The shape of the utility curve, beyond the bare fact that it tends to bend downward as the money increases, is impossible to pin down precisely,[†] though contemporary economists and psychologists are constantly devising ever-more-intricate experiments to refine our understanding of its properties. ("Now just get your head settled comfortably at the center of the fMRI, if you don't mind, and I'm going to ask you to rank the following six poker strategies in order from most enticing to least enticing, and after that, if you wouldn't mind just holding still while my postdoc takes this cheek swab . . . ?")

We know, at least, that there is no universal curve; different people in different contexts assign different utilities to money. This fact is important. It gives us pause, or it ought to, when we start making generalizations about economic behavior. Greg Mankiw, the Harvard economist whom we last saw in chapter 1 faintly praising Reaganomics, wrote a widely circulated blog post in 2008 explaining that increased income taxes proposed by presidential candidate Barack Obama would lead him to slack off at work. After all, Mankiw was already at an equilibrium, where the utility of the dollars he'd earn from another hour of work would be exactly canceled by the negative utility imposed by the loss of an hour with his kids. Diminish the number of dollars Mankiw makes per hour, and that trade stops being worth it; he cuts back on work until he drops to the income level where an hour with his kids is worth the same to him as an hour spent working for his Obama-diminished pay. He agrees with Reagan's view of the economy as seen from the standpoint of

---

[*] Although, as Karl Menger—Abraham Wald's PhD advisor—pointed out in 1934, there are variants of the St. Petersburg game so generous that even Bernoulli's logarithmic players would seem to be bound to pay arbitrarily many ducats to play. What if the kth prize is 2^{2^k} ducats?

[†] Indeed, most people would say the utility curve does not even literally *exist*, as such—it should be thought of a loose guideline, not as a real thing with a precise shape we haven't yet measured exactly.

a cowboy-movie star; when the tax rate goes up, you make fewer cowboy movies.

But not everybody is Greg Mankiw. In particular, not everybody has the same utility curve he has. The comic essayist Fran Lebowitz tells a story about her youth in Manhattan, driving a cab. She started driving at the beginning of the month, she said, and kept driving every day until she'd made enough money to pay for rent and food. Then she stopped driving and wrote for the rest of the month. For Lebowitz, all money above a certain threshold contributes essentially zero further utility; she has a different-looking curve than Mankiw does. Hers goes flat once her rent is paid. What happens to Fran Lebowitz if income taxes go up? She works *more*, not less, to bring herself back up to the threshold.*

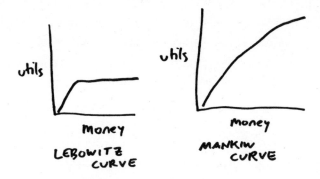

Bernoulli was not the only mathematician to arrive at the idea of utility and its nonlinear relation with money. He'd been anticipated by at least two other researchers. One was Gabriel Cramer of Geneva; the other was a young correspondent of Cramer's, none other than the needle thrower Georges-Louis LeClerc, Comte de Buffon. Buffon's interest in probability was not restricted to parlor games. Late in life, he reminisced about his encounter with the vexing St. Petersburg paradox: "I dreamed about this problem some time without finding the knot; I could not see that it was possible to make mathematical calculations agree with common sense without introducing some moral considerations; and hav-

---

* Lebowitz wrote in her book *Social Studies*, "Stand firm in your refusal to remain conscious during algebra. In real life, I assure you, there is no such thing as algebra." I claim this example shows there is mathematics in Lebowitz's life, whether she refers to it as such or not!

ing expressed my ideas to Mr. Cramer, he told me that I was right, and
that he had also resolved this question by a similar approach."

Buffon's conclusion mirrored Bernoulli's, and he perceives the nonlinearity especially clearly:

> Money must not be estimated by its numerical quantity: if the metal,
> that is merely the sign of wealth, was wealth itself, that is, if the hap-
> piness or the benefits that result from wealth were proportional to
> the quantity of money, men would have reason to estimate it nu-
> merically and by its quantity, but it is barely necessary that the ben-
> efits that one derives from money are in just proportion with its
> quantity; a rich man of one hundred thousand *ecus* income is not ten
> times happier than the man of only ten thousand *ecus*; there is more
> than that what money is, as soon as one passes certain limits it has
> almost no real value, and cannot increase the well-being of its pos-
> sessor; a man that discovered a mountain of gold would not be richer
> than the one that found only one cubic fathom.

The doctrine of expected utility is appealingly straightforward and
simple: presented with a set of choices, pick the one with the highest
expected utility. It is perhaps the closest thing we have to a simple mathematical theory of individual decision making. And it captures many
features of the way humans make choices, which is why it remains a
central part of the quantitative social scientist's tool kit. Pierre-Simon
Laplace, on the last page of his 1814 treatise *A Philosophical Essay on
Probabilities*, writes, "We see, in this Essay, that the theory of probabilities is, in the end, only common sense boiled down to 'calculus'; it points
out in a precise way what rational minds understand by means of a sort
of instinct, without necessarily being aware of it. It leaves nothing to
doubt, in the choice of opinions and decisions; by its use one can always
determine the most advantageous choice."

Again we see it: mathematics is the extension of common sense by
other means.

But expected utility doesn't get at everything. Once again, the troubling complications enter in the form of a puzzle. This time, the
puzzle-bearer was Daniel Ellsberg, who later became famous as the

whistle-blower who leaked the Pentagon Papers to the civilian press. (In mathematical circles, which can be parochial at times, it would not be outlandish to hear it said of Ellsberg, "You know, before he got involved in politics, he did some really important work.")

In 1961, a decade before his explosion into public view, Ellsberg was a brilliant young analyst at the RAND Corporation, consulting with the U.S. government on strategic matters surrounding nuclear war—how it could be prevented, or, barring that, effectively conducted. At the same time, he was working toward a Harvard PhD in economics. On both tracks, he was thinking deeply about the process by which human beings made decisions in the face of the unknown. At the time, the theory of expected utility held a supreme position in the mathematical analysis of decisions. Von Neumann and Morgenstern,* in their foundational book *The Theory of Games and Economic Behavior,* had proven that all people who obeyed a certain set of behavior rules, or axioms, *had* to act as if their choices were governed by the drive to maximize some utility function. These axioms—later refined by Leonard Jimmie Savage, a member of the wartime Statistical Research Group with Abraham Wald—were the standard model of behavior under uncertainty at the time.

Game theory and expected utility theory still play a great role in the study of negotiations among people and states, but never more so than at RAND at the height of the Cold War, where the writings of von Neumann and Morgenstern were the subject of Pentateuchal levels of reverence and analysis. The researchers at RAND were studying something fundamental to human life: the process of choice and competition. And the games they studied, like Pascal's wager, were played for very high stakes.

Ellsberg, the young superstar, had a taste for crossing up established expectations. After graduating third from his class at Harvard, he had startled his intellectual comrades by enlisting in the Marine Corps, where he served for three years as an infantryman. In 1959, as a Harvard Junior Fellow, he delivered a lecture on strategy in foreign policy at the Boston Public Library, in which he famously contemplated the effective-

---

* The same Oskar Morgenstern who got Abraham Wald out of pure math and eventually out of occupied Austria.

ness of Adolf Hitler as a geopolitical tactician: "There is the artist to study, to learn what *can* be hoped for, what can be done with the threat of violence." (Ellsberg always insisted that he didn't recommend that the United States adopt Hitler-style strategies, but only wanted to make dispassionate study of their effectiveness—maybe so, but it's hard to doubt he was trying to get a rise out of his audience.)

So it's perhaps no surprise that Ellsberg was not content to accept the prevailing views. In fact, he'd been picking at the foundations of game theory since his undergraduate senior thesis. At RAND, he devised a famous experiment now known as Ellsberg's paradox.

Suppose there's an urn* with ninety balls inside. You know that thirty of the balls are red; concerning the other sixty balls, you know only that some are black and some are yellow. The experimenter describes to you the following four bets.

> RED: You get $100 if the next ball pulled from the urn is red; otherwise, you get nothing.
> BLACK: You get $100 if the next ball is black, otherwise nothing.
> NOT-RED: You get $100 if the next ball is either black or yellow, otherwise nothing.
> NOT-BLACK: You get $100 if the next ball is either red or yellow, otherwise nothing.

Which bet do you prefer; RED or BLACK? What about NOT-RED versus NOT-BLACK?

Ellsberg quizzed his subjects to find out which of these bets they preferred, given the choice. What he found was that the people he polled tended to prefer RED to BLACK. With RED, you know where you stand: you've got a 1-in-3 chance of getting the money. With BLACK, you have no idea what odds to expect. As for NOT-RED and NOT-BLACK, the situation is just the same; Ellsberg's subjects liked NOT-RED better, preferring the state of knowing that their chance of a payoff is exactly 2/3.

---

* I have never even seen an urn, but it's some kind of iron law of probability theory that if randomly colored balls are to be chosen, it must be an urn that holds them.

Now suppose you have a more complicated choice: you have to pick *two* of the bets. And not any two you like: you have to take either "RED and NOT-RED" or "BLACK and NOT-BLACK." If you prefer RED to BLACK and NOT-RED to NOT-BLACK, it seems reasonable that you prefer "RED and NOT-RED" to "BLACK and NOT-BLACK."

But now here's the problem. Picking RED and NOT-RED is the same thing as giving yourself $100. But so is BLACK and NOT-BLACK! How can one be preferable to the other when they're *the same thing?*

For a proponent of expected utility theory, Ellsberg's results looked very strange. Each bet must be worth a certain number of utils, and if RED has more utility than BLACK, and NOT-RED more than NOT-BLACK, it just has to be the case that RED + NOT-RED is worth more utils than BLACK + NOT-BLACK; but the two are the same. If you want to believe in utils, you have to believe that the participants in Ellsberg's study are just plain wrong in their preferences; they are bad at calculating, or they're not paying close attention to the question, or they're simply crazy. Since the people Ellsberg asked were in fact well-known economists and decision theorists, this conclusion presents its own problems for the status quo.

For Ellsberg, the answer to the paradox is simply that expected utility theory is incorrect. As Donald Rumsfeld would later put it, there are known unknowns and there are unknown unknowns, and the two are to be processed differently. The "known unknowns" are like RED—we don't know which ball we'll get, but we can quantify the probability that the ball will be the color we want. BLACK, on the other hand, subjects the player to an "unknown unknown"—not only are we not sure whether the ball will be black, we don't have any knowledge of how likely it is to be black. In the decision-theory literature, the former kind of unknown is called *risk*, the latter *uncertainty*. Risky strategies can be analyzed numerically; uncertain strategies, Ellsberg suggested, were beyond the bounds of formal mathematical analysis, or at least beyond the bounds of the flavor of mathematical analysis beloved at RAND.

None of which is to deny the incredible utility of utility theory. There are many situations, lotteries being one, where the mystery we're subject to is all risk, governed by well-defined probabilities; and there are many more circumstances where "unknown unknowns" are present but play only a small role. We see here the characteristic push and pull of the mathematical approach to science. Mathematicians like Bernoulli and von Neumann construct formalisms that apply a penetrating light to a sphere of inquiry only dimly understood before; mathematically fluent scientists like Ellsberg work to understand the limits of those formalisms, to refine and improve them where it's possible to do so, and to post strongly worded warning signs where it's not.

Ellsberg's paper is written in a vivid, literary style uncharacteristic of technical economics. In his concluding paragraph, he writes of his experimental subjects that "the Bayesian or Savage approach gives wrong predictions and, by their lights, bad advice. They act in conflict with the axioms deliberately, without apology, because it seems to them the sensible way to behave. Are they clearly mistaken?"

In the world of cold war Washington and RAND, decision theory and game theory were held in the highest intellectual esteem, seen as the scientific tools that would win the next world war, as the atom bomb had won the last one. That those tools might actually be limited in their application, especially in contexts for which there was no precedent and thus no means of estimating probabilities—like, say, *the instantaneous reduction of the human race to radioactive dust*—must have been at least a little troubling for Ellsberg. Was it here, over a disagreement about math, that his doubts about the military establishment really began?

THIRTEEN

WHERE THE TRAIN
TRACKS MEET

The notion of utility helps make sense of a puzzling feature of the Cash WinFall story. When Gerald Selbee's betting group bought massive quantities of tickets, they used Quic Pic, letting the lottery's computers pick the numbers on their slips at random. Random Strategies, on the other hand, picked their numbers themselves; this meant they had to fill out hundreds of thousands of slips by hand, then feed them through the machines at their chosen convenience stores one by one, a massive and incredibly dull undertaking.

The winning numbers are completely random, so every lottery ticket has the same expected value; Selbee's 100,000 Quic Pics would bring in the same amount of prize money, on average, as Harvey and Lu's 100,000 artisanally marked tickets. As far as expected value is concerned, Random Strategies did a lot of painful work for no reward. Why?

Consider this case, which is simpler but of the same nature. Would you rather have $50,000, or would you rather have a 50/50 bet between losing $100,000 and gaining $200,000? The expected value of the bet is

$$(1/2) \times (-\$100,000) + (1/2) \times (\$200,000) = \$50,000,$$

the same as the cash. And there is indeed some reason to feel indifferent between the two choices; if you made that bet time after time after time, you'd almost certainly make $200,000 about half the time and lose $100,000 the other half. Imagine you alternated winning and losing: after two bets you've won $200,000 and lost $100,000 for a net gain of $100,000, after four bets you're up $200,000, after six bets $300,000, and so on: a profit of $50,000 per bet on average, just the same as if you'd gone the safe route.

But now pretend for a moment that you're not a character in a word problem in an economics textbook, but rather an actual person—an actual person who does not have $100,000 cash on hand. When you lose that first bet and your bookie—let us say your big, angry, bald, power-lifting bookie—comes to collect, do you say, "An expected value calculation shows that it's very likely I'll be able to pay you back in the long run"? You do not. That argument, while mathematically sound, will not achieve its goals.

If you're an actual person, you should take the $50,000.

This reasoning is well captured by utility theory. If I'm a corporation with limitless funds, losing $100,000 might not be so bad—let's say it's worth –100 utils—while winning $200,000 brings me 200 utils. In that case, dollars and utils might match up to be nicely linear; a util is just another name for a grand.

But if I'm an actual person with meager savings, the calculus is rather different. Winning $200,000 would change my life more than it would the corporation's, so maybe it's worth more to me—say 400 utils. But losing $100,000 doesn't just clean out my bank account, it puts me in hock to the angry bald power lifter. That's not just a bad day for the balance sheet, it's a serious injury hazard. Maybe we rate it at –1,000 utils. In which case the expected utility of the bet is

$$(1/2) \times (-1000) + (1/2) \times (400) = -300$$

The negative utility of this bet means this is not only worse than a sure $50,000, *it's worse than doing nothing at all.* The 50% chance of being totally wiped out is a risk you just can't afford—at least, not without the promise of a much bigger reward.

This is a mathematical way of formalizing a principle you already know: the richer you are, the more risks you can afford to take. Bets like the one above are like risky stock investments with a positive expected dollar payoff; if you make a lot of these investments, you might sometimes lose a bunch of cash at once, but in the long run you'll come out ahead. The rich person, who has enough reserves to absorb those occasional losses, invests and gets richer; the nonrich people stay right where they are.

A risky investment can make sense even if you don't have the money to cover your losses—as long as you have a backup plan. A certain market move might come with a 99% chance of making a million dollars and a 1% chance of losing $50 million. Should you make that move? It has a positive expected value, so it seems like a good strategy. But you might also balk at the risk of absorbing such a big loss—especially because small probabilities are notoriously hard to be certain about.* The pros call moves like this "picking up pennies in front of a steamroller"—most of the time you make a little money, but one small slip and you're squashed.

So what do you do? One strategy is to leverage yourself up to the eyeballs until you've got enough paper assets to make the risky move, but scaled up by a factor of one hundred. Now you're very likely to make $100 million per transaction—great! And if the steamroller gets you? You're out $5 billion. Except you're not—because the world economy, in these interconnected times, is a big rickety tree house held together with rusty nails and string. An epic collapse of one part of the structure runs a serious risk of pulling down the whole shebang. The Federal Reserve has a strong disposition not to let that happen. As the old saying goes, if you're down a million bucks, it's your problem; but if you're down five billion bucks, it's the government's problem.

This financial strategy is cynical, but it often works—it worked for Long-Term Capital Management in the 1990s, as chronicled in Roger Lowenstein's superb book *When Genius Failed*, and it worked for the firms that survived, and even profited from, the financial collapse of

---

* Analysts like Nassim Nicholas Taleb argue, persuasively in my opinion, that it's a fatal error to assign numerical probabilities to rare financial events at all.

2008. Absent fundamental changes that seem nowhere in sight, it will work again.*

Financial firms are not human, and most humans, even rich humans, don't like uncertainty. The rich investor might happily take the 50-50 bet with an expected value of $50,000, but would probably prefer to take the $50,000 outright. The relevant term of art is *variance*, a measure of how widely spread out the possible outcomes of a decision are, and how likely one is to encounter the extremes on either end. Among bets with the same expected dollar value, most people, especially people without limitless liquid assets, prefer the one with lower variance. That's why some people invest in municipal bonds, even though stocks offer higher rates of return in the long run. With bonds, you're *sure* you're going to get your money. Invest in stocks, with their greater variance, and you're likely to do better—but you might end up much worse.

Battling variance is one of the main challenges of managing money, whether you call it that or not. It's because of variance that retirement funds diversify their holdings. If you have all your money in oil and gas stocks, one big shock to the energy sector can torch your whole portfolio. But if you're half in gas and half in tech, a big move in one batch of stocks needn't be accompanied by any action in the others; it's a lower-variance portfolio. You want to have your eggs in different baskets, *lots* of different baskets; this is exactly what you do when you stash your savings in a giant index fund, which distributes its investments across the entire economy. The more mathematically minded financial self-help books, like Burton Malkiel's *A Random Walk down Wall Street*, are fond of this strategy; it's dull, but it works. *If retirement planning is exciting . . .*

Stocks, at least in the long run, tend to get more valuable on average; investing in the stock market, in other words, is a positive expected-value move. For bets that have *negative* expected value, the calculus flips; people hate a sure loss as much as they like a sure win. So you go for bigger variance, not smaller. You don't see people swagger up to the rou-

---

* Of course there's ample reason to believe that some people inside the banks knew their investments were pretty likely to founder and that they lied about this; the point is that *even when bankers are honest* the incentives push them toward taking stupid risks at the public's eventual expense.

lette wheel and lay one chip on every number; that's just an unnecessarily elaborate way of handing chips to the dealer.

What does all this have to do with Cash WinFall? As we said at the top, the expected dollar value of 100,000 lottery tickets is what it is, no matter which tickets you buy. But the variance is a different story. Suppose, for instance, I decide to go into the high-volume betting game, but I take a different approach; I buy 100,000 copies of the same ticket.

If that ticket happens to match 4 out of the 6 numbers in the lottery drawing, then I'm the lucky holder of 100,000 pick-4 winners, and I'm basically going to sweep up the entire $1.4 million prize pool, for a tidy 600% profit. But if my set of numbers is a loser, I lose my whole $200,000 pile. That's a high-variance bet, with a big chance of a big loss and a small chance of an even bigger win.

So "don't put all your money on one number" is pretty good advice—much better to spread your bets around. But wasn't that exactly what Selbee's gang was doing by using the Quic Pic machine, which chooses numbers at random?

Not quite. First of all, while Selbee wasn't putting all his money on one ticket, he *was* buying the same ticket multiple times. At first, that seems strange. At his most active, he was buying 300,000 tickets per drawing, letting the computer pick his numbers randomly from almost 10 million choices. So his purchases amounted to a mere 3% of the possible tickets; what are the odds he'd buy the same ticket twice?

Actually, they're really, really good. Old chestnut: bet the guests at a party that two people in the room have the same birthday. It had better be a good-sized party—say there are thirty people there. Thirty birthdays out of 365 options* isn't very many, so you might think it pretty unlikely that two of those birthdays would land on the same day. But the relevant quantity isn't the number of people: it's the number of *pairs* of people. It's not hard to check that there are 435 pairs of people,† and each pair has a 1 in 365 chance of sharing a birthday; so in a party that size you'd expect to see a pair sharing a birthday, or maybe even two

---

* 366 if you count leap days, but we're not going for precision here.
† The first person in the pair can be any of the 30 people, and the second any of the 29 who remain, giving 30×29 choices; but this counts each pair twice, since it counts {Ernie, Bert} and {Bert, Ernie} separately; so the right number of pairs is $(30 \times 29)/2 = 435$.

pairs. In fact, the chance that two people out of thirty share a birthday turns out to be a little over 70%—pretty good odds. And if you buy 300,000 randomly chosen lottery tickets out of 10 million options, the chance of buying the same ticket twice is so close to 1 that I'd rather just say "it's a certainty" than figure out how many more 9s I'd need after "99.9%" to specify the probability on the nose.

And it's not just repeated tickets that cause the trouble. As always, it can be easier to see what's going on with the math if we make the numbers small enough that we can draw pictures. So let's posit a lottery draw with just seven balls, of which the state picks three as the jackpot combination. There are thirty-five possible jackpot combos, corresponding to the thirty-five different ways that three numbers can be chosen from the set 1, 2, 3, 4, 5, 6, 7. (Mathematicians like to say, for short, "7 choose 3 is 35.") Here they are, in numerical order:

123 124 125 126 127
134 135 136 137
145 146 147
156 157
167
234 235 236 237
245 246 247
256 257
267
345 346 347
356 357
367
456 457
467
567

Say Gerald Selbee goes to the store and uses the Quic Pic to buy seven tickets at random. His chance of winning the jackpot remains pretty small. But in this lottery, you also get a prize for hitting two out of three numbers. (This particular lottery structure is sometimes called the *Tran-*

sylvanian lottery, though I could find no evidence that such a game has ever been played in Transylvania, or by vampires.)

Two out of three is a pretty easy win. So I don't have to keep typing "two out of three," let's call a ticket that wins this lesser prize a *deuce*. If the jackpot drawing is 1, 4 and 7, for example, the four tickets with a 1, a 4, and some number *other* than 7 are all deuces. And besides those four, there are the four tickets that hit 1-7 and the four that hit 4-7. So twelve out of thirty-five, just over a third of the possible tickets, are deuces. Which suggests there are probably at least a couple of deuces among Gerald Selbee's seven tickets. To be precise, you can compute that Selbee has

5.3% chance of no deuces
19.3% chance of exactly one deuce
30.3% chance of two deuces
26.3% chance of three deuces
13.7% chance of four deuces
4.3% chance of five deuces
0.7% chance of six deuces
0.1% chance of all seven tickets being deuces.

The expected number of deuces is thus

$$5.3\% \times 0 + 19.3\% \times 1 + 30.3\% \times 2 + 26.3\% \times 3 + 13.7\% \times 4 \\ + 4.3\% \times 5 + 0.7\% \times 6 + 0.1\% \times 7 = 2.4$$

The Transylvanian version of James Harvey, on the other hand, doesn't use the Quic Pic; he fills out his seven tickets by hand, and here they are:

124
135
167
257
347
236
456

Suppose the lottery draws 1, 3, and 7. Then Harvey's holding three deuces: 135, 167, and 347. What if the lottery draws 3, 5, 6? Then Harvey once again has three deuces among his tickets, with 135, 236, and 456. Keep trying possible combinations and you'll quickly see that Harvey's choices have a very special property: either he wins the jackpot, or he wins *exactly* three deuces. The chance that the jackpot is one of Harvey's seven tickets is 7 out of 35, or 20%. So he has a

20% chance of no deuces
80% chance of three deuces.

His expected number of deuces is

$$20\% \times 0 + 80\% \times 3 = 2.4$$

the same as Selbee's, as it must be. But the variance is much smaller; Harvey has very little uncertainty about how many deuces he's going to get. That makes Harvey's portfolio a lot more attractive to potential cartel members. Note especially: whenever Harvey doesn't get three deuces, he wins the jackpot. That means that Harvey's strategy *guarantees* a substantial minimum payoff, something the Quic-Pickers like Selbee can never do. Picking the numbers yourself can get rid of your risk while maintaining the reward—if you pick the numbers right.

And how do you do that? That is—literally, for once!—the million-dollar question.

First try: just ask your computer to do it. Harvey and his team were MIT students, presumably able to knock off a few dozen lines of code before their morning coffee. Why not just write a program to run through all combinations of 300,000 WinFall tickets to see which one provided the lowest-variance strategy?

That wouldn't be a hard program to write. The one small problem would be the way all matter and energy in the universe decayed into heat death by the time your program had handled the first tiny fragment of a microsliver of the data it was trying to analyze. From the point of view of a modern computer, 300,000 is not, a very large number. But the objects that the proposed program has to pick through are not the 300,000 tickets—they are the possible sets of 300,000 tickets to be purchased from the 10 million possible Cash WinFall tickets. How many of those sets are there? More than 300,000. More than the number of subatomic particles that exist or have ever existed. A lot more. You've probably never even *heard* of a number as big as the number of ways to select your 300,000 tickets.*

What we're up against here is the dreaded phenomenon known by computer-science types as "the combinatorial explosion." Put simply: very simple operations can change manageably large numbers into absolutely impossible ones. If you want to know which of the fifty states is the most advantageous place to site your business, that's easy; you just have to compare fifty different things. But if you want to know which *route* through the fifty states is the most efficient—the so-called traveling salesman problem—the combinatorial explosion goes off, and you face difficulty on a totally different scale. There are about 30 vigintillion routes to choose from. In more familiar terms, that's 30 thousand trillion trillion trillion trillion trillion.

Boom!

So there'd better be another way to choose our lottery tickets to tamp down variance. Would you believe me if I told you it all came down to plane geometry?

WHERE THE TRAIN TRACKS MEET

Parallel lines don't meet. That's what makes them parallel.

But parallel lines sometimes *appear* to meet—think of a pair of train tracks, alone in an empty landscape, the two rails seeming to converge as your eyes follow them closer and closer to the horizon. (I find it helps to have some country music playing if you want a really vivid mental image here.) This is the phenomenon of *perspective*; when you try to depict the three-dimensional world on your two-dimensional field of vision, something has to give.

The people who first figured out what was going on here were the

* Unless you've heard of a googolplex. Now *that* is a big number, boy howdy.

people who needed to understand both how things are and how things look, and the difference between the two: namely, painters. The moment, early in the Italian Renaissance, at which painters understood perspective was the moment visual representation changed forever, the moment when European paintings stopped looking like your kid's drawings on the refrigerator door (if your kid mostly drew Jesus dead on the cross) and started looking like the things they were paintings of.*

How exactly Florentine artists like Filippo Brunelleschi came to develop the modern theory of perspective has occasioned a hundred quarrels among art historians, into which we won't enter here. What we know for sure is that the breakthrough joined aesthetic concerns with new ideas from mathematics and optics. A central point was the understanding that the images we see are produced by rays of light that bounce off objects and subsequently strike our eye. This sounds obvious to a modern ear, but believe me, it wasn't obvious then. Many of the ancient scientists, most famously Plato, argued that vision must involve a kind of fire that emanated from the eye. This view goes at least as far back as Alcmaeon of Croton, one of the Pythagorean weirdos we met in chapter 2. The eye must generate light, Alcmaeon argued: what other source could there be for the *phosphene*, the stars you see when you shut your eyes and press down on your eyeball? The theory of vision by reflected rays was worked out in great detail by the eleventh-century Cairene mathematician Abu 'Ali al-Hasan ibn al-Haytham (but let's call him Alhazen, as most Western writers do). His treatise on optics, the *Kitab al-Manazir*, was translated into Latin and taken up eagerly by philosophers and artists seeking a more systematic understanding of the relation between sight and the thing seen. The main point is this: a point P on your canvas represents a *line* in three-dimensional space. Thanks to Euclid, we know there's a unique line containing any two specified points. In this case, the line is the one containing P and your eye. Any object in the world that lies on that line gets painted at point P.

Now imagine you're Filippo Brunelleschi standing out on the flat

* Or at least they looked like certain kinds of optical representations of the things they were paintings of, which over the years we've come to think of as realistic; what counts as "realism" has been the subject of hot contention among art critics for about as long as there's been art criticism.

prairie, the canvas on an easel in front of you, painting the train tracks.*
The track consists of two rails, which we call R_1 and R_2. Each one of
these rails, drawn on the canvas, is going to look like a line. And just as a
point on the canvas corresponds to a line in space, a line on the canvas
corresponds to a plane. The plane P_1 corresponding to R_1 is the one swept
out by the lines joining each point on the rail to your eye. In other words,
it's the unique plane containing both your eye and the rail R_1. Similarly,
the plane P_2 corresponding to R_2 is the one containing your eye and R_2.
Each of the two planes cuts the canvas in a line, and we call these lines
L_1 and L_2.

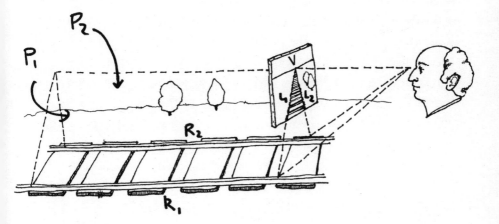

The two rails are parallel. *But the two planes are not.* How could they
be? They meet at your eye, and parallel planes do not meet anywhere.
But planes that aren't parallel have to intersect in a line. In this case, the
line is horizontal, emanating from your eye and proceeding parallel to
the train tracks. The line, being horizontal, does not meet the prairie—it
shoots out toward the horizon, never touching the ground. But—and
here is the point—it meets the canvas, at some point V. Since V is on the
plane R_1, it must be on the line L_1 where R_1 cuts the canvas. And since V
is also on R_2, it must be on L_2. In other words, V is the point on the canvas where the painted train tracks meet. In fact, any straight path on the
prairie that runs parallel to the train tracks will look, on the canvas, like

* Anachronistic, okay, but just go with it.

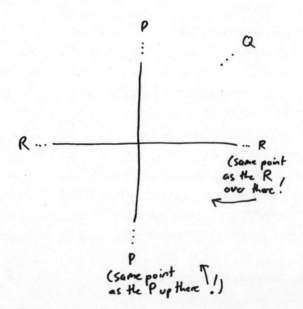

a line through V. V is the so-called vanishing point, the point through which the paintings of all lines parallel to the tracks must pass. In fact, every pair of parallel tracks determines some vanishing point on the canvas; where the vanishing point is depends on which direction the parallel lines are going. (The only exceptions are pairs of lines parallel to the canvas itself, like the slats between the rails—they'll still look parallel in your painting.)

The conceptual shift that Brunelleschi made here is the heart of what mathematicians call projective geometry. Instead of points in the landscape, we think of lines through our eye. At first glance, the distinction might seem purely semantic; each point on the ground determines one and only one line between the point and our eye, so what does it matter whether we think about the point or think about the line? The difference is just this: there are more lines through our eye than there are points on the ground, because there are *horizontal* lines, which don't intersect the ground at all. These correspond to the vanishing points on our canvas, the places where train tracks meet. You might think of this line as a point on the ground that is "infinitely far away" in the direction of the tracks. And indeed, mathematicians usually call them *points at infinity*. When you take the plane Euclid knew and glue on the points at infinity, you get the *projective plane*. Here's a picture of it:

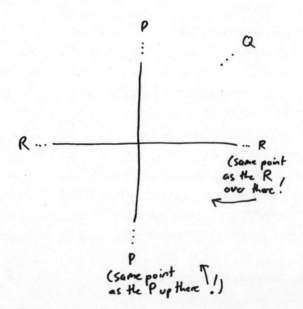

Most of the projective plane looks just like the regular flat plane you're used to. But the projective plane has more points, those so-called points at infinity: one for each possible direction along which a line can be oriented in the plane. You should think of the point P, which corresponds to the vertical direction, as being infinitely high up along the vertical axis—but also infinitely *low down* along the vertical axis. In the projective plane, the two ends of the y-axis *meet* at the point at infinity, and the axis is revealed to be not really a line but a circle. In the same way, Q is the point that's infinitely far northeast (or southwest!) and R is the point at the end of the horizontal axis. Or rather, at *both* ends. If you travel infinitely far to the right, until you arrive at R, and then keep on going, you find yourself still traveling rightward but now heading back toward the center from the left edge of the picture.

This kind of leaving-one-way-and-coming-back-the-other enthralled the young Winston Churchill, who recalled vividly the one mathematical epiphany of his life:

> I had a feeling once about Mathematics, that I saw it all—Depth beyond depth was revealed to me—the Byss and the Abyss. I saw, as one might see the transit of Venus—or even the Lord Mayor's Show, a quantity passing through infinity and changing its sign from plus to minus. I saw exactly how it happened and why the tergiversation was inevitable: and how the one step involved all the others. It was like politics. But it was after dinner and I let it go!

In fact, point R is not just the endpoint of the horizontal axis, but of *any* horizontal line. If two different lines are both horizontal, they are parallel; and yet, in projective geometry, they meet, at the point at infinity. David Foster Wallace was asked in a 1996 interview about the ending of *Infinite Jest*, which many people found abrupt: Did he, the interviewer asked, avoid writing an ending because he "just got tired of writing it"? Wallace replied, rather testily: "There is an ending as far as I'm concerned. Certain kinds of parallel lines are supposed to start converging in such a way that an 'end' can be projected by the reader somewhere beyond the right frame. If no such convergence or projection occurred to you, then the book's failed for you."

The projective plane has the defect that it's kind of hard to draw, but the advantage that it makes the rules of geometry much more agreeable. In Euclid's plane, two different points determine a single line, and two different lines determine a single intersection point—unless they're parallel, in which case they don't meet at all. In mathematics, we like rules, and we don't like exceptions. In the projective plane, you don't have to make any exceptions to the rule that two lines meet at a point, because parallel lines meet too. Any two vertical lines for instance, meet at P, and any two lines pointing northeast to southwest meet at Q. Two points determine a single line, two lines meet at a single point, end of story.* It's perfectly symmetrical and elegant in a way that classical plane geometry is not. And it's not coincidence that projective geometry arose naturally from attempts to solve the practical problem of depicting the three-dimensional world on a flat canvas. Mathematical elegance and practical utility are close companions, as the history of science has shown again and again. Sometimes scientists discover the theory and leave it to mathematicians to figure out why it's elegant, and other times mathematicians develop an elegant theory and leave it to scientists to figure out what it's good for.

One thing the projective plane is good for is representational painting. Another is picking lottery numbers.

A TINY GEOMETRY

The geometry of the projective plane is governed by two axioms:

> Every pair of points is contained in exactly one common line.
> Every pair of lines contains exactly one common point.

Once mathematicians had found *one* kind of geometry that satisfied these two perfectly tuned axioms, it was natural to ask whether there

---

* But if the lines containing R are all horizontal, and the lines containing P are all vertical, what is the line through R and P? It is a line we haven't drawn, the *line at infinity*, which contains all the points at infinity and none of the points of the Euclidean plane.

were any more. It turns out there are a lot. Some are big, some are small. The very tiniest is called the *Fano plane*, after its creator, Gino Fano, who in the late nineteenth century was one of the first mathematicians to take seriously the idea of finite geometries. It looks like this:

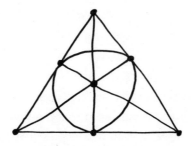

This is a small geometry indeed, consisting of only seven points! The "lines" in this geometry are the curves shown in the diagram; they're small, too, possessing only three points each. There are seven of them, six of which *look* like lines and the other of which looks like a circle. And yet this so-called geometry, exotic as it is, satisfies axioms 1 and 2 just as well as Brunelleschi's plane did.

Fano had an admirably modern approach—he had, to use Hardy's phrase, "the habit of definition," avoiding the unanswerable question of what geometry *really was*, and asking, instead: Which phenomena behave like geometry? In Fano's own words:

A base del nostro studio noi mettiamo una *varietà* qualsiasi di enti di qualunque natura; enti che chiameremo, per brevità, punti indipendentemente però, ben inteso, dalla loro stessa natura.

That is:

As a basis for our study we assume an arbitrary *collection* of entities of an arbitrary nature, entities which, for brevity, we shall call points, but this is quite independent of their nature.

For Fano and his intellectual heirs, it doesn't matter whether a line "looks like" a line, a circle, a mallard duck, or anything else—all that

matters is that lines *obey the laws* of lines, set down by Euclid and his successors. If it walks like geometry, and it quacks like geometry, we call it
geometry. To one way of thinking, this move constitutes a rupture between mathematics and reality, and is to be resisted. But that view is too
conservative. The bold idea that we can think geometrically about systems that don't look like Euclidean space,* and even call these systems
"geometries" with head held high, turned out to be critical to understanding the geometry of the relativistic space-time we live in; and nowadays we use generalized geometric ideas to map Internet landscapes,
which are even further removed from anything Euclid would recognize.
That's part of the glory of math; we develop a body of ideas, and once
they're correct, *they're correct*, even when applied far, far outside the context in which they were first conceived.

For example: here's Fano's plane again, but with the points labeled by
the numbers 1 through 7:

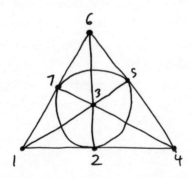

Look familiar? If we list the seven lines, recording for each the set of
three points that constitute it, we get:

124
135

---

* To be fair, there is another sense in which the Fano plane really does look like more traditional
geometry. Descartes taught us how to think of points on the plane as pairs of *coordinates* x and y,
which are real numbers; if you use Descartes's construction but draw your coordinates from
number systems other than the real numbers, you get other geometries. If you do Cartesian geometry using the Boolean number system beloved of computer scientists, which has only two
numbers, the bits 0 and 1, you get the Fano plane. That's a beautiful story, but it's not the story
we're telling just now. See the endnotes for a little more of it.

167
257
347
236
456

This is none other than the seven-ticket combo we saw in the last section, the one that hits each pair of numbers exactly once, guaranteeing a minimum payoff. At the time, that property seemed impressive and mystical. How could anyone have come up with such a perfectly arranged set of tickets?

But now I've opened the box and revealed the trick: it's simple geometry. Each pair of numbers appears on exactly one ticket, because each pair of points appears on exactly one line. It's just Euclid, even though we're speaking now of points and lines Euclid would not have recognized as such.

I'M SORRY, DID YOU SAY "BOFAB"?

The Fano plane tells you how to play the seven-number Transylvanian lottery without taking on any risk, but what about the Massachusetts lottery? There are lots of finite geometries with more than seven points, but none, unfortunately, that precisely meet the requirements of Cash WinFall. Something more general is needed. The answer doesn't come directly from Renaissance painting or Euclidean geometry, but from another unlikely source—the theory of digital signal processing.

Suppose I want to send an important message to a satellite, like "Turn on right thruster." Satellites don't speak English, so what I'm actually sending is a sequence of 1s and 0s, what computer scientists call *bits*:

1110101 . . .

This message seems crisp and unambiguous. But in real life, communication channels are noisy. Maybe a cosmic ray strikes the satellite just as it's receiving your transmission and garbles one bit of the message, so that the satellite receives

1010101 . . .

That message doesn't seem very different; but if the changing of the
bit switches the instruction from "right thruster" to "left thruster," the
satellite might be in for serious trouble.

Satellites are expensive, so this is trouble you really want to avoid. If
you were trying to talk to a friend at a noisy party, you might have to
repeat yourself to keep the noise from drowning out your message. The
same trick works with the satellite; in our original message, we can repeat each bit twice, sending 00 instead of 0 and 11 instead of 1:

11 11 11 00 11 00 11 . . .

Now, when the cosmic ray strikes the second bit of the message, the
satellite sees

10 11 11 00 11 00 11 . . .

The satellite *knows* that each two-bit segment is supposed to be either 00 or 11, so that initial "10" is a red flag; something has gone wrong.
But what? That's tough for the satellite to figure out: since it doesn't
know exactly where the noise corrupted the signal, there's no way to tell
whether the original message started with 00 or 11.

This problem, too, can be fixed. Just repeat three times instead of
twice:

111 111 111 000 111 000 111 . . .

The message comes through corrupted, like this:

101 111 111 000 111 000 111 . . .

But now the satellite is in good shape. That first three-bit segment, it
knows, is supposed to be 000 or 111, so the presence of 101 means something's gone awry. But if the original message was 000, *two* bits must

have been corrupted in very close proximity, an unlikely event as long as the frequency of message-zapping cosmic rays is pretty small. So the satellite has good reason to let majority rule: if two out of the three bits are 1, the odds are very good that the original message was 111.

What you've just witnessed is an example of an *error-correcting code*, a communications protocol that allows the receiver to cancel out the errors in a noisy signal.* The idea, like basically everything else in information theory, comes from Claude Shannon's monumental 1948 paper, "A Mathematical Theory of Communication."

A mathematical theory of communication! Doesn't that sound a little grandiose? Isn't communication a fundamentally human activity that can't be reduced to cold numbers and formulas?

Understand this: I warmly endorse, in fact *highly recommend*, a bristly skepticism in the face of all claims that such-and-such an entity can be explained, or tamed, or fully understood, by mathematical means.

And yet the history of mathematics is a history of aggressive territorial expansion, as mathematical techniques get broader and richer, and mathematicians find ways to address questions previously thought of as outside their domain. "A mathematical theory of probability" sounds unexceptional now, but once it would have seemed a massive overreach; math was about the certain and the true, not the random and the maybeso! All that changed when Pascal, Bernoulli, and others found mathematical laws that governed the workings of chance.† A mathematical theory of infinity? Before the work of Georg Cantor in the nineteenth century, the study of the infinite was as much theology as science; now, we understand Cantor's theory of multiple infinities, each one infinitely larger than the last, well enough to teach it to first-year math majors. (To be fair, it does kind of blow their minds.)

These mathematical formalisms don't capture every detail of the phenomena they describe, and aren't intended to. There are questions about randomness, for instance, about which probability theory is silent. To some people, the questions that stay outside the reach of math are the

---

* And *every* signal is noisy, to some degree or another.
† Ian Hacking's *The Emergence of Probability* covers the story superbly.

most interesting ones. But to think carefully about chance, nowadays, *without* having probability theory somewhere in mind is a mistake. If you don't believe me, ask James Harvey. Or, better yet, ask the people whose money he won.

Will there be a mathematical theory of consciousness? Of society? Of aesthetics? People are trying, that's for sure, with only limited success so far. You should distrust all such claims on instinct. But you should also keep in mind that they might end up getting some important things right.

The error-correcting code does not at first seem like revolutionary mathematics. If you're at a noisy party, repeat yourself—problem solved! But that solution has a cost. If you repeat every bit of your message three times, your message takes three times as long to transmit. That might not be a problem at a party, but it could be a problem if you need the satellite to turn on its right thruster *right this second*. Shannon, in the paper that launched the theory of information, identified the basic trade-off that engineers still grapple with today: the more resistant to noise you want your signal to be, the slower your bits are transmitted. The presence of noise places a cap on the length of a message your channel can reliably convey in a given amount of time; this limit was what Shannon called the *capacity* of the channel. Just as a pipe can only handle so much water, a channel can only handle so much information.

But fixing errors doesn't require you to make your channel three times skinnier, as the "repeat three times" protocol does. You can do better—and Shannon knew this perfectly well, because one of his colleagues at Bell Labs, Richard Hamming, had already figured out how.

Hamming, a young veteran of the Manhattan Project, had low-priority access to Bell's ten-ton Model V mechanical relay computer; he was only allowed to run his programs on the weekends. The problem was that any mechanical error could halt his computation, with no one available to start the machine running again until Monday morning. This was annoying. And annoyance, as we know, is one of the great spurs to technical progress. Wouldn't it be better, Hamming thought, if the machine could correct its own errors and keep on plugging? And so he developed a plan. The input to the Model V can be thought of as a string of 0s and 1s, just like the transmission to the satellite—the math doesn't care whether those digits are bits in a digital stream, the states of an electrical relay, or holes in a strip of tape (at the time, a state-of-the-art data interface).

Hamming's first step was to break the message up into blocks of three symbols:

111 010 101 . . .

The *Hamming code** is a rule that transforms each of these three-digit blocks into a seven-digit string. Here's the codebook:

    000 -> 0000000
    001 -> 0010111
    010 -> 0101011
    011 -> 0111100
    101 -> 1011010
    110 -> 1100110
    100 -> 1001101
    111 -> 1110001

So the encoded message would look like

1110001 0101011 1011010. . . .

Those seven-bit blocks are called *code words*. The eight code words are the only blocks the code allows; if the receiver sees anything else coming over the wire, something has gone wrong for sure. Say you receive 1010001. You know this can't be right, because 1010001 isn't a code word. What's more, the message you received differs in only one place from the code word 1110001. And there's no *other* code word that's so close to the messed-up transmission you actually saw. So you can feel pretty safe in guessing that the code word your correspondent meant to send was 1110001, which means that the corresponding 3-digit block in the original message was 111.

---

* For the technical sticklers, what I'm describing here is actually the dual of the usual Hamming code; in this case, it's an example of a *punctured Hadamard code*.

You might think we just got lucky. What if the mystery transmission had been close to two different code words? Then we wouldn't be able to make a confident judgment. But that can't happen, and here's why. Look again at the lines in the Fano plane:

124
135
167
257
347
236
456

How would you describe this geometry to a computer? Computers like to be talked to in 0s and 1s, so write each line as a string of 0s and 1s, where a 0 in place n stands for "point n is on the line" and a 1 in place n means "point n is not on the line." So that first line, 124, gets represented as

0010111

and the second line, 135, is

0101011

You'll notice that both strings are code words in the Hamming code. In fact, the seven nonzero code words in the Hamming code match up exactly to the seven lines in the Fano plane. The Hamming code and the Fano plane (and, for that matter, the optimal ticket combo for the Transylvanian lottery) are exactly the same mathematical object in two different outfits!

This is the secret geometry of the Hamming code. A code word is a set of three points in the Fano plane that form a line. Flipping a bit in the string is the same thing as adding or deleting a point, so as long as the original code word wasn't 0000000, the bollixed transmission you get corresponds to a set with either two or four points.* If you receive a two-point set, you know how to figure out the missing point; it's just the third point on the unique line that joins the two points you received. What if you receive a four-point set of the form "line plus one extra point?" Then you can infer that the correct message consists of those three points in your set that form a line. A subtlety presents itself: how do you know there's only *one* way of choosing such a set of three points? It helps if we give our points names: call them A, B, C, and D. If A, B, and C all lie on a line, then A, B, and C must be the set of points your correspondent meant to send. But what if A, C, and D also lie along a line? No worries: this is impossible, because the line containing A, B, and C and the line containing A, C, and D would then have the two points A and C in common. But two lines can only intersect in *one* point; that's the rule.† In other words, thanks to the axioms of geometry, the Hamming code has the same magical error-correcting property as "repeat three times"; if a message gets modified by a single bit en route, the receiver can always figure out what message the transmitter meant to send. But instead of multiplying your transmission time by three, your new improved code sends just seven bits for every three bits of your original message, a more efficient ratio of 2.33.

The discovery of error-correcting codes, both Hamming's first codes and the more powerful ones that came later, transformed the engineering of information. The goal no longer had to be building systems so heavily shielded and double-checked that no error could ever arise. After Hamming and Shannon, it sufficed to make errors *rare enough* that the flexibility of the error-correcting code could counteract whatever noise got through. Error-correcting codes are now found wherever data needs to be communicated quickly and reliably. The Mars orbiter *Mariner 9* sent pictures of the Martian surface back to Earth using one such code,

---

* If the original codeword *is* 0000000, then the version with one bit messed up has six 0s and only one 1, making the receiver quite confident that 0000000 was the intended signal.
† If you haven't thought about this before, you have probably found that the argument in this paragraph is hard to follow. The reason it's hard to follow is that you can't get an argument of this kind into your brain by sitting and reading about it—you have to get a pen out and try to write down a set of four points which contains two different lines in the Fano plane, and then fail to do that, and then understand why you failed. There is no other way. I encourage you to write directly in the book, if it's not borrowed from the library or displayed on a screen.

the Hadamard code. Compact discs are encoded with the Reed-Solomon code, which is why you can scratch them and they still sound perfect. (Readers born after, say, 1990 who are unfamiliar with compact discs can just think of flash drives, which use among other things the similar Bose-Chaudhuri-Hocquenghem codes to avoid data corruption.) Your bank's routing number is encoded using a simple code called a *checksum*. This one is not quite an error-correcting code, but merely an error-*detecting* code like the "repeat each bit twice" protocol; if you type one digit wrong, the computer executing the transfer may not be able to puzzle out what number you actually meant, but it can at least figure out something's wrong and avoid sending your money to the wrong bank.

It's not clear whether Hamming understood the full range of applications of his new technique, but his bosses at Bell certainly had some idea, as Hamming found out when he tried to publish his work:

The Patent Department would not release the thing until they had patent coverage. . . . I didn't believe that they could patent a bunch of mathematical formulas. I said they couldn't. They said, "Watch us." They were right. And since then I have known that I have a very weak understanding of patent laws because, regularly, things that you shouldn't be able to patent—it's outrageous—you can patent.

Math moves faster than the patent office: The Swiss mathematician and physicist Marcel Golay learned about Hamming's ideas from Shannon, and developed many new codes of his own, not knowing that Hamming himself had worked out many of the same codes behind the patent curtain. Golay published first, leading to a confusion about credit that persists to the present day. As for the patent, Bell got it, but lost the right to charge for the license as part of an antitrust settlement in 1956.

What makes the Hamming code work? To understand this, you have to come at it from the other direction, asking: What would make it fail?

Remember, the bête noire of an error-correcting code is a block of digits that's simultaneously close to two *different* code words. A recipient presented with the offending string of bits would be flummoxed, having no principled way to determine which of the near-miss code words appeared in the original transmission.

It sounds like we're using a metaphor here: Blocks of binary digits don't have locations, so what can we mean by saying one is "close" to another? One of Hamming's great conceptual contributions was to insist that this wasn't merely a metaphor, or didn't have to be. He introduced a new notion of distance, now called the *Hamming distance*, which was adapted to the new mathematics of information just as the distance Euclid and Pythagoras understood was adapted to the geometry of the plane. Hamming's definition was simple: the distance between two blocks is the number of bits you need to alter in order to change one block into the other. So the distance between the code words 0010111 and 0101011 is 4; in order to get from the former code word to the latter, you have to change the bits in the second, third, fourth, and fifth places.

Hamming's eight code words are a good code because no block of seven bits is within Hamming distance 1 of two different code words. If it were, the two code words would be within Hamming distance 2 of each other.[*] But you can check for yourself and see that no two of those code words differ in just two places; in fact, any two different code words are at a Hamming distance of at least 4 from each other. You can think of the code words as something like electrons in a box, or antisocial people in an elevator. They've got a confined space to live in, and within those constraints, they try to make as much mutual distance from each other as they possibly can.

This same principle underlies all manner of communications that are robust to noise. Natural language works this way: if I write **lanvuage** instead of **language**, you can figure out what it was I meant to say, because there's no other word in English that's one letter substitution away from **lanvuage**. This breaks down, of course, when you start looking at shorter words: a dog, a cog, a bog, and a log are all perfectly good things to which one might refer in English, and a burst of noise wiping out the first phoneme makes it impossible to tell which one was meant. Even in

---

[*] For the experts: that Hamming distance satisfies the *triangle inequality*.

this case, though, you can use the semantic distance between those words to help you correct errors. If it bit you, it was probably a dog; if you fell off it, it was probably a log. And so on.

You can make language more efficient—but when you do, you hit the same hard tradeoff Shannon discovered. Many people of a nerdy and/or mathematical persuasion* have labored to create languages that would convey information compactly and precisely, without any of the redundancy, synonymy, and ambiguity that languages like English indulge themselves in. Ro was an artificial language created in 1906 by the Reverend Edward Powell Foster, who aimed to replace the thicket of English vocabulary with a lexicon in which the meaning of each word could be derived logically from its sound. It's perhaps no surprise that among the Ro enthusiasts was Melvil Dewey, whose Dewey Decimal System imposed on the stacks of the public library a similarly rigid organization. Ro is indeed admirably compact; lots of long English words, like **ingredient**, come out much shorter in Ro, where you just say **cegab**. But the compactness comes at a cost; you lose the error correction English offers as a built-in feature. Small elevator, very crowded, the passengers don't have much personal space; which is to say, each word in Ro is very close to lots of others, creating opportunities for confusion. The word for "color" in Ro is **bofab**. But if you change one letter, to make it **bogab**, you have the word for "sound." **Bokab** means "electricity" and **bolab** means "flavor." Worse still, the logical structure of Ro leads similar-sounding words to have similar meanings too, making it impossible to figure out what's going on from context. **Bofoc**, **bofof**, **bofog**, and **bofol** mean "red," "yellow," "green," and "blue" respectively. It makes a certain kind of sense to have conceptual similarity represented in sound; but it also makes it very difficult to talk about color in Ro at a crowded party. "I'm sorry, was that 'bofoc' or 'bofog'?"[†]

Some modern constructed languages, on the other hand, go the other way, making explicit use of the principles Hamming and Shannon laid

---

* Not the same thing!

† I would like to think the fact that **bebop** is Ro for "elastic" is an undiscovered fragment of the secret history of jazz, but it's probably just a coincidence.

out; Lojban, one of the most successful contemporary examples,* has a strict rule that no two of the basic roots, or *ginsu*, are allowed to be too phonetically close.

Hamming's notion of "distance" follows Fano's philosophy—a quantity that quacks like distance has the right to behave like distance. But why stop there? The set of points at distance less than or equal to 1 from a given central point has a name in Euclidean geometry; it is called a circle, or, if we are in higher dimensions, a sphere.† So we're compelled to call the set of strings at Hamming distance at most 1‡ from a code word a "Hamming sphere," with the code word at the center. For a code to be an error-correcting code, no string—no *point*, if we're to take this geometric analogy seriously—can be within distance 1 of two different code words; in other words, we ask that no two of the Hamming spheres centered at the code words share any points.

So the problem of constructing error-correcting codes has the same structure as a classical geometric problem, that of *sphere packing*: how do we fit a bunch of equal-sized spheres as tightly as possible into a small space, in such a way that no two spheres overlap? More succinctly, how many oranges can you stuff into a box?

The sphere-packing problem is a lot older than error-correcting codes; it goes back to the astronomer Johannes Kepler, who wrote a short booklet in 1611 called *Strena Seu De Nive Sexangula*, or "The Six-Cornered Snowflake." Despite the rather specific title, Kepler's book contemplates the general question of the origin of natural form. Why do snowflakes and the chambers of honeycombs form hexagons, while the seed chambers of an apple are more apt to come in fives? Most relevantly for us right now: why do the seeds of pomegranates tend to have twelve flat sides?

Here's Kepler's explanation. The pomegranate wants to fit as many

* According to the FAQ at lojban.org, the number of people who can speak Lojban conversationally "ranges beyond what can be counted on the fingers of one hand," which in this business is indeed pretty good.
† To be more precise, a sphere is the set of points at distance *exactly* 1 from the center; the space described here, a filled-in sphere, is usually called a *ball*.
‡ Which is to say, at distance either 0 or 1, since Hamming distances, unlike the usual distances in geometry, have to be whole numbers.

seeds as possible inside its skin; in other words, it is carrying out a sphere-packing problem. If we believe nature does as good a job as can be done, then these spheres ought to be arranged in the densest possible fashion. Kepler argued that the tightest possible packing was obtained as follows. Start with a flat layer of seeds, arranged in a regular pattern like so:

The next layer is going to look just the same as this one, but cunningly placed so that each seed sits in the little triangular divot formed by three seeds below it. Then just keep adding more layers in the same way. It's best to be a little careful here: only half the divots are going to support spheres in the next layer up, and at each stage you have a choice of *which* half of the divots you want to fill. The most customary choice, called the *face-centered cubic lattice*, has the nice property that every layer has the spheres placed directly over the spheres three layers below. According to Kepler, there is no denser way to pack spheres in space. And in the face-centered cubic lattice, each sphere touches exactly twelve others. As the pomegranate seeds grew, Kepler reasoned, each one would press against its twelve neighbors, flattening its surface near the point of contact and producing the twelve-sided figures he observed.

Whether Kepler was right about pomegranates I have no idea,* but his claim that the face-centered cubic lattice was the densest possible sphere packing became a topic of intense mathematical interest for centuries. Kepler offered no proof of his statement; apparently it just seemed right to him that the face-centered cubic lattice couldn't be beat. Generations of grocers, who stack oranges in a face-centered cubic configuration without any worry as to whether their method is the absolute best possible, agree with him. Mathematicians, that demanding tribe, wanted absolute confirmation. And not just about circles and spheres; once you're

* We do know, though, that the atoms in the solid forms of aluminum, copper, gold, iridium, lead, nickel, platinum, and silver arrange themselves in face-centered cubic form. One more example of a mathematical theory finding applications its creators could not have contemplated.

in the realm of pure mathematics, nothing stops you from going beyond circles and spheres to yet higher dimensions, packing the so-called hyperspheres of dimension greater than 3. Does the geometric story of high-dimensional sphere packings give insight into the theory of error-correcting codes, as the geometric story of the projective plane did? In this case, the flow has mostly been in the other direction;* the insights from coding theory have instigated progress in sphere packings. John Leech, in the 1960s, used one of Golay's codes to build an incredibly dense packing of twenty-four-dimensional spheres, in a configuration now known as the Leech lattice. It's a crowded place, the Leech lattice, where each of the twenty-four-dimensional spheres touches 196,560 of its neighbors. We still don't know whether it's the tightest possible twenty-four-dimensional packing, but in 2003, Henry Cohn[†] and Abhinav Kumar proved that if a denser lattice exists, it beats Leech by a factor of at most

1.00000000000000000000000000000165.

In other words: close enough.

You can be forgiven for not caring about twenty-four-dimensional spheres and how to smoosh them together, but here's the thing; any mathematical object as startling as the Leech lattice is bound to be important. It turned out that the Leech lattice was very rich in symmetries of a truly exotic kind. The master group theorist John Conway, upon encountering the lattice in 1968, worked out all its symmetries in a twelve-hour spree of computation on a single giant roll of paper. These symmetries ended up forming some of the final pieces of the general theory of finite symmetry groups that preoccupied algebraists for much of the twentieth century.[‡]

---

* Though in contexts where signals are modeled as sequences of real numbers, not sequences of 0s and 1s, the sphere-packing problem is precisely what's needed to design good error-correcting codes.

† Cohn works at Microsoft Research, which is in a way a continuation of the Bell Labs model of pure math supported by high-tech industry, hopefully to the benefit of both.

‡ Yet another great story too long and twisty to wander into here, but see Mark Ronan's *Symmetry and the Monster*.

As for good old three-dimensional oranges, it turns out Kepler was right that his packing was the best possible—but that wasn't proved for almost four hundred years, finally falling in 1998 to Thomas Hales, then a professor at the University of Michigan. Hales settled the matter by means of a difficult and delicate argument that reduced the problem to an analysis of a mere few thousand configurations of spheres, which he dealt with by means of a massive computer calculation. The difficult and delicate argument posed no problem for the math community; we're used to those, and this part of Hales's work was quickly judged and found correct. The massive computer calculation, on the other hand, was trickier. A proof can be checked down to the last detail, but a computer program is a different sort of thing. In principle, a human can check every line of code; but even having done so, how can you be sure the code ran correctly?

Mathematicians have almost universally accepted Hales's proof, but Hales himself seems to have been stung by the initial discomfort with the proof's reliance on computation. Since the resolution of the Kepler conjecture, he's moved away from the geometry that made him famous and turned instead to the project of formal verification of proofs. Hales envisions, and is working to create, a future mathematics that looks very different from our own. In his view, mathematical proofs, whether computer-aided or carried out by humans with pencils, have gotten so complicated and interdependent that we can no longer reasonably have full confidence in their correctness. The classification of finite simple groups, the now-completed program of which Conway's analysis of the Leech lattice formed a crucial part, is distributed over hundreds of papers by hundreds of authors, totaling some ten thousand pages. No human alive can be said to understand it all. So how can we be sure it's really right?

Hales thinks we have no choice but to start over again, rebuilding the vast corpus of mathematical knowledge within a formal structure that can be verified by machine. If the code that checks the formal proofs is itself checkable (and this, Hales convincingly argues, is a feasible goal) then we can free ourselves forevermore from controversies like the one Hales endured over whether a proof is really a proof. And from there?

The next step, maybe, is computers that can construct proofs, or even *have ideas*, without any human intervention at all.

If this actually happens, is mathematics over? Of course, if machines overtake and then surpass humans in all mental dimensions, using us as slaves or livestock or playthings, as some of the most extravagant futurists predict, then yeah, math is over, along with everything else. But short of that, I think mathematics will probably survive. After all, math has already been computer-aided for decades. Many calculations that once would have counted as "research" are now considered no more creative or praiseworthy than adding a series of ten-digit numbers; once your laptop can do it, it's not mathematics anymore. But this hasn't put mathematicians out of work. We've managed to stay just ahead of the ever-increasing sphere of computer dominance, like action heroes outracing a fireball.

And if machine intelligences of the future can take over from us much of the work we know as research now? We'll reclassify that research as "computation." And whatever we quantitatively minded humans are doing with our newly freed-up time, that's what we'll call "mathematics."

The Hamming code is pretty good, but one might hope to do better still. After all, there's something wasteful about Hamming's code: even in the days of punched tape and mechanical relays, computers were reliable enough that almost all seven-bit blocks would come through unscathed. The code seems too conservative; surely we could get away with adding fewer failsafe bits to our message. And so we can; that's what Shannon's famous theorem proves. For example, if errors come at a rate of one per thousand bits, Shannon tells you there are codes that make each message only 1.2% longer than its unencoded form. And better yet, by making the basic blocks longer and longer, you can find codes that achieve this speed and satisfy any desired degree of reliability, however strict.

How did Shannon construct these excellent codes? Well, here's the thing—he didn't. When you encounter an intricate construction like Hamming's, you're naturally inclined to think an error-correcting code is a very special thing, designed and engineered and tweaked and retweaked until every pair of code words has been gingerly nudged apart without any other pair being forced together. Shannon's genius was to see that this vision was totally wrong. Error-correcting codes are the opposite of special. What Shannon proved—and once he understood *what* to prove, it was really not so hard—was that *almost all* sets of code words exhibited the error-correcting property; in other words, a completely random code, with no design at all, was extremely likely to be an error-correcting code.

This was a startling development, to say the least. Imagine you were tasked with building a hovercraft; would your first approach be to throw a bunch of engine parts and rubber tubing on the ground at random, figuring the result would probably float?

Hamming, still impressed forty years later, said of Shannon's proof in 1986:

> Courage is one of the things that Shannon had supremely. You have only to think of his major theorem. He wants to create a method of coding, but he doesn't know what to do so he makes a random code. Then he is stuck. And then he asks the impossible question, "What would the average random code do?" He then proves that the average code is arbitrarily good, and that therefore there must be at least one good code. Who but a man of infinite courage could have dared to think those thoughts? That is the characteristic of great scientists; they have courage. They will go forward under incredible circumstances; they think and continue to think.

If a random code was very likely to be an error-correcting code, what's the point of Hamming? Why not just choose code words completely at random, secure in the knowledge that Shannon's theorem makes it very likely your code corrects errors? Here's one problem with that plan. It's not enough for a code to be able to correct errors in principle; it has to be practical. If one of Shannon's codes uses blocks of size 50, then the number of code words is the number of 0-1 strings fifty bits long, which is 2 to the 50th power, a little over a quadrillion. Big number. Your spacecraft receives a signal, which is supposed to be one of these quadrillion code words, or at least close to one. But *which* one? If you have to flip through the quadrillion code words one by one, you're in big trouble. It's the combinatorial explosion again, and in this context it forces on us another tradeoff. Codes that have a lot of structure, like the Hamming codes, tend to be easy to decode. But these very special codes, it turns out, are usually not as efficient as the completely random ones that Shannon studied! And in the decades between then and now, mathematicians have tried to ride that conceptual boundary between structure and randomness, laboring to construct codes random enough to be fast, but structured enough to be decodable.

The Hamming code is great for the Transylvanian lottery, but not so effective at Cash WinFall. The Transylvanian lottery has just seven numbers; Massachusetts offered forty-six. We're going to need a bigger code. The best one I could find for the purpose was discovered by R. H. F. Denniston of the University of Leicester in 1976. And it's a beauty.

Denniston wrote down a list of 285,384 six-number combinations from a choice of forty-eight numbers. The list starts like this:

    1 2 48 3 4 8
    2 3 48 4 5 9
    1 2 48 3 6 32 . . .

The first two tickets have four numbers in common: 2, 3, 4, and 48. But—and here is the miracle of the Denniston system—you will never find any two of those 285,384 tickets that have *five* numbers in common. You can translate the Denniston system into a code, much as we did with the Fano plane: replace each ticket with a string of 48 1s and 0s, with a 0 in the places corresponding to the numbers on your ticket, and a 1 in the places corresponding to the numbers not on your ticket. So the first ticket above would translate into the codeword:

    000011101111111111111111111111111111111111111110

Check for yourself: the fact that no two tickets agree on five out of six numbers means that this code, like the Hamming code, has no two code words separated by a Hamming distance of less than four.[*]

---

[*] What's the point, when Shannon proved that a totally random choice of code should work just as well? Yes, in a sense, but his theorem in its strongest form requires that code words be allowed to get arbitrarily long. In a case like this, where the code words are fixed to have length 48, you can beat the random code with a little extra care, and this is exactly what Denniston did.

Another way to say this is that every five-number combination appears on at most one of Denniston's tickets. And it gets better: in fact, every five-number combination appears on *exactly* one ticket.*

As you can imagine, a lot of care is required in choosing the tickets on Denniston's list. Denniston includes in his paper a computer program in ALGOL that verifies that his list really does have the magical property he claims, a rather advanced gesture for the 1970s. Still, he insists that the computer's role in the collaboration is to be understood as strictly subordinate to his own: "I should like, indeed, to make it clear that all the results announced here were found without recourse to computers, even though I suggest that computers may be used to verify them."

Cash WinFall has only forty-six numbers, so to play it Denniston-style, you have to destroy the beautiful symmetry a bit by throwing out all the tickets in Denniston's system containing a 47 or a 48. This still leaves you with 217,833 tickets. Suppose you get together $435,666 out of the couch cushions and decide to play these numbers. What happens?

The lottery draws six numbers—say, 4, 7, 10, 11, 34, 46. In the unlikely event these match one of your tickets exactly, you win the jackpot. But even if not, you're still in line to win a healthy pile of money for matching five of the six numbers. Do you have a ticket with 4, 7, 10, 11, 34 on it? *One* of Denniston's tickets does, so the only way you can miss out is if the Denniston ticket with those five numbers was 4, 7, 10, 11, 34, 47 or 4, 7, 10, 11, 34, 48, and thus got trashed.

But what about a different five-number combination, like 4, 7, 10, 11, 46? Maybe you had bad luck the first time, because 4, 7, 10, 11, 34, 47 was one of Denniston's tickets. But then 4, 7, 10, 11, 46, 47 cannot be on Denniston's list, because it would agree in five places with a ticket you already know to be there. In other words, if evil 47 makes you miss out on one five-out-of-six prize, it can't make you miss out on any others. The same goes for 48. So of the six possible five-number wins:

4, 7, 10, 11, 34

---

* In math terms, this is because Denniston's list of tickets forms what's called a *Steiner system*. Added in press: In January 2014, Peter Keevash, a young mathematician at Oxford, announced a major breakthrough, proving the existence of more or less *all* possible Steiner systems that mathematicians had wondered about.

4, 7, 10, 11, 46
4, 7, 10, 34, 46
4, 7, 11, 34, 46
4, 10, 11, 34, 46
7, 10, 11, 34, 46

you are *guaranteed* to have at least four of them among your tickets. In fact, if you buy the 217,833 Denniston tickets, you have a

2% chance of hitting the jackpot
72% chance of winning six of the five-out-of-six prizes
24% chance of winning five of the five-out-of-six prizes
2% chance of winning four of the five-out-of-six prizes

Compare this with the Selbee Quick Pick strategy of choosing tickets randomly. In that case, there's a small chance, 0.3%, of getting shut out of the five-out-of-six prize tier entirely. Worse, there's a 2% chance of getting just one of those prizes, a 6% chance of getting two, an 11% chance of getting three, and a 15% chance of getting four. The guaranteed returns of the Denniston strategy are replaced by risk. Naturally, that risk comes with an upside, too—team Selbee has a 32% chance of getting more than six of those prizes, impossible if you pick your tickets according to Denniston. The expected value of Selbee's tickets is the same as that of Denniston's, or anyone else's. But the Denniston method shields the player from the winds of chance. In order to play the lottery without risk, it's not enough to play hundreds of thousands of tickets; you have to play the *right* hundreds of thousands of tickets.

Is this strategy the reason Random Strategies spent the time to fill out hundreds of thousands of tickets by hand? Were they using Denniston's system, developed in the spirit of utterly pure mathematics, to siphon money from the Lottery at no risk to themselves? Here's where my reporting hit a wall. I was able to get in touch with Yuran Lu, but he didn't know exactly how those tickets had been chosen; he told me only that they had a "go-to guy" in the dorm who handled all such algorithmic matters. I can't be sure whether the go-to guy used the Denniston system, or something like it. But if he didn't, I think he probably should have.

OKAY, FINE, YOU CAN PLAY POWERBALL

At this point, we've documented exhaustively how the choice to play the lottery is almost always a poor one in terms of expected number of dollars, and how, even in those rare cases where the expected monetary value of a lottery ticket exceeds its cost, great care is required in order to squeeze as much expected utility as possible out of the tickets you buy.

This leaves mathematically minded economists with one inconvenient fact to explain, the same one that baffled Adam Smith more than two hundred years ago: lotteries are very, very popular. The lottery is not the kind of situation Ellsberg studied, in which people face decisions against unknown and unknowable odds. The minuscule chance of winning the lottery is posted for all to see. The principle that people tend to make choices that more or less maximize their utility is a pillar of economics, and does an adequate job modeling behavior in everything from business practice to romantic choices. But not Powerball. This kind of irrational behavior is as unacceptable to a certain species of economist as the irrational magnitude of the hypotenuse was to the Pythagoreans. It doesn't fit their model of what can be; and yet it is.

Economists are more flexible than Pythagoreans. Rather than angrily drowning the bearers of bad news, they adjust their models to fit reality. One popular account was offered by our old buddies Milton Friedman and Leonard Savage, who proposed that lottery players follow a squiggly utility curve, reflecting that people think about wealth in terms of classes, not numerical amounts. If you're a middle-class worker who spends five bucks a week on the lottery, and you lose, that choice costs you a little money but doesn't change your class position; despite the loss of money, the negative utility is pretty close to zero. But if you *win*, well, that moves you into a different stratum of society. You can think of this as the "deathbed" model—on your deathbed, will you care that you died with a little less money because you played the lottery? Probably not at all. Will you care that you retired at thirty-five and spent the rest of your life snorkeling off Cabo because you hit the Powerball? Yes. Yes you will.

In a bigger departure from classical theory, Daniel Kahnemann and Amos Tversky suggested that people in general tend to follow a different path from the one the utility curve demands, not just when Daniel Ells-berg sticks an urn in front of them, but in the general course of life. Their "prospect theory," for which Kahnemann later won the Nobel Prize, is now seen as the founding document of behavioral economics, which aims to model with the greatest possible fidelity the way people *do* act, not the way that, according to an abstract notion of rationality, they *should*. In the Kahnemann-Twersky theory, people tend to place more weight on low-probability events than a person obedient to the von Neumann-Morgenstern axioms would; so the allure of the jackpot exceeds what a strict expected utility calculation would license.

But the simplest explanation doesn't require much theoretical heavy lifting at all. Simply: buying a lottery ticket, whether you win or not, is, in some small way, fun. Not Caribbean vacation fun, not all-night dance party fun, but one or two dollars' worth of fun? Quite possibly so. There are reasons to doubt this explanation (for instance, lottery players them-selves tend to cite the prospect of winning as their primary reason for playing), but it does an admirable job of explaining the behavior we see.

Economics isn't like physics and utility isn't like energy. It is not con-served, and an interaction between two beings can leave both with more utility than they started with. This is the sunny free-marketeer's view of the lottery. It's not a regressive tax, it's a *game*, where people pay the state a small fee for a few minutes of entertainment the state can provide very cheaply, and the proceeds keep the libraries open and the street-lights on. Just as when two countries trade with each other, both parties to the transaction come out ahead.

So yes—play Powerball, if Powerball is fun for you. Math gives you permission!

There are problems with this view, to be sure. Here's Pascal again, delivering a typically morose take on the excitement of gambling:

> This man spends his life without weariness in playing every day for
> a small stake. Give him each morning the money he can win each
> day, on condition he does not play; you make him miserable. It will
> perhaps be said that he seeks the amusement of play and not the
> winnings. Make him then play for nothing; he will not become ex-
> cited over it, and will feel bored. It is then not the amusement alone

that he seeks; a languid and passionless amusement will weary him. He must get excited over it, and deceive himself by the fancy that he will be happy to win what he would not have as a gift on condition of not playing.

Pascal sees the pleasures of gambling as contemptible. And enjoyed to excess, they can of course be harmful. The reasoning that endorses lotteries also suggests that methamphetamine dealers and their clients enjoy a similar win-win relationship. Say what you want about meth, you can't deny it is broadly and sincerely enjoyed.*

But what about another comparison? Instead of strung-out tweakers, think about small-business owners, the pride of America. Opening a store or selling a service isn't the same thing as buying a lottery ticket; you have some measure of control over your success. But the two enterprises have something in common: for most people, opening a business is a bad bet. It doesn't matter how delicious you believe your barbecue sauce to be, how disruptively innovative you expect your app to be, how ruthless and borderline felonious you intend your business practices to be—you are much more likely to fail than to succeed. That's the nature of entrepreneurship: you balance a very, very small probability of making a fortune against a modest probability of eking out a living against a substantially larger probability of losing your pile, and for a large proportion of potential entrepreneurs, when you crunch the numbers, the expected financial value, like that of a lottery ticket, is less than zero. Typical entrepreneurs (like typical lottery customers) overrate their chance of success. Even businesses that survive typically make their proprietors less money than they'd have drawn in salary from an existing company. And yet society benefits from a world in which people, against their wiser judgment, launch businesses. We want restaurants, we want barbers, we want smartphone games. Is entrepreneurship "a tax on the stupid"? You'd be called crazy if you said so. Part of that is because we esteem a business owner more highly than we do a gambler; it's hard to

* I am not making this argument up; if you want to see it pushed through in full, see Gary Becker and Kevin Murphy's theory of rational addiction.

separate our moral feelings about an activity from the judgments we make about its rationality. But part of it—the *biggest* part—is that the utility of running a business, like the utility of buying a lottery ticket, is not measured only in expected dollars. The very act of realizing a dream, or even trying to realize it, is part of its own reward.

That, at any rate, is what James Harvey and Yuran Lu decided. After the downfall of WinFall, they moved west and founded a Silicon Valley startup that sells an online chat system for businesses. (Harvey's profile page coyly lists "non-traditional investment strategies" among his interests.) As I write, they're still looking for venture capital funding. Maybe they'll get it. But if not, I'll bet you'll quickly find them starting again, expected value or no expected value, hoping the next ticket they try is a winner.

PART IV

· · · · ·

Regression

*Includes: Hereditary genius, the curse of the Home
Run Derby, arranging elephants in rows and columns,
Bertillonage, the invention of the scatterplot, Galton's
ellipse, rich states vote for Democrats but rich people vote
for Republicans, "Is it possible, then, that lung cancer is
one of the causes of smoking cigarettes?," why handsome
men are such jerks.*

FOURTEEN

THE TRIUMPH OF
MEDIOCRITY

T he early 1930s, like the present period, was a time of soul-searching for the American business community. Something had gone wrong, that much was plain. But what kind of thing? Was the great crash of 1929, and the subsequent depression, an unpredictable catastrophe? Or was the American economy systemically flawed?

Horace Secrist was as well placed as anyone could be to answer this question. Secrist was a professor of statistics and director of the Bureau for Business Research at Northwestern, an expert in the application of quantitative methods to business, and the author of a widely used statistics textbook for students and business executives. Since 1920, years before the crash, he had been compiling meticulously detailed statistics on hundreds of business concerns, from hardware stores to railroads to banks. Secrist tabulated expenses, total sales, outlays on wages and rent, and every other piece of data he could get, trying to locate and taxonomize the mysterious variations which made some businesses thrive and others falter.

So in 1933, when Secrist was ready to reveal the results of his analysis, people in both academia and business were inclined to listen. All the more so when he revealed the striking nature of his results in a 468-page

volume, thickly marbled with tables and graphs. Secrist pulled no pun-
ches: he called his book *The Triumph of Mediocrity in Business*.

"Mediocrity tends to prevail in the conduct of competitive business,"
Secrist wrote. "This is the conclusion to which this study of the costs
(expenses) and profits of thousands of firms unmistakably points. Such
is the price which industrial (trade) freedom brings."

How did Secrist arrive at such a doomy conclusion? First, he strati-
fied the businesses in each sector, carefully segregating the winners
(high income, low expenses) from the inefficient duds. The 120 clothing
stores Secrist studied, for instance, were first ranked by ratio of sales to
expenses in 1916, then divided into six groups, or "sextiles," of twenty
shops each. Secrist expected to see the shops in the top sextile consoli-
date their gains over time, growing ever more superior as they honed
their already top-of-market skills. What he found was precisely the opposite. By 1922, the clothing stores in the highest sextile had lost most
of their advantage over the typical shop; they were still better-than-
average stores, but by and large they were no longer the standouts. What's
more, the lowest sextile—the *worst* stores—experienced the same effect
in the opposite direction, improving their performance toward the average. Whatever genius had propelled the top-sextile stores to excel had
mostly exhaused itself in a mere six years. Mediocrity had triumphed.

Secrist found the same phenomenon in every kind of business. Hardware stores regressed to mediocrity; so did grocery stores. And it didn't
matter what metric you used. Secrist tried measuring his companies by
the ratio of wages to sales, the ratio of rent to sales, and whatever other
economic stat he could put his hands on. It didn't matter. With time, the
top performers started to look and behave just like the members of the
common mass.

Secrist's book arrived as a bucket of cold water to the face of an already uncomfortable business elite. Many reviewers saw in Secrist's
graphs and tables a numerical disproof of the mythology that sustained
entrepreneurship. Robert Riegel of the University of Buffalo wrote, "The
results confront the business man and the economist with an insistent
and to some degree tragic problem. While there are exceptions to the
general rule, the conception of an early struggle, crowned with success

for the able and efficient, followed by a long period of harvesting the rewards, is thoroughly dissipated."

What force was pushing the outliers toward the middle? It had to have something to do with human behavior, because the phenomenon didn't seem to show up in the natural world. Secrist, with characteristic thoroughness, had carried out a similar test on the average July temperature for 191 U.S. cities. Here there was no regression. The cities that were hottest in 1922 were just as hot in 1931.

After decades of recording statistics and studying the operation of American business, Secrist thought he knew the answer. It was built into the nature of competition itself to push down successful businesses and promote their incompetent rivals. Secrist wrote:

> Complete freedom to enter trade and the continuance of competition mean the perpetuation of mediocrity. New firms are recruited from the relatively "unfit"—at least from the inexperienced. If some succeed, they must meet the competitive practices of the class, the market, to which they belong. Superior judgment, merchandising sense, and honesty, however, are always at the mercy of the unscrupulous, the unwise, the misinformed and the injudicious. The results are that retail trade is over-crowded, shops are small and inefficient, volume of business inadequate, expenses relatively high, and profits small. So long as the field of activity is freely entered, and it is; and so long as competition is "free," and within the limits suggested above, it is; neither superiority nor inferiority will tend to persist. Rather, mediocrity tends to become the rule. The average level of the intelligence of those conducting business holds sway, and the practices common to such trade mentality become the rule.

Can you imagine a business school professor saying something like this today? It's unthinkable. In modern discourse, free-market competition is the cleansing blade that cuts down the incompetent and the 10%-less-than-maximally-competent alike. Inferior firms are at the mercy of their betters, not the other way around.

But Secrist saw the free market, with its firms of different sizes and skill levels jostling against each other, as something like the one-room schoolhouse that was, by 1933, already well on its way to disuse. As Secrist describes it: "Pupils of all ages, of different mentality, and of training, grouped together in a single room, were to be educated. Pandemonium, discouragement, and inefficiency, of course, resulted. Common sense later pointed to the desirability of classification, grading, special treatment—corrections which opened the way for native ability to assert itself, and for superiority to withstand being watered down and diluted by inferiority."

That last part sounds a little—well, how should I put it—can you think of anybody *else* in 1933 who was talking about the importance of superior beings withstanding dilution by inferior ones?

Given the flavor of Secrist's views on education, it may come as no surprise that his ideas about regression to mediocrity descend from those of the nineteenth-century British scientist and pioneering eugenicist Francis Galton. Galton was the youngest of seven children and a sort of child prodigy. Galton's bedridden older sister Adèle took on his education as her chief amusement; he could sign his name at two, and by four was writing her letters like this: "I can cast up any sum in addition and can multiply by 2, 3, 4, 5, 6, 7, 8, 10. I can also say the pence table. I read French a little and I know the Clock." Galton started medical studies at eighteen, but after his father died, leaving him a substantial fortune, he found himself suddenly less motivated to pursue a traditional career. For a while Galton was an explorer, leading expeditions into the African interior. But the epochal publication of *The Origin of Species* in 1859 catalyzed a drastic shift in his interests: Galton recalls that he "devoured its contents and assimilated them as fast as they were devoured," and from then on, the greater share of Galton's work was devoted to the heredity of human characteristics, both physical and mental. This work led him to a suite of policy preferences that are decidedly unsavory from a modern point of view. The opening of his 1869 book *Hereditary Genius* gives the flavor:

> I propose to show in this book that a man's natural abilities are derived by inheritance, under exactly the same limitations as are the form and physical features of the whole organic world. Consequently,

as it is easy, notwithstanding those limitations, to obtain by careful
selection a permanent breed of dogs or horses gifted with peculiar
powers of running, or of doing anything else, so it would be quite
practicable to produce a highly-gifted race of men by judicious marriages during several consecutive generations.

Galton made his case by means of a detailed study of British men of
achievement, from clerics to wrestlers, arguing that notable Englishmen*
tend to have disproportionately notable relatives. *Hereditary Genius* met
with a great deal of resistance, particularly from the clergy; Galton's
purely naturalistic view of worldly success left little room for a more
traditional view of Providence. Especially irksome was Galton's claim
that success in ecclesiastical pursuits was itself subject to hereditary influence: that, as one reviewer complained, "a pious man owes his piety
not so much (as we had ever believed) to the direct action of the Holy
Ghost on his soul, blowing like the wind where it listeth, but rather to
his earthly father's physical bequest of a constitution adapted to the religious emotions." Whatever friends Galton had among the religious establishment were surely lost three years later, when he published a short
article titled "Statistical Inquiries into the Efficacy of Prayer." (Executive
summary: prayer not so efficacious.)
      By contrast, Galton's book was received with great excitement, if not
uncritical acceptance, by the Victorian scientific community. Charles
Darwin wrote Galton in a kind of intellectual frenzy, not even waiting
until he'd finished the book:

DOWN, BECKENHAM, KENT, S.E.
December 23rd

MY DEAR GALTON,
      *—I have only read about 50 pages of your book (to Judges), but I
must exhale myself, else something will go wrong in my inside. I do not
think I ever in all my life read anything more interesting and original—*

* He apologizes in the introduction for the omission of foreigners, remarking, "I should have especially liked to investigate the biographies of Italians and Jews, both of whom appear to be rich
in families of high intellectual breeds."

and how Well and clearly you put every point! George, who has
finished the book, and who expressed himself in just the same terms,
tells me that the earlier chapters are nothing in interest to the later ones!
It will take me some time to get to these latter chapters, as it is read
aloud to me by my wife, who is also much interested. You have made a
convert of an opponent in one sense, for I have always maintained that,
excepting fools, men did not differ much in intellect, only in zeal and
hard work; and I still think this is an eminently important difference.
I congratulate you on producing what I am convinced will prove a
memorable work. I look forward with intense interest to each reading,
but it sets me thinking so much that I find it very hard work; but that is
wholly the fault of my brain and not of your beautifully clear style.
    —Yours most sincerely,
      (Signed) CH. DARWIN

To be fair, Darwin might have been biased, being Galton's first cousin. What's more, Darwin truly believed that mathematical methods offered scientists an enriched view of the world, even though his own work was far less quantitative than Galton's. He wrote in his memoirs, reflecting on his early education,

I attempted mathematics, and even went during the summer of 1828 with a private tutor (a very dull man) to Barmouth, but I got on very slowly. The work was repugnant to me, chiefly from my not being able to see any meaning in the early steps in algebra. This impatience was very foolish, and in after years I have deeply regretted that I did not proceed far enough at least to understand something of the great leading principles of mathematics, for men thus endowed seem to have an extra sense.

In Galton, Darwin may have felt he was finally seeing the outset of the extrasensory biology he was mathematically unequipped to launch on his own.

The critics of *Hereditary Genius* contended that, while heredity of intellectual tendencies was real, Galton was overstating its strength relative to other factors affecting achievement. So Galton set out to understand the *extent* to which our parental inheritance determined our fate. But quantifying the heredity of "genius" wasn't so easy: how, exactly, was one to measure just how notable his notable Englishmen were? Undeterred, Galton turned to human characteristics that could be more easily placed on a numerical scale, like height. As Galton and everyone else already knew, tall parents tend to have tall children. When a six-foot-two man and a five-foot-ten woman get married, their sons and daughters are likely to be taller than average.

But now here is Galton's remarkable discovery: those children are *not* likely to be as tall as their parents. The same goes for short parents, in the opposite direction; their kids will be tend to be short, but not as short as they themselves are. Galton had discovered the phenomenon now called *regression to the mean*. His data left no doubt that it was real.

"However paradoxical it may appear at first sight," Galton wrote in his 1889 book *Natural Inheritance*, "it is theoretically a necessary fact,* and one that is clearly confirmed by observation, that the Stature of the adult offspring must on the whole, be more *mediocre* than the stature of their Parents."

So, too, Galton reasoned, must it be for mental achievement. And this conforms with common experience; the children of a great composer, or scientist, or political leader, often excel in the same field, but seldom so much so as their illustrious parent. Galton was observing the same phenomenon that Secrist would uncover in the operations of business. Excellence doesn't persist; time passes, and mediocrity asserts itself.†

But there's one big difference between Galton and Secrist. Galton was, in his heart, a mathematician, and Secrist was not. And so Galton understood *why* regression was taking place, while Secrist remained in the dark.

Height, Galton understood, was determined by some combination of inborn characteristics and external forces; the latter might include envi-

---

* Technical but important note: When Galton says "necessary," he is making use of the biological fact that the distribution of human height is roughly the same from generation to generation. It's theoretically possible for there to be no regression, but this would force an increase in variation, so that each generation would have more gigantic giants and more diminutive pipsqueaks.

† It's hard to understand how Secrist, who was familiar with Galton's work on human height, managed to convince himself that regression to the mean was found only in variables under human control. When a theory really has got your brain in its grip, contradictory evidence—even evidence you already know—sometimes becomes invisible.

ronment, childhood health, or just plain chance. I am six foot one, and in part that's because my father is six foot one and I share some of his height-promoting genetic material. But it's also because I ate reasonably nutritious food as a child and didn't undergo any unusual stresses that would have stunted my growth. And my height was no doubt bumped up and down by who knows how many other experiences I underwent, in utero and ex. Tall people are tall because their heredity predisposes them to be tall, or because external forces encourage them to be tall, or both. And the taller a person is, the likelier it is that *both* factors are pointing in the upward direction.

In other words, people drawn from the tallest segment of the population are almost certain to be taller than their genetic predisposition would suggest. They were born with good genes, but they also got a boost from environment and chance. Their children will share their genes, but there's no reason the external factors will once again conspire to boost their height over and above what heredity accounts for. And so, on average, they'll be taller than the average person, but not quite so exceedingly tall as their beanpole parents. *That's* what causes regression to the mean: not a mysterious mediocrity-loving force, but the simple workings of heredity intermingled with chance. That's why Galton writes that regression to the mean is "theoretically a necessary fact." At first, it came to him as a surprising feature of his data, but once he understood what was going on, he saw it couldn't possibly have come out any other way.

It's just the same for businesses. Secrist wasn't wrong about the firms that had the fattest profits in 1922; it's likely that they ranked among the most well managed companies in their sectors. But they were lucky, too. As time went by, their management might well have remained superior in wisdom and judgment. But the companies that were lucky in 1922 were no more likely than any other companies to be lucky ten years later. And so those top-sextile companies start slipping in the rankings as the years go by.

In fact, almost any condition in life that involves random fluctuations in time is potentially subject to the regression effect. Did you try a new apricot-and-cream-cheese diet and find you lost three pounds? Think back to the moment you decided to slim down. More than likely it was a moment at which the normal up-and-down of your weight had you at the top of your usual range, because those are the kinds of moments when you look down at the scale, or just at your midsection, and say, jeez, I've gotta *do* something. But if that's the case, you might well have lost three pounds anyway, apricots or no apricots, when you trended back toward your normal weight. You've learned very little about the efficacy of the diet.

You might try to address this problem by random sampling: choose two hundred patients at random, check which ones are overweight, and then try the diet on the overweight folks. But then you'd be doing just what Secrist did. The heaviest segment of the population is a lot like the top sextile of businesses. They are certainly more likely than the average person to have a consistent weight problem. But they are *also* more likely to be at the top of their weight range on the day you happened to weigh them. Just as Secrist's high performers degraded toward mediocrity with time, so will your heavy patients lose weight, whether the diet is effective or not. That's why the better sort of diet studies don't just study the effects of one diet; they compare two candidate diets to see which induces more weight loss. Regression to the mean should affect each group of dieters equally, so *that* comparison is fair.

Why is the second novel by a breakout debut writer, or the second album by an explosively popular band, so seldom as good as the first? It's not, or not entirely, because most artists only have one thing to say. It's because artistic success is an amalgam of talent and fortune, like everything else in life, and thus subject to regression to the mean.[*]

Running backs who sign big multiyear contracts tend to record fewer yards per carry in the season following.[†] Some people claim that's because they no longer have a financial incentive to stretch for that extra yard, and that psychological factor probably does play a role. But just as important is that they signed the big contract as a result of having a mas-

---

[*] These cases are complicated by the fact that novelists and musicians tend to get better with more practice. F. Scott Fitzgerald's second novel (can you even name it?) is pretty bad compared to his debut, *This Side of Paradise*, but when his style matured he turned out to have a little bit left in the tank.

[†] This fact, along with its interpretation, comes from Brian Burke at Advanced NFL Stats, whose clear exposition and rigorous attention to statistical good sense should be a model for all serious sports analysts.

sively good year. It would be bizarre if they *didn't* return to a more ordinary level of performance the following season.

"ON PACE"

As I write, it's April, the beginning of baseball season, when every year we're treated to a bouquet of news stories about which players are "on pace" to perform which unimaginable record-shattering feat. Today on ESPN I learn that "Matt Kemp is off to a blazing start, hitting .460 and on pace for 86 home runs, 210 RBIs, and 172 runs scored." These eye-popping numbers (no one in the history of major-league baseball has ever hit more than 73 home runs in a season) are a typical example of false linearity. It's like a word problem: "If Marcia can paint 9 houses in 17 days, and she has 162 days to paint as many houses as she can . . ."

Kemp hit nine home runs in the Dodgers' first seventeen games, a rate of 9/17 runs per game. So an amateur algebraist might write down the following linear equation:

$$H = G \times (9 / 17)$$

where H is the number of home runs Kemp hits for the full season, and G is the number of games his team plays. A baseball season is 162 games long. And when you plug in 162 for G, you get 86 (or rather 85.7647, but 86 is the closest whole number).

But not all curves are lines. Matt Kemp will not hit eighty-six home runs this year. And it's regression to the mean that explains why. At any point in the season, it's pretty likely that the league leader in home runs is a good home run hitter. Indeed, it's clear from Kemp's history that there are intrinsic Matt Kemp qualities that enable him regularly to club a baseball with awe-inspiring force. But the league leader in home runs is also very likely to have been lucky. Which means that, whatever his league-leading pace is, you can expect it to drop as the season goes on.

No one at ESPN, to be fair, thinks Matt Kemp is going to hit eighty-six home runs. These "on pace" statements, when made in April, are usually delivered in a half-joking tone: "Of course he won't, but what *if* he kept this up?" But as the summer goes on, the tongue draws farther and farther out of the cheek, until by midseason people are quite serious about using a linear equation to project a player's statistics to the end of the year.

But it's still wrong. If there's regression to the mean in April, there's regression to the mean in July.

Ballplayers get this. Derek Jeter, when bugged about being on pace to break Pete Rose's career hit record, told the *New York Times*, "One of the worst phrases in sports is 'on pace for.'" Wise words!

Let's make this less theoretical. If I'm leading the American League in home runs at the All-Star break, how many home runs should I expect to hit the rest of the way?

The All-Star break divides the baseball season into a "first half" and a "second half," but the second half is actually a bit shorter: in recent years, between 80% and 90% as long as the first half. So you might expect me to hit about 85% as many home runs in the second half as I did in the first.*

But history says this is the wrong thing to expect. To get a sense of what really goes on, I looked at first-half American League home run leaders in nineteen seasons between 1976 and 2000 (excluding years shortened by strikes and those where there was a tie for first-half leader). Only three (Jim Rice in 1978, Ben Oglivie in 1980, and Mark McGwire in 1997) hit as many as 85% of their first-half total after the break. And for every one of those, there's a hitter like Mickey Tettleton, who led the AL with twenty-four homers at the 1993 all-star break and managed only eight the rest of the way. The sluggers, on average, hit only 60% as many home runs in the second half as they had in their league-leading first. This decline isn't due to fatigue, or the August heat; if it were, you'd see a similarly large decline in home run production around the league. It's simple regression to the mean.

And it's not restricted to the very best home run hitter in the league. The Home Run Derby, held during the All-Star break each year, is a

---

* Actually, the overall home run rate appears to dip slightly in the second half; but this may be because late-season call-ups are getting more at bats. In a data set consisting of elite home run hitters, the second-half home run rate and the first-half home run rate were the same (J. McCollum and M. Jaiclin, *Baseball Research Journal*, Fall 2010).

competition where baseball's top mashers compete to hit as many moon shots as they can against a batting-practice pitcher. Some batters complain that the artificial conditions of the derby throw off their timing and make it harder to hit home runs in the weeks after the break: the Home Run Derby Curse. The *Wall Street Journal* ran a breathless story, "The Mysterious Curse of the Home Run Derby," in 2009, which was vigorously rebutted by the statistically minded baseball blogs. That didn't stop the *Journal* from revisiting the same ground in 2011, with "The Great Derby Curse Strikes Once Again." But there is no curse. The participants in the derby are there because they had an awfully good start to the season. Regression demands that their later production, on average, won't keep up with the pace they've set.

As for Matt Kemp, he injured a hamstring in May, missed a month, and was a different player when he returned. He finished the 2012 season not with the eighty-six home runs he was "on pace" for, but twenty-three.

There's something the mind resists about regression to the mean. We want to believe in a force that brings down the mighty. It's not satisfying enough to accept what Galton knew in 1889: the apparently mighty are seldom quite as mighty as they look.

SECRIST MEETS HIS MATCH

This crucial point, invisible to Secrist, was not so obscure to more mathematically minded researchers. In contrast to Secrist's generally respectful reviews was the famous statistical smackdown delivered by Harold Hotelling in the *Journal of the American Statistical Association*. Hotelling was a Minnesotan, the son of a hay dealer, who went to college to study journalism and there discovered an extraordinary talent for mathematics. (Francis Galton, had he gone on to study the heredity of notable Americans, would have been pleased to know that despite Hotelling's humble upbringing his ancestors included a secretary of the Massachusetts Bay Colony and an Archbishop of Canterbury.) Like Abraham Wald, Hotelling started in pure math, writing a PhD dissertation in algebraic topology at Princeton. He would go on to lead the wartime Statistical Research Group in New York—the same place Wald explained to the army how to put the armor where the bullet holes weren't. In 1933, when Secrist's book came out, Hotelling was a young professor at Columbia who had already made major contributions to theoretical statistics, especially in relation to economic problems. He was said to enjoy playing Monopoly in his head; having memorized the board and the frequencies of the various Chance and Community Chest cards, this was a simple exercise in random number generation and mental book-keeping. This should give some impression both of Hotelling's mental powers and of the sort of thing he enjoyed.

Hotelling was totally devoted to research and the generation of knowledge, and in Secrist he may have seen something of a kindred soul. "The labor of compilation and of direct collection of data," he wrote sympathetically, "must have been gigantic."

Then the hammer drops. The triumph of mediocrity observed by Secrist, Hotelling points out, is more or less automatic whenever we study a variable that's affected by both stable factors and the influence of chance. Secrist's hundreds of tables and graphs "prove nothing more than that the ratios in question have a tendency to wander about." The result of Secrist's exhaustive investigation is "mathematically obvious from general considerations, and does not need the vast accumulation of data adduced to prove it." Hotelling drives his point home with a single, decisive observation. Secrist believed the regression to mediocrity resulted from the corrosive effect of competitive forces over time; com-petition was what caused the top stores in 1916 to be hardly above aver-age in 1922. But what happens if you select the stores with the highest performance in 1922? As in Galton's analysis, these stores are likely to have been both lucky and good. If you turn back the clock to 1916, what-ever intrinsic good management they possess should still be in force, but their luck may be totally different. Those stores will typically be closer to mediocre in 1916 than in 1922. In other words, if regression to the mean were, as Secrist thought, the natural result of competitive forces, those forces would have to *work backward in time as well as forward.*

Hotelling's review is polite but firm, distinctly more in sorrow than in anger: he is trying to explain to a distinguished colleague, in the kind-est way possible, that he has wasted ten years of his life. But Secrist didn't take the hint. The issue after next of *JASA* ran his contentious letter of response, pointing out a few misapprehensions in Hotelling's review, but otherwise spectacularly missing the point. Secrist insisted once again that the regression to mediocrity was not a mere statistical generality, but rather was particular to "data affected by competitive pressure and managerial control." At this point Hotelling stops being nice and lays it out straight. "The thesis of the book," he writes in response, "when correctly interpreted, is essentially trivial. . . . To 'prove' such a mathematical result by a costly and prolonged numerical study of many kinds of business profit and expense ratios is analogous to proving the multiplication table by arranging elephants in rows and columns, and then doing the same for numerous other kinds of animals. The performance, though perhaps entertaining, and having a certain pedagogical value, is not an important contribution either to zoölogy or mathematics."

THE TRIUMPH OF MEDIOCRITY IN ORAL-ANAL TRANSIT TIME

It's hard to blame Secrist too much. It took Galton himself some twenty years to fully grasp the meaning of regression to the mean, and many subsequent scientists misunderstood Galton exactly as Secrist had. The biometrician Walter F. R. Weldon, who had made his name by showing that Galton's findings about the variation in human traits held equally well for shrimp, said in a 1905 lecture about Galton's work:

> Very few of those biologists who have tried to use his methods have taken the trouble to understand the process by which he was led to adopt them, and we constantly find regression spoken of as a peculiar property of living things, by virtue of which variations are diminished in intensity during their transmission from parent to child, and the species is kept true to type. This view may seem plausible to those who simply consider that the mean deviation of children is less than that of their fathers: but if such persons would remember the equally obvious fact that there is also a regression of fathers on children, so that the fathers of abnormal children are on the whole less

abnormal than their children, they would either have to attribute this feature of regression to a vital property by which children are able to reduce the abnormality of their parents, or else to recognize the real nature of the phenomenon they are trying to discuss.

Biologists are eager to think regression stems from biology, management theorists like Secrist want it to come from competition, literary critics ascribe it to creative exhaustion—but it is none of these. It is mathematics.

And still, despite the entreaties of Hotelling, Weldon, and Galton himself, the message hasn't totally sunk in. It's not just the *Wall Street Journal* sports page that gets this wrong; it happens to scientists, too. One particularly vivid example comes from a 1976 *British Medical Journal* paper on the treatment of diverticular disease with bran. (I am just old enough to remember 1976, when bran was spoken of by health enthusiasts with the kind of reverence that omega-3 fatty acids and antioxidants enjoy today.) The authors recorded each patient's "oral-anal transit time"—that is, the length of time a meal spent in the body between entrance and exit—before and after the bran treatment. They found that bran has a remarkable regularizing effect. "All those with rapid times slowed down towards 48 hours . . . those with medium length transits showed no change . . . and those with slow transit times tended to speed up towards 48 hours. Thus bran tended to modify both slow and fast initial transit times towards a 48-hour mean." This, of course, is precisely what you'd expect if bran had no effect at all. To put it delicately, we all have our fast days and our slow days, whatever our underlying level of intestinal health. And an unusually quick transit on Monday is likely to be followed by a more average transit time on Tuesday, bran or no bran.*

Then there's the rise and fall of Scared Straight. The program took juvenile offenders on tours of prisons, where inmates warned them about the horrors that awaited them on the inside if they didn't drop their criminal ways pronto. The original program, held in New Jersey's Rahway

---

* The authors do gesture at the existence of regression: "While this phenomenon could merely be a regression towards the mean, we conclude that increasing the fibre intake does have a genuine physiological action in slowing fast transit times and accelerating slow transit times in patients with diverticular disease." Where this conclusion comes from, apart from faith in bran, is hard to say.

State Prison, was featured in an Oscar-winning documentary in 1978 and quickly spawned imitations across the United States and as far away as Norway. Teenagers raved about the moral kick in the pants they got from Scared Straight, and wardens and prisoners liked the opportunity to contribute something positive to society. The program resonated with a popular, deep-seated sense that overindulgence by parents and society were to blame for youth crime. Most important, Scared Straight worked. One representative program, in New Orleans, reported that participants were arrested less than half as often after Scared Straight than before.

Except it didn't work. The juvenile offenders are like Secrist's low-performing stores: selected, not at random, but by virtue of being the worst of their kind. Regression tells you that the very worst-behaved kids this year will likely still be behavior problems next year; *but not as much so.* The decline in arrest rate is just what you'd expect even if Scared Straight had no effect.

Which isn't to say Scared Straight was completely ineffective. When the program was put through randomized trials, where a randomly selected subgroup of juvenile offenders were put through Scared Straight and then compared to the remaining kids, who didn't participate, researchers found that the program actually *increased* antisocial behavior. Maybe it should have been called Scared Stupid.

FIFTEEN

GALTON'S ELLIPSE

Galton had shown that regression to the mean was in effect whenever the phenomenon being studied was influenced by the play of chance forces. But how strong were those forces, by comparison with the effect of heredity?

In order to hear what the data was telling him, Galton had to put it in a form more graphically revealing than a column of numbers. He later recalled, "I began with a sheet of paper, ruled crossways, with a scale across the top to refer to the statures of the sons, and another down the side for the statures of their fathers, and there also I had put a pencil mark at the spot appropriate to the stature of each son and to that of his father."

This method of visualizing the data is the spiritual descendant of René Descartes's analytic geometry, which asks us to think about points in the plane as pairs of numbers, an x-coordinate and a y-coordinate, joining algebra and geometry in a tight clasp they've been locked in ever since.

Each father-son pair has an associated pair of numbers: namely, the height of the father followed by the height of the son. My father is six foot one and so am I—seventy-three inches each—so if we'd been in Galton's data set we would have been recorded as (73,73). And Galton would have recorded our existence by making a mark on his sheet of paper with x-coordinate 73 and y-coordinate 73. Each parent and child in Galton's voluminous records required another mark on the paper, until in the end his sheet bore a vast spray of dots, representing the whole range of variation in stature. Galton had invented the type of graph we now call a *scatterplot.*[*]

Scatterplots are spectacularly good at revealing the relationship between two variables; look in just about any contemporary scientific journal and you'll see a raft of them. The late nineteenth century was a kind of golden age of data visualization. In 1869 Charles Minard made his famous chart showing the dwindling of Napoleon's army on its path into Russia and its subsequent retreat, often called the greatest data graphic ever made; this, in turn, was a descendant of Florence Nightingale's coxcomb graph[†] showing in stark visual terms that most of the British soldiers lost in the Crimean War had been killed by infections, not Russians.

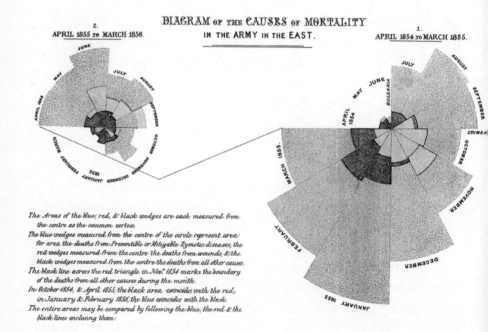

* Or at least reinvented it: the astronomer John Herschel constructed a sort of scatterplot in 1833 to study the orbits of binary stars. This isn't the same Herschel who discovered Uranus, by the way; that was his dad, William Herschel. Notable Englishmen and their notable relatives!
† What Nightingale actually called the *coxcomb* was the booklet containing the graph, not the graph itself, but everybody calls it the coxcomb and it's too late to change it now.

The coxcomb and scatterplot play to our cognitive strengths: our brains are sort of bad at looking at columns of numbers, but absolutely ace at locating patterns and information in a two-dimensional field of vision.

In some cases, that's easy. For instance, suppose that every son and father had *equal* height, the way my dad and I do. This represents a situation where chance plays no role at all, and your stature is completely determined by your patrimony. Then all the points in our scatterplot would have x and y coordinates equal; in other words, they'd be stuck to the diagonal line whose equation is x = y:

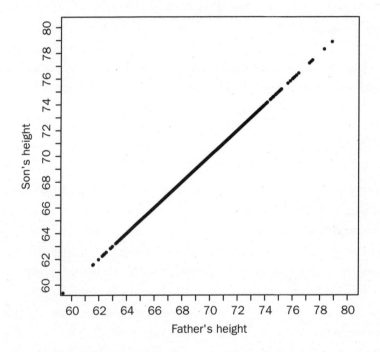

Note that the density of the dots is greater near the middle and less near the extremes; more men are five feet nine inches tall (sixty-nine inches) than are six foot one or five foot four.

Now what about the opposite extreme, where the heights of fathers and sons are totally independent? In that case, the scatterplot would look something like this:

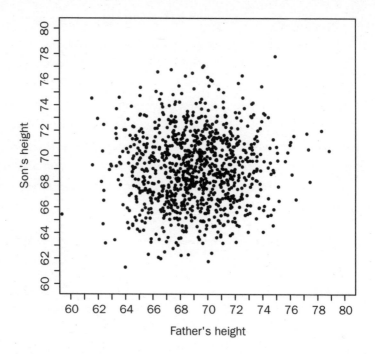

This picture, unlike the first one, shows no bias toward the diagonal. If you restrict your attention to sons whose fathers were six foot one (seventy-three inches), corresponding to a vertical slice in the right half of the scatterplot, the dots measuring the height of the sons are still centered on five foot nine. We say that the *conditional expectation* of the son's height (that is, how tall the son will be on average given that his father stands six foot one) is the same as the *unconditional expectation* (the average height of sons computed without any restriction on the father). This is what Galton's sheet of paper would have looked like if there were no heritable differences at all affecting height. It's regression to the mean in its most intense form; the sons of tall fathers regress *all the way* to the mean, ending up no taller than the sons of shorties.

But Galton's scatterplot didn't look like either of those two extreme cases. Instead, it was intermediate between them:

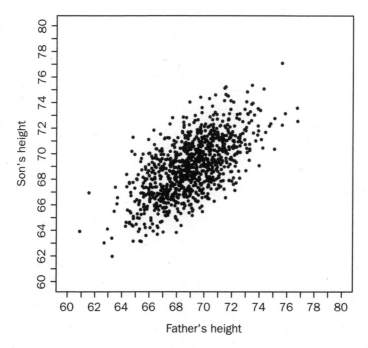

What does the average son of a six-foot-one-inch father look like in this plot? I've drawn a vertical slice to show you which points on the scatterplot correspond to those father-son pairs.

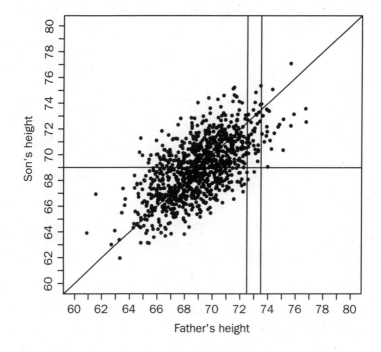

NUMBER OF ADULT CHILDREN OF VARIOUS STATURES BORN OF 205 MID-PARENTS OF VARIOUS STATURES.
(All Female heights have been multiplied by 1·08).

Heights of the Mid-parents in inches.	Heights of the Adult Children.														Total Number of		Medians.
	Below	62·2	63·2	64·2	65·2	66·2	67·2	68·2	69·2	70·2	71·2	72·2	73·2	Above	Adult Children.	Mid-parents.	
Above	..	..	..	..	..	..	..	..	..	..	..	1	3	..	4	5	..
72·5	..	..	..	..	..	..	..	1	2	1	2	7	2	4	19	6	72·2
71·5	..	..	..	..	1	3	4	3	5	10	4	9	2	2	43	11	69·9
70·5	1	..	2	..	1	1	3	12	18	14	7	4	3	2	68	22	69·5
69·5	..	..	3	16	4	17	27	20	33	25	20	11	4	3	183	41	68·9
68·5	1	..	4	11	16	25	31	34	48	21	18	4	3	3	219	49	68·2
67·5	..	3	5	14	15	36	38	28	38	19	11	4	..	..	211	33	67·6
66·5	..	3	3	5	2	17	17	14	13	4	..	..	..	..	78	20	67·2
65·5	1	..	9	5	7	11	11	7	7	5	2	1	..	..	66	12	66·7
64·5	1	1	4	4	1	5	5	..	2	..	..	..	..	..	23	5	65·8
Below	1	..	2	4	1	2	2	1	1	..	..	..	..	..	14	1	..
Totals	5	7	32	59	48	117	138	120	167	99	64	41	17	14	928	205	..
Medians	..	..	66·3	67·8	67·9	67·7	67·9	68·3	68·5	69·0	69·0	70·0	..	..	..	..	..

NOTE.—In calculating the Medians, the entries have been taken as referring to the middle of the squares in which they stand. The reason why the headings run 62·2, 63·2, &c., instead of 62·5, 63·5, &c., is that the observations are unequally distributed between 62 and 63, 63 and 64, &c., there being a strong bias in favour of integral inches. After careful consideration, I concluded that the headings, as adopted, best satisfied the conditions. This inequality was not apparent in the case of the Mid-parents.

You can see that the dots near the "six-foot-one-inch father" slice are more heavily concentrated below the diagonal than above, so that the sons are on average shorter than the father. On the other hand, they are plainly biased to lie mostly above sixty-nine inches, the height of the average man. In the data set I plotted, the average height of those sons turns out to be just under six feet: taller than average, but not as tall as Dad. You are looking at a *picture* of regression to the mean.

Galton noticed very quickly that his scatterplots, generated by the interplay of heredity and chance, had a geometric structure that was anything but random. They seemed to be enclosed, more or less, by an ellipse, centered on the point where both parents and child were of exactly average height.

The tilted elliptical shape of the data is quite clear even in the raw data in the table reproduced on page 316, from Galton's 1886 paper "Regression Towards Mediocrity in Hereditary Stature"; look at the figure formed by the nonzero entries in the table. The table also makes clear that I haven't told the whole story of Galton's data set; for instance, his y-coordinate is not "height of the father," but "average of the father's height with 1.08 times the mother's height,"* what Galton calls the "mid-parent."

In fact, Galton did more—he carefully drew curves on his scatterplot along which the density of points was roughly constant. Curves of this kind are called *isopleths*, and they're very familiar to you, if not under that tongue-twisting name. If we start from a map of the United States, and draw a curve through all the cities where today's high temperature is exactly 75 degrees, 50 degrees, or any other fixed value, you get the familiar curves of the weather map; these are called *isotherms*. A really hardcore weather map might also include *isobars*, joining areas of equal barometric pressure, or *isonephs*, areas of equal cloud cover. If we measure elevation instead of temperature, the isopleths are the contour lines, sometimes called isohypses, you find on topographic maps. This isopleth map shows the average annual number of snowstorms per year across the continental U.S.:

---

* The 1.08 is to make the average heights of the mothers roughly match those of the fathers, so that male and female height are measured on the same scale.

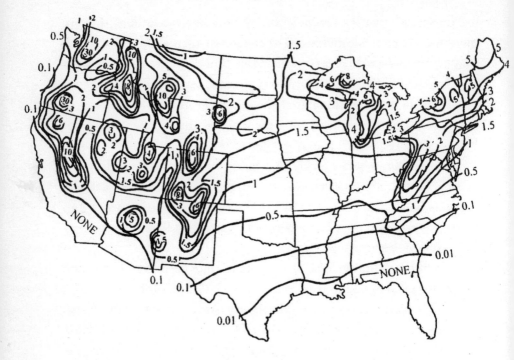

The isopleth wasn't Galton's invention; the first published isoplethic map was produced in 1701 by Edmond Halley, the British Astronomer Royal we last saw explaining to the king how to price annuities correctly.* Navigators already knew that magnetic north and true north didn't always agree; understanding exactly how and where the disagreement appeared was obviously critical for successful ocean travel. The curves on Halley's map were *isogons*, showing sailors where the discrepancy between magnetic north and true north were constant. The data was based on measurements Halley made aboard the *Paramore*, which crossed the Atlantic several times with Halley himself at the helm. (This guy really knew how to keep busy between comets.)

Galton found an amazing regularity: his isopleths were all ellipses, one contained within the next, each one with the same center. It was like the contour map of a perfectly elliptical mountain, with its peak at the pair of heights most frequently observed in Galton's sample: average

---

* Isopleths go back even further than this. The first ones we know about were isobaths (curves of constant depth) drawn on maps of rivers and harbors, which go back at least as far as 1584; but Halley seems to have invented the technique independently, and certainly popularized it.

height for both parents and children. The mountain is none other than the three-dimensional version of the gendarme's hat that de Moivre had studied; in modern language we call it the bivariate normal distribution.

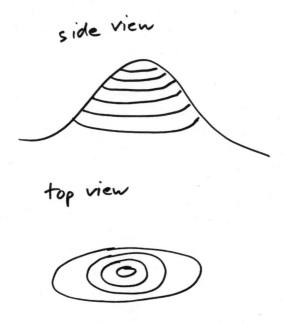

side view

top view

When the son's height is completely unrelated to those of the parents, as in the second scatterplot above, Galton's ellipses are all circles, and the scatterplot looks roughly round. When the son's height is completely determined by heredity, with no chance element involved, as in the first scatterplot, the data lies along a straight line, which one might think of as an ellipse that has gotten as elliptical as it possibly can. In between, we have ellipses of various levels of skinniness. That skinniness, which the classical geometers called the *eccentricity* of the ellipse, is a measure of the extent to which the height of the father determines that of the son. High eccentricity means that heredity is powerful and regression to the mean is weak; low eccentricity means the opposite, that regression to the mean holds sway. Galton called his measure *correlation*, the term we still use today. If Galton's ellipse is almost round, the correlation is near 0; when the ellipse is skinny, lined up along the northeast-southwest axis, the correlation comes close to 1. By means of the eccentricity—a geometric quantity at least as old as the work of Apollonius of Perga in the third

century BCE—Galton had found a way to measure the association between two variables, and in so doing had solved a problem at the cutting
edge of nineteenth-century biology: the quantification of heredity.

A proper skeptical attitude now requires you to ask: What if your
scatterplot *doesn't* look like an ellipse? What then? There's a pragmatic
answer: in practice, the scatterplots of real-life data sets often do array
themselves in rough ellipses. Not always, but often enough to make the
technique widely applicable. Here's what it looks like when you plot the
share of voters who voted for John Kerry in 2004 against the share Barack
Obama got in 2008. Each dot represents a single House district:

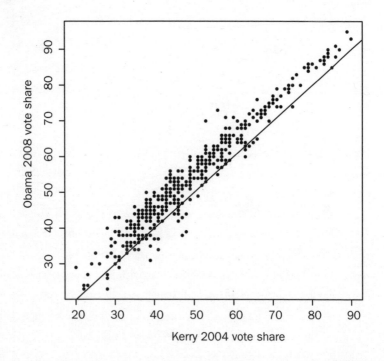

The ellipse is plain to see; and it's very skinny; vote share for Kerry is
highly correlated with vote share for Obama. The plot floats noticeably
above the diagonal, reflecting the fact that Obama generally did better
than Kerry.

Here's a plot of several years of daily stock price changes for Google
and GE:

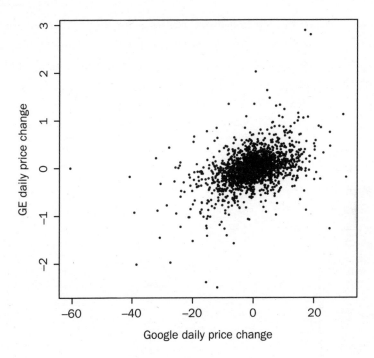

Here's a picture we've already seen, average SAT score plotted against tuition for a group of North Carolina colleges:

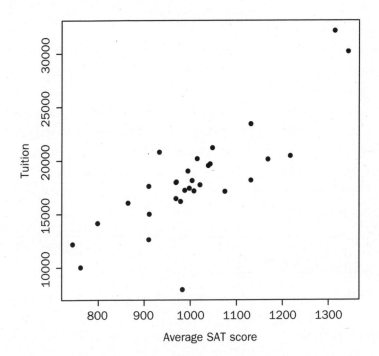

And here are the 50 U.S. states arranged in a scatterplot by average income and George W. Bush's share of the 2004 presidential vote, with wealthy liberal states like Connecticut down in the lower right and Republican states of more modest means, like Idaho, in the upper left.

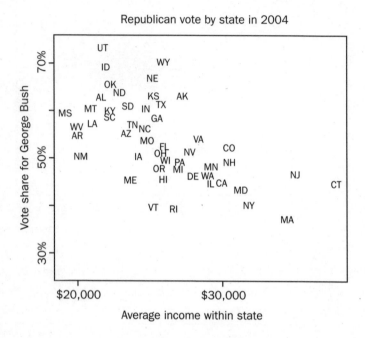

Republican vote by state in 2004

These data sets come from very different sources, but all four scatterplots arrange themselves in the same vaguely elliptical shape that the heights of parents and children did. In the first three cases, the correlation is *positive*; an increase in one variable is associated with an increase in the other, and the ellipse points northeast to southwest. In the last picture, the correlation is negative: In general, the richer states tend to skew more Democratic, and the ellipse points northwest to southeast.

THE UNREASONABLE EFFECTIVENESS OF CLASSICAL GEOMETRY

For Apollonius and the Greek geometers, ellipses were *conic sections*: surfaces obtained by slicing a cone along a plane. Kepler showed (although it took the astronomical community some decades to catch on) that the planets traveled in elliptical orbits, not circular ones as had been previously thought. Now, the very same curve arises as the natural shape enclosing heights of parents and children. Why? It's not because there's some hidden cone governing heredity which, when lopped off at just the right angle, gives Galton's ellipses. Nor is it that some form of genetic gravity enforces the elliptical form of Galton's charts via Newtonian laws of mechanics.

The answer lies in a fundamental property of mathematics—in a sense, the very property that has made mathematics so magnificently useful to scientists. In math there are many, many complicated objects, but only a few simple ones. So if you have a problem whose solution admits a simple mathematical description, *there are only a few possibilities for the solution.* The simplest mathematical entities are thus ubiquitous, forced into multiple duty as solutions to all kinds of scientific problems.

The simplest curves are lines. And it's clear that lines are everywhere in nature, from the edges of crystals to the paths of moving bodies in the absence of force. The *next* simplest curves are those cut out by quadratic equations,* in which no more than two variables are ever multiplied together. So squaring a variable, or multiplying two different variables, is allowed, but cubing a variable, or multiplying one variable by the square of another, is strictly forbidden. Curves in this class, including ellipses, are still called conic sections out of deference to history; but more forward-looking algebraic geometers call them *quadrics.*[†] Now there are lots of quadratic equations: any such is of the form

$$A\, x^2 + B\, xy + C\, y^2 + D\, x + E\, y + F = 0$$

for some values of the six constants A, B, C, D, E, and F. (The reader who feels so inclined can check that no other type of algebraic expression is allowed, subject to our requirement that we are only allowed to multiply two variables together, never three.) That seems like a lot of choices—infinitely many, in fact! But these quadrics turn out to fall into

---

* You could also make a case for curves of exponential growth and decay, which are just as ubiquitous as conic sections.
† Why they are called *quadrics* as opposed to *quadratics* is a nomenclatural mystery I have not managed to penetrate.

three main classes: ellipses, parabolas, and hyperbolas.[*] Here's what they look like:

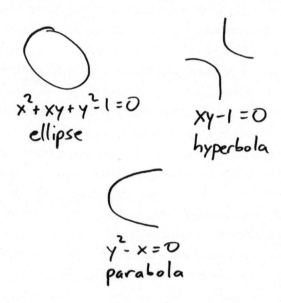

$$x^2 + xy + y^2 - 1 = 0$$
ellipse

$$xy - 1 = 0$$
hyperbola

$$y^2 - x = 0$$
parabola

We encounter these three curves again and again as the solution to scientific problems; not only the orbits of planets, but the optimal designs of curved mirrors, the arcs of projectiles, and the shapes of rainbows.

Or even beyond science. My colleague Michael Harris, a distinguished number theorist at the Institut de Mathématiques de Jussieu in Paris, has a theory that three of Thomas Pynchon's major novels are governed by the three conic sections: *Gravity's Rainbow* is about parabolas (all those rockets, launching and dropping!), *Mason & Dixon* about ellipses, and *Against the Day* about hyperbolas. This seems as good to me as any other organizing theory of these novels I've encountered; certainly Pynchon, a former physics major who likes to drop references to Möbius strips and the quaternions in his novels, knows very well what the conic sections are.

Galton observed that the curves he drew by hand looked like ellipses, but was not quite geometer enough to be sure that this precise curve, and not some other more or less ovoid figure, was actually in charge. Was

---

[*] There are actually a few extra cases, like the curve with the equation $xy = 0$, which is a pair of lines crossing at the point (0,0); these are considered "degenerate" and we will not speak of them here.

he letting his desire for an elegant and universal theory affect his perception of the data he'd collected? He wouldn't be the first or last scientist to make that mistake. Galton, careful as always, sought the advice of J. D. Hamilton Dickson, a mathematician at Cambridge. He even went so far as to conceal the origin of his data, presenting it as a problem arising from physics, to avoid prejudicing Dickson toward any particular conclusion. To Galton's delight, Dickson quickly confirmed that the ellipse was not only the curve that the data suggested, but the curve that theory demanded.

"The problem may not be difficult to an accomplished mathematician," Galton wrote, "but I certainly never felt such a glow of loyalty and respect towards the sovereignty and wide sway of mathematical analysis as when his answer arrived, confirming, by purely mathematical reasoning, my various and laborious statistical conclusions with far more minuteness than I had dared to hope, because the data ran somewhat roughly, and I had to smooth them with tender caution."

BERTILLONAGE

Galton understood quickly that the idea of correlation wasn't limited to the study of heredity; it applied to *any* pair of qualities that might bear some relation to one another.

As it happened, Galton was in possession of a massive database of anatomical measurements, of the sort that were enjoying a vogue in the late nineteenth century, thanks to the work of Alphonse Bertillon. Bertillon was a French criminologist with a spirit very much like Galton's; he was devoted to a rigorously quantitative view of human life and confident about the benefits such an approach would bring.* In particular, Bertillon was appalled by the unsystematic and haphazard way in which French police identified criminal suspects. How much better and more modern it would be, Bertillon reasoned, to attach to each miscreant

---

* For all his enthusiasm for data, though, Bertillon blew it in the biggest case he ever handled; he helped convict Alfred Dreyfus of treason with a bogus "geometric proof" that a letter offering to sell French military documents was written in Dreyfus's handwriting. See L. Schneps and C. Schneps, *Math on Trial*, for a full account of the case and Bertillon's unfortunate involvement.

Frenchman a series of numerical measurements: the length and breadth of the head, the length of fingers and feet, and so on. In Bertillon's system, each arrested suspect was measured and his data filed on cards and stored away for future use. Now, if the same man were nabbed again, identifying him was a simple matter of getting out the calipers, taking his numbers, and comparing them with the cards on file. "Aha, Mr. 15-6-56-42, thought you'd get away, didn't you?" You can replace your name by an alias, but there's no alias for the shape of your head.

Bertillon's system, so in keeping with the analytic spirit of the time, was adopted by the Paris Prefecture of Police in 1883, and quickly spread throughout the world. At its height, *bertillonage* held sway in police departments from Bucharest to Buenos Aires. "The Bertillon cabinet," Raymond Fosdick wrote in 1915, "became the distinguishing mark of the modern police organization." In its time, the practice was so common and uncontroversial in the United States that Justice Anthony Kennedy brought it up in his majority opinion in the 2013 case *Maryland vs. King*, allowing states to take DNA samples from felony arrestees: in Kennedy's view, a DNA sequence was just another sequence of data points attached to a suspect, a sort of twenty-first-century Bertillon card.

Galton asked himself: Was Bertillon's choice of measurements the best possible? Or could you identify suspects more accurately if you took even more measurements? The problem, Galton realized, is that bodily measurements aren't entirely independent. If you've already measured a suspect's hands, do you really need to measure his feet, too? You know what they say about men with big hands: their feet are, statistically speaking, also likely to be of greater than average size. So the addition of the foot length doesn't add as much information to the Bertillon card as one might initially hope. Adding more and more measurements—if they are poorly chosen—may provide steadily diminishing returns.

To study this phenomenon, Galton made another scatterplot, this one of height versus "cubit," the distance from the elbow to the tip of the middle finger. To his astonishment, he saw the same elliptical pattern that had emerged from the heights of fathers and sons. Once again, he had graphically demonstrated that the two variables, height and cubit, were *correlated*, even though one didn't strictly determine the other. If two measurements are highly correlated (like the length of the left foot and the length of the right) there's little point in taking the time to record both numbers. The best measurements to take are the ones that are uncorrelated with each of the others. And the relevant correlations could be computed from the vast array of anthropometric data Galton had already gathered.

As it happens, Galton's invention of correlation didn't lead to the institution of a vastly improved Bertillon system. That was largely thanks to Galton himself, who championed a competing system, dactyloscopy— what we now call fingerprinting. Like Bertillon's system, fingerprinting reduced a suspect to a list of numbers or symbols that could be marked on a card, sorted, and filed. But fingerprinting enjoyed certain obvious advantages, most notably that a criminal's fingerprints were often available for measurement in circumstances where the criminal himself was not. This point was made vividly by the case of Vincenzo Peruggia, who stole the *Mona Lisa* from the Louvre in a daring daylight theft in 1911. Peruggia had been arrested in Paris before, but his dutifully recorded Bertillon card, filed in its cabinet according to the lengths and widths of his various physical features, was not of much use. Had the cards contained dactyloscopic information, the fingerprint Peruggia left on the *Mona Lisa*'s discarded frame would have identified him at once.[*]

ASIDE: CORRELATION, INFORMATION, COMPRESSION, BEETHOVEN

I lied a little about the Bertillon system. In fact, he didn't record the exact numerical value of each physical characteristic, but only whether it was small, medium, or large. When you measure the length of the finger, you divide the criminals into three groups: small-fingered, medium-fingered, large-fingered. And then when you measure the cubit, you

---

[*] That's how Fosdick tells the story, at any rate, in "The Passing of the Bertillon System of Identification." As with any famous crime of yesteryear, there's a huge accretion of uncertainty and conspiracy theory around the *Mona Lisa* theft, and other sources tell different stories about the role of the fingerprints.

divide each of these three groups into three subgroups, so that the criminals are divided ninefold in all. Making all five measurements in the basic Bertillon system divides the criminals into

$$3 \times 3 \times 3 \times 3 \times 3 = 3^5 = 243$$

groups; and for each of these 243, there are seven options for eye and hair color. So, in the end, Bertillon classified suspects into $3^5 \times 7 = 1701$ tiny categories. Once you've arrested more than 1701 people, some categories will inevitably contain more than one suspect; but the number of people in any one category is likely to be rather small, small enough that a gendarme can easily flip through the cards to find a photograph matching the man in chains before him. And if you cared to add more measurements, tripling the number of categories each time you did so, you could easily make categories so small that no two criminals—for that matter, no two Frenchmen of any kind—would share the same Bertillon code.

It's a neat trick, keeping track of something complicated like the shape of a human being with a short string of symbols. And the trick isn't limited to human physiognomy. A similar system, called the Parsons code,* is used to classify musical melodies. Here's how it goes. Take a melody—one we all know, like Beethoven's "Ode to Joy," the glorious finale of the Ninth Symphony. We mark the first note with a *. And for each note thereafter, you mark down one of three symbols: **u** if the note at hand goes up from the previous note, **d** if it goes down, or **r** if it repeats the note that came before. The first two notes of Ode to Joy are the same, so you start out with *r. Then a higher note followed by a still higher one: *ruu. Next you repeat the top note, and then follow with a string of four descents: so the code for the whole opening segment is *ruurdddd.

You can't reproduce the sound of Beethoven's masterpiece from the Parsons code, any more than you can sketch a picture of a bank robber from his Bertillon measurements. But if you have a cabinet full of music

---

* Readers of a certain age may enjoy knowing that the Parsons who invented the Parsons code was the father of Alan Parsons, who recorded "Eye in the Sky."

categorized by Parsons code, the string of symbols does a pretty good job of identifying any given tune. If, for instance, you have the "Ode to Joy" in your head but can't remember what it's called, you can go to a website like Musipedia and type in *ruurdddd. That short string is enough to cut the possibilities down to "Ode to Joy" or Mozart's Piano Concerto No. 12. If you whistle to yourself a mere seventeen notes, there are

$$3^{16} = 3 \times 3 \times 3 \times 3 \times 3 \times 3 \times 3 \times 3 \times 3 \times 3 \times 3 \times 3 \times 3 \times 3 \times 3$$
$$\times 3 = 43,046,721$$

different Parsons codes; that's surely greater than the number of melodies ever recorded, and makes it pretty rare for two songs to have the same code. Each time you add a new symbol, you're multiplying the number of codes by three; and thanks to the miracle of exponential growth, a very short code gives you an astonishingly high capacity for discriminating between two songs.

But there's a problem. Back to Bertillon: What if we found that the men who came into the police station always had cubits in the same size category as their fingers? Then what look like nine choices for the first two measurements are really only three: small finger/small cubit, medium finger/medium cubit, and long finger/long cubit; two-thirds of the drawers in our Bertillon cabinet sit empty. The total number of categories is not really 1701, but a mere 567, with a corresponding diminution of our ability to distinguish one criminal from another. Another way to think of this: we thought that we were taking five measurements, but given that the cubit conveys exactly the same information as the finger, we were effectively taking only four. That's why the number of possible cards is cut down from $7 \times 3^5 = 1701$ to $7 \times 3^4 = 567$. (The 7 is counting the possibilities for eye and hair color.) More relationships between the measurements would make the effective number of categories still smaller and the Bertillon system still less powerful.

Galton's great insight was that the same thing applies even if finger length and cubit length aren't *identical*, but only *correlated*. Correlations between the measurements make the Bertillon code less informative. Once again, Galton's keen wisdom provided him a kind of intellectual prescience. What he'd captured was, in embryonic form, a way of thinking that would become fully formalized only a half-century later, by Claude Shannon in his theory of information. As we saw in chapter 13, Shannon's formal measure of information was able to provide bounds on how quickly bits could flow through a noisy channel; in much the same way, Shannon's theory provides a way of capturing the extent to which correlation between variables reduces the informativeness of a card. In modern terms we would say that the more strongly correlated the measurements, the less information, in Shannon's precise sense, a Bertillon card conveys.

Nowadays, though Bertillonage is gone, the idea that the best way to keep track of identity is by a sequence of numbers has achieved total dominance; we live in a world of digital information. And the insight that correlation reduces the effective amount of information has emerged as a central organizing principle. A photograph, which used to be a pattern of pigment on a sheet of chemically coated paper, is now a string of numbers, each one representing the brightness and color of a pixel. An image captured on a 4-megapixel camera is a list of 4 million numbers—no small commitment of memory for the device shooting the picture. But these numbers are highly correlated with each other. If one pixel is bright green, the next one over is likely to be as well. The actual information contained in the image is much less than 4 million numbers' worth—and it's precisely this fact that makes it possible* to have *compression*, the critical mathematical technology that allows images, videos, music, and text to be stored in much smaller spaces than you'd think. The presence of correlation makes compression possible; actually *doing* it involves much more modern ideas, like the theory of wavelets developed in the 1970s and '80s by Jean Morlet, Stéphane Mallat, Yves Meyer, Ingrid Daubechies, and others; and the rapidly developing area of compressed sensing, which started with a 2005 paper by Emmanuel Candès, Justin Romberg, and Terry Tao, and has quickly become its own active subfield of applied math.

---

* Okay, it's not literally just a matter of correlations between pairs of pixels, but it does come down to the amount of information (in Shannon's sense) conveyed by an image.

THE TRIUMPH OF MEDIOCRITY IN WEATHER

There's one thread we still need to tie off. We've seen how regression to the mean explains the "triumph of mediocrity" that Secrist discovered. But what about the triumph of mediocrity that Secrist *didn't* observe? When he tracked the temperatures of U.S. cities, he found that the hottest ones in 1922 were still hottest in 1931. This observation is crucial to his argument that the regression of business enterprises was something specific to human endeavor. If regression to the mean is a universal phenomenon, why don't temperatures do it too?

The answer is simple: they do.

The table below shows the average January temperatures in degrees Fahrenheit at thirteen weather stations in southern Wisconsin, no two of which are farther than an hour's drive apart:

	Jan 2011	Jan 2012
Clinton	15.9	23.5
Cottage Grove	15.2	24.8
Fort Atkinson	16.5	24.2
Jefferson	16.5	23.4
Lake Mills	16.7	24.4
Lodi	15.3	23.3
Madison airport	16.8	25.5
Madison arboretum	16.6	24.7
Madison, Charmany	17.0	23.8
Mazomanie	16.6	25.3
Portage	15.7	23.8
Richland Center	16.0	22.5
Stoughton	16.9	23.9

When you make a Galton-style scatterplot of these temperatures you see that, in general, the cities that were warmer in 2011 tended to be warmer in 2012.

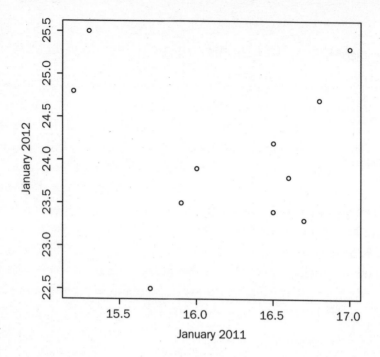

But the three warmest stations in 2011 (Charmany, Madison airport, and Stoughton) ended up the warmest, seventh warmest, and eighth warmest in 2012. Meanwhile, the three coldest 2011 stations (Cottage Grove, Lodi, and Portage) got relatively warmer: Portage was tied for fourth coldest, Lodi was second coldest, and Cottage Grove was actually warmer in 2012 than most of the other cities. In other words, both the hottest and the coldest groups moved toward the middle of the rankings, just as with Secrist's hardware stores.

Why didn't Secrist see this effect? Because he chose his weather stations in a different way. His cities weren't restricted to a small chunk of the upper Midwest, but were spread out much more widely. Suppose we look at the January temperatures as you range around California instead of Wisconsin:

	Jan 2011	Jan 2012
Eureka	48.5	46.6
Fresno	46.6	49.3
Los Angeles	59.2	59.4
Riverside	57.8	58.9

	Jan 2011	Jan 2012
San Diego	60.1	58.2
San Francisco	51.7	51.6
San Jose	51.2	51.4
San Luis Obispo	54.5	54.4
Stockton	45.2	46.7
Truckee	27.1	30.2

No regression to be seen. The cold places, like Truckee up in the Sierra Nevadas, stay cold, and the hot places, like San Diego and LA, stay hot. Plotting these temperatures gives you a very different-looking picture:

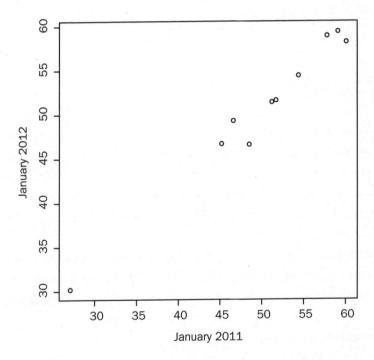

The Galtonian ellipse around these ten points would be very narrow indeed. The differences you see in the temperatures in the table reflect the fact that some places in California are just plain colder than others, and the underlying differences between the cities swamp the chance fluctuation from year to year. In Shannon's language, we'd say there's lots of signal and not so much noise. For the cities in south-central Wisconsin, it's just the opposite. Climatically speaking, Mazomanie and Fort Atkinson are not very different. In any given year, the ranking of these cities by temperature is going to have a lot to do with chance. There's lots of noise, not so much signal.

Secrist thought the regression he painstakingly documented was a new law of business physics, something that would bring more certainty and rigor to the scientific study of commerce. But it was just the opposite. If businesses were like cities in California—some really hot, some really not, reflecting inherent differences in business practice—you'd see correspondingly less regression to the mean. What Secrist's findings really show is that businesses are much more like the cities in Wisconsin. Superior management and business insight play a role, but so does plain luck, in roughly equal measure.

EUGENICS, ORIGINAL SIN, AND THIS BOOK'S MISLEADING TITLE

In a book called *How Not to Be Wrong* it's a bit strange to write about Galton without saying much about his greatest fame among non-mathematicians: the theory of eugenics, of which he's usually called the father. If, as I claim, an attention to the mathematical side of life is helpful in avoiding mistakes, how could a scientist like Galton, so clear-eyed with regard to mathematical questions, be so wrong about the merits of breeding human beings for desirable properties? Galton saw his own opinions on this subject as modest and sensible, but they shock the contemporary ear:

> As in most other cases of novel views, the wrong-headedness of objectors to Eugenics has been curious. The most common misrepresentations now are that its methods must be altogether those of compulsory unions, as in breeding animals. It is not so. I think that stern compulsion ought to be exerted to prevent the free propagation of the stock of those who are seriously afflicted by lunacy, feeble-mindedness, habitual criminality, and pauperism, but that is quite

different from compulsory marriage. How to restrain ill-omened marriages is a question by itself, whether it should be effected by seclusion, or in other ways yet to be devised that are consistent with a humane and well-informed public opinion. I cannot doubt that our democracy will ultimately refuse consent to that liberty of prop-agating children which is now allowed to the undesirable classes, but the populace has yet to be taught the true state of these things. A democracy cannot endure unless it be composed of able citizens; therefore it must in self-defence withstand the free introduction of degenerate stock.

What can I say? Mathematics is a way not to be wrong, but it isn't a way not to be wrong about *everything*. (Sorry, no refunds!) Wrongness is like original sin; we are born to it and it remains always with us, and constant vigilance is necessary if we mean to restrict its sphere of influ-ence over our actions. There is real danger that, by strengthening our abilities to analyze some questions mathematically, we acquire a general confidence in our beliefs, which extends unjustifiably to those things we're still wrong about. We become like those pious people who, over time, accumulate a sense of their own virtuousness so powerful as to make them believe the bad things they do are virtuous too.

I'll do my best to resist that temptation. But watch me carefully.

THE ADVENTURES OF KARL PEARSON ACROSS THE TENTH DIMENSION

It is difficult to overstate the impact of Galton's creation of correlation on the conceptual world we now inhabit—not only in statistics, but in every precinct of the scientific enterprise. If you know one thing about the word *correlation* it's that "correlation does not imply causation"—two phenomena can be correlated, in Galton's sense, even if one doesn't cause the other. This, by itself, was not news. People certainly understood that siblings are more likely than other pairs of people to share physical char-acteristics, and that this isn't because tall brothers cause their younger sisters to be tall. But there's still a causal relationship lurking in the background: the tall parents whose genetic contribution aids in causing both children to be tall. In the post-Galton world, you could talk about an association between two variables while remaining completely agnostic about the existence of *any* particular causal relationship, direct or indirect. In its way, the conceptual revolution Galton engendered has something in common with the insight of his more famous cousin, Charles Darwin. Darwin showed that one could meaningfully talk about progress without any need to invoke purpose. Galton showed that one could meaningfully talk about association without any need to invoke underlying cause.

Galton's original definition of correlation was somewhat limited, applying only to those variables whose distribution followed the bell curve law we saw in chapter 4. But the notion was quickly adapted and generalized by Karl Pearson* to apply to any variables whatsoever.

Were I to write down Pearson's formula right now, or were you to go look it up, you would see a mess of square roots and ratios, which, unless you have Cartesian geometry at your fingertips, would not be very illuminating. But in fact, Pearson's formula has a very simple geometric description. Mathematicians ever since Descartes have enjoyed the wonderful freedom to flip back and forth between algebraic and geometric descriptions of the world. The advantage of algebra is that it's easier to formalize and to type into a computer. The advantage of geometry is that it allows us to bring our physical intuition to bear on the situation, particularly when you can draw a picture. I seldom feel I *really* understand a piece of mathematics until I know what it's all about in geometric language.

So what, for a geometer, is correlation all about? It'll help to have an example at hand. Look again at the table on pages 332–333, which lists average January temperatures in ten California cities in 2011 and 2012. As we saw, the 2011 and 2012 temperatures have a strong positive correlation; in fact, Pearson's formula yields a sky-high value of 0.989.

If we want to study the relation between temperature measurements in two different years, it doesn't matter if you modify each entry in the table by the same amount. If 2011 temperature is correlated with 2012

---

* Dad of Egon Pearson, who battled with R. A. Fisher in an earlier chapter.

temperature, it's just as correlated with "2012 temperature + 5 degrees." Another way to put it: if you take all the points in the diagram above and move them up five inches, it doesn't change the shape of Galton's ellipse, merely its location. It turns out to be useful to shift the temperatures by a uniform amount to make the average value equal to *zero* in both 2011 and 2012. If you do that, you get a table that looks like this:

	Jan 2011	Jan 2012
Eureka	−1.7	−4.1
Fresno	−3.6	−1.4
Los Angeles	9.0	8.7
Riverside	7.6	8.2
San Diego	9.9	7.5
San Francisco	1.5	0.9
San Jose	1.0	0.7
San Luis Obispo	4.3	3.7
Stockton	−5.0	−4.0
Truckee	−23.1	−20.5

The rows of the table have negative entries for cold cities like Truckee and positive entries for balmier places like San Diego.

Now here's the trick. That column of ten numbers keeping track of the January 2011 temperatures is a list of numbers, yes. But it's also a *point*. How so? This goes back to our hero, Descartes. You can think of a pair of numbers (x,y) as a point in the plane, x units to the right and y units upward from the origin. In fact, we can draw a little arrow pointing from the origin to our point (x,y), an arrow called a *vector*.

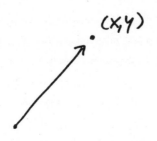

In the same way, a point in three-dimensional space is described by a list of three coordinates (x,y,z). And nothing except habit and craven fear keeps us from pushing this further. A list of four numbers can be thought of as a point in four-dimensional space, and a list of ten numbers, like the California temperatures in our table, is a point in ten-dimensional space. Better yet, think of it as a ten-dimensional vector.

Wait, you may rightfully ask: How am I supposed to think about that? What does a ten-dimensional vector look like?

It looks like this:

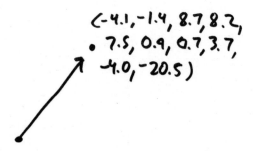

That's the dirty little secret of advanced geometry. It may sound impressive that we can do geometry in ten dimensions (or a hundred, or a million . . .), but the mental pictures we keep in our mind are two- or at most three-dimensional. That's all our brains can handle. Fortunately, this impoverished vision is usually enough.

High-dimensional geometry can seem a little arcane, especially since the world we live in is three-dimensional (or four-dimensional, if you count time, or maybe twenty-six-dimensional, if you're a certain kind of string theorist, but even then, you think the universe doesn't extend very far along most of those dimensions). Why study geometry that isn't realized in the universe?

One answer comes from the study of data, currently in extreme vogue. Remember the digital photo from the four-megapixel camera: it's described by 4 million numbers, one for each pixel. (And that's before we take color into account!) So that image is a 4-million-dimensional vector; or, if you like, a point in 4-million-dimensional space. And an image that changes with time is represented by a point that's *moving around* in a 4-million-dimensional space, which traces out a curve in 4-million-dimensional space, and before you know it you're doing 4-million-dimensional calculus, and then the fun can really start.

Back to temperature. There are two columns in our table, each of which provides us with a ten-dimensional vector. They look like this:

The two vectors point in roughly the same direction, which reflects the fact that the two columns are not in fact so different; as we've already seen, the coldest cities in 2011 stayed cold in 2012, and ditto for the warm ones.

And this is Pearson's formula, in geometric language. The correlation between the two variables is determined by the *angle* between the two vectors. If you want to get all trigonometric about it, the correlation is the cosine of the angle. It doesn't matter if you remember what cosine means; you just need to know that the cosine of an angle is 1 when the angle is 0 (i.e. when the two vectors are pointing in the same direction) and −1 when the angle is 180 degrees (vectors pointing in opposite directions). Two variables are positively correlated when the corresponding vectors are separated by an acute angle—that is, an angle smaller than 90 degrees—and negatively correlated when the angle between the vectors is larger than 90 degrees, or obtuse. It makes sense: vectors at an acute angle to one another are, in some loose sense, "pointed in the same direction," while vectors that form an obtuse angle seem to be working at cross purposes.

When the angle is a *right* angle, neither acute nor obtuse, the two variables have a correlation of zero; they are, at least as far as correlation goes, unrelated to each other. In geometry, we call a pair of vectors that form a right angle perpendicular, or orthogonal. And by extension, it's common practice among mathematicians and other trig aficionados to use the word "orthogonal" to refer to something unrelated to the issue at hand—"You might expect that mathematical skills are associated with magnificent popularity, but in my experience, the two are orthogonal." Slowly this usage is creeping out of the geekolect into the wider language. You can just about see it happening in a recent Supreme Court oral argument:

> MR. FRIEDMAN: I think that issue is entirely orthogonal to the
> issue here because the Commonwealth is acknowledging—
> CHIEF JUSTICE ROBERTS: I'm sorry. Entirely what?
> MR. FRIEDMAN: Orthogonal. Right angle. Unrelated. Irrelevant.
> CHIEF JUSTICE ROBERTS: Oh.
> JUSTICE SCALIA: What was that adjective? I like that.
> MR. FRIEDMAN: Orthogonal.
> JUSTICE SCALIA: Orthogonal?
> MR. FRIEDMAN: Right, right.
> JUSTICE SCALIA: Ooh.
> (Laughter.)

I'm rooting for *orthogonal* to catch on. It's been a while since a mathy word really broke out into demotic English. *Lowest common denominator* has by now lost its mathematical flavor almost entirely, and *exponentially*—just don't get me started on *exponentially.*[*]

The application of trigonometry to high-dimensional vectors in order to quantify correlation is not, to put it mildly, what the developers of the cosine had in mind. The Nicaean astronomer Hipparchus, who wrote down the first trigonometric tables in the second century BCE, was trying to compute the time lapse between eclipses; the vectors he dealt with described objects in the sky, and were solidly three-dimensional. But a mathematical tool that's just right for one purpose tends to make itself useful again and again.

The geometric understanding of correlation clarifies aspects of statistics that might otherwise be murky. Consider the case of the wealthy liberal elitist. For a while now, this slightly disreputable fellow has been a familiar character in political punditry. Perhaps his most devoted

---

[*] Though perhaps it's best not to complain too loudly about the incorrect use of *exponential* to mean simply "fast"—I recently saw a sportswriter, who had no doubt been scolded at some point about exponential, refer to sprinter Usain Bolt's "astonishing, logarithmic rise in speed," which is even worse.

chronicler is the political writer David Brooks, who wrote a whole book about the group he called the Bohemian Bourgeoisie, or Bobos. In 2001, contemplating the difference between suburban, affluent Montgomery County, Maryland (my birthplace!), and middle-class Franklin County, Pennsylvania, he speculated that the old political stratification by economic class, with the GOP standing up for the moneybags and the Democrats for the working man, was badly out of date.

> Like upscale areas everywhere, from Silicon Valley to Chicago's North Shore to suburban Connecticut, Montgomery County supported the Democratic ticket in last year's presidential election, by a margin of 63 percent to 34 percent. Meanwhile, Franklin County went Republican, by 67 percent to 30 percent.

First of all, this "everywhere" is a little strong. Wisconsin's richest county is Waukesha, centered on the tony suburbs west of Milwaukee. Bush crushed Gore there, 65–31, while Gore narrowly won statewide.

Still, Brooks is pointing to a real phenomenon, one we saw depicted quite plainly in a scatterplot a few pages back. In the contemporary U.S. electoral landscape, rich states are more likely than poor states to vote for the Democrats. Mississippi and Oklahoma are Republican strongholds, while the GOP doesn't even bother to contest New York and California. In other words, being from a rich state is positively correlated with voting Democratic.

But statistician Andrew Gelman found that the story is more complicated than the Brooksian portrait of a new breed of latte-sipping, Prius-driving liberals with big tasteful houses and NPR tote bags full of cash. In fact, rich people are still more likely to vote Republican than poor people are, an effect that's been consistently present for decades. Gelman and his collaborators, digging deeper into the state-by-state data, find a very interesting pattern. In some states, like Texas and Wisconsin, richer counties tend to vote more Republican. In others, like Maryland, California, and New York, the richer counties are more Democratic. Those last states happen to be the ones where many political pundits live. In their limited worlds, the rich neighborhoods *are* loaded with rich liberals, and it's natural for them to generalize this experience to the rest of the country. Natural, but when you look at the overall numbers, plainly wrong.

But there seems to be a paradox here. Being rich is positively correlated with being from a rich state, more or less by definition. And being from a rich state is positively correlated with voting for Democrats. Doesn't that mean being rich *has* to be correlated with voting Democratic? Geometrically: if vector 1 is at an acute angle to vector 2, and vector 2 is at an acute angle to vector 3, does vector 1 have to be at an acute angle to vector 3?

No! Proof by picture:

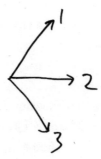

Some relationships, like "bigger than," are *transitive*; if I weigh more than my son and my son weighs more than my daughter, it's an absolute certainty that I weigh more than my daughter. "Lives in the same city as" is transitive, too—if I live in the same city as Bill, who lives in the same city as Bob, then I live in the same city as Bob.

Correlation is not transitive. It's more like "blood relation"—I'm related to my son, who's related to my wife, but my wife and I aren't blood relatives to each other. In fact, it's not a terrible idea to think of correlated variables as "sharing part of their DNA." Suppose I run a boutique money management firm with just three investors, Laura, Sara, and Tim. Their stock positions are pretty simple: Laura's fund is split 50-50 between Facebook and Google, Tim's is half General Motors and half Honda, and Sara, poised between old economy and new, goes half Honda, half Facebook. It's pretty obvious that Laura's returns will be positively correlated with Sara's; they have half their portfolio in common. And the correlation between Sara's returns and Tim's will be equally strong. But there's no reason to think Tim's performance has to be correlated with

Laura's.* Those two funds are like the parents, each contributing half its "genetic material" to form Sara's hybrid fund.

The non-transitivity of correlation is somehow obvious and mysterious at the same time. In the mutual-fund example, you'd never be fooled into thinking that a rise in Tim's performance gives much information about how Laura's doing. But our intuition does less well in other domains. Consider, for instance, the case of "good cholesterol," the common name for cholesterol conveyed around the bloodstream by high-density lipoproteins, or HDL. It's been known for decades that high levels of HDL cholesterol in the blood are associated with a lower risk of "cardiovascular events." If you're not a native speaker of medicalese, that means people with plenty of good cholesterol are less likely on average to clutch their hearts and keel over dead.

We also know that certain drugs reliably increase HDL levels. A popular one is niacin, a form of vitamin B. If niacin increases HDL, and more HDL is associated with lower risk of cardiovascular events, then it seems like popping niacin is a wise idea; that's why my physician recommended it to me, as yours probably did too, unless you're a teenager or a marathon runner or a member of some other metabolically privileged caste.

The problem is, it's not clear it works. Niacin supplementation recorded promising results in small clinical trials. But a large-scale trial carried out by the National Heart, Lung, and Blood Institute was halted in 2011, a year and a half before the scheduled finish, because the results were so weak it didn't seem worth it to continue. Patients who got niacin had higher HDL levels, all right, but they had just as many heart attacks and strokes as everybody else. How can this be? Because correlation isn't transitive. Niacin is correlated with high HDL, and high HDL is correlated with low risk of heart attack, but that doesn't mean that niacin prevents heart attacks.

Which isn't to say that manipulating HDL cholesterol is a dead end. Every drug is different, and it might be clinically relevant *how* you boost that HDL number. Back to the investment firm: we know that Tim's returns are correlated with Sara's, so you might try to improve Sara's earnings by taking measures to improve Tim's. If your approach were to

* Except insofar as the whole stock market tends to move in concert, of course.

issue a falsely optimistic stock tip to goose GM's stock price, you'd find that you improved Tim's performance, while Sara got no benefit. But if you did the same thing to Honda, Tim's and Sara's numbers would both improve.

If correlation were transitive, medical research would be a lot easier than it actually is. Decades of observation and data collection have given us lots of known correlations to work with. If we had transitivity, doctors could just chain these together into reliable interventions. We know that women's estrogen levels are correlated with lower risk of heart disease, and we know that hormone replacement therapy can raise those levels, so you might expect hormone replacement therapy to be protective against heart disease. And, indeed, that used to be conventional clinical wisdom. But the truth, as you've probably heard, is a lot more complicated. In the early 2000s, the Women's Health Initiative, a long-term study involving a gigantic randomized clinical trial, reported that hormone replacement therapy with estrogen and progestin appeared actually to *increase* the risk of heart disease in the population they studied. More recent results suggest that the effect of hormone replacement therapy might be different in different groups of women, or that estrogen alone might be better for your heart than the estrogen-progestin combo, and so on.

In the real world, it's next to impossible to predict what effect a drug will have on a disease, even if you know a lot about how it affects biomarkers like HDL or estrogen level. The human body is an immensely complex system, and there are only a few of its features we can measure, let alone manipulate. Based on the correlations we can observe, there are lots of drugs that might plausibly have a desired health effect. And so you try them out in experiments, and most of them fail dismally. To work in drug development requires a resilient psyche, not to mention a vast pool of capital.

UNCORRELATED DOESN'T MEAN UNRELATED

When two variables are correlated, we've seen that they're somehow related to each other. So what if they're not? Does that mean the variables are completely unrelated, neither one affecting the other? Far from it.

Galton's notion of correlation is limited in a very important way: it detects *linear* relations between variables, where an increase in one variable tends to coincide with a proportionally large increase (or decrease) in the other. But just as not all curves are lines, not all relationships are linear relationships.

Take this one:

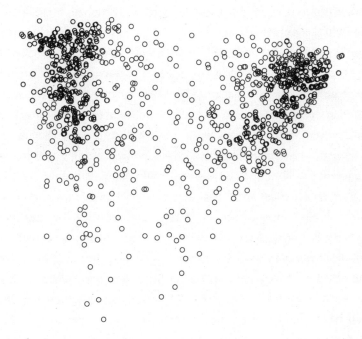

You're looking at a picture I made of a political survey taken by Public Policy Polling on December 15, 2011; there are one thousand dots, each representing a voter who responded to a twenty-three-question poll. The position of a point on the left-right axis represents, well, left and right: people who said they supported President Obama, approved of the Democratic Party, and opposed the Tea Party tend to be on the left-hand side, while those who favored the GOP, disliked Harry Reid, and believed there is a "War on Christmas" are over on the right. The vertical axis stands roughly for "informedness"—voters toward the bottom of the graph tended to answer "don't know" to more insidery questions like "Do you approve or disapprove of the job [Senate Minority Leader] Mitch McConnell is doing?" and to express little or no excitement about the 2012 presidential election.

One can check that the variables measured by two axes are uncorrelated,* just as eyeballing the graph suggests; it doesn't look like the points tend to be farther left or right as you move up the page. But that doesn't mean that the two variables aren't related to each other. In fact, the relation is quite clear from the picture. The plot is "heart-shaped," with a lobe on either side and a point at the bottom. As the voters get more informed, they don't get more Democratic or more Republican, but they do get *more polarized*: lefties go farther left, right-wingers get farther right, and the sparsely populated space in the middle gets even sparser. In the lower half of the graph, the less-informed voters tend to adopt a more centrist stance. The graph reflects a sobering social fact, which is by now commonplace in the political science literature. Undecided voters, by and large, aren't undecided because they're carefully weighing the merits of each candidate, unprejudiced by political dogma. They're undecided because they're barely paying attention.

A mathematical tool, like any scientific instrument, detects some kinds of phenomena but not others; a correlation computation can't see the heart-shapedness (cardiomorphism?) of this scatterplot any more than your camera can detect gamma rays. Keep this in mind when you're told that two phenomena in nature or society were found to be uncorrelated. It doesn't mean there's no relationship, only that there's no relationship of the sort that correlation is designed to detect.

---

* Technical note for those who care: in fact, this is the two-dimensional projection provided by a principal component analysis on the poll answers, so the uncorrelatedness of the two axes is automatic. The interpretation of the axes is my own. This example is meant merely to illustrate a point about correlation, and should not under any circumstances be taken as actual social science!

SIXTEEN

DOES LUNG CANCER MAKE YOU SMOKE CIGARETTES?

A nd what about when two variables *are* correlated? What does that really mean?

To make this simple, let's start with the simplest kind of variable, a *binary* variable with only two possible values. Oftentimes a binary variable is the answer to a yes-or-no question: "Are you married?" "Do you smoke?" "Are you now, or have you ever been, a member of the Communist Party?"

When you're comparing two binary variables, correlation takes on a particularly simple form. To say that marital status and smoking status are negatively correlated, for example, is simply to say that married people are less likely than the average person to smoke. Or, to put it another way, smokers are less likely than the average person to be married. It's worth taking a moment to persuade yourself that those two things are indeed the same! The first statement can be written as an inequality

married smokers / all married people < all smokers / all people

and the second as

married smokers / all smokers < all married people / all
people

If you multiply both sides of each inequality by the common denominator (all people) × (all smokers) you can see that the two statements are different ways of saying the same thing:

(married smokers) × (all people) < (all smokers) × (all married
people)

In the same way, if smoking and marriage were *positively* correlated,
it would mean that married people were more likely than average to
smoke and smokers more likely than average to be married.

One problem presents itself immediately. Surely the chance is very
small that the proportion of smokers among married people is *exactly the
same* as the proportion of smokers in the whole population. So, absent a
crazy coincidence, marriage and smoking will be correlated, either positively or negatively. And so will sexual orientation and smoking, U.S.
citizenship and smoking, first-initial-in-the-last-half-of-the-alphabet
and smoking, and so on. *Everything* will be correlated with smoking, in
one direction or the other. It's the same issue we encountered in chapter
7; the null hypothesis, strictly speaking, is just about always false.

To throw up our hands and say, "Everything is correlated with everything else!" would be fairly uninformative. So we don't report on all of
these correlations. When you read a report that one thing is correlated
with another, you're implicitly being told that the correlation is "strong
enough" to be worth reporting—usually because it passed a test of statistical significance. As we've seen, the statistical significance test brings
with it many dangers, but it is, at least, a signal that makes a statistician
sit up, take notice, and say, "Something must be going on."

But what? Here we come to the really sticky part. Marriage is negatively correlated with smoking; that's a fact. A typical way to express
that fact is to say

"If you're a smoker, you're less likely to be married."

But one small change makes the meaning very different:

"If you were a smoker, you'd be less likely to be married."

It seems strange that changing the sentence from the indicative to the subjunctive mood can change what it says so drastically. But the first sentence is merely a statement about what is the case. The second concerns a much more delicate question: What *would* be the case if we changed something about the world? The first sentence expresses a correlation; the second suggests a causation. As we've already mentioned, the two are not the same. That smokers are less frequently married than others doesn't mean that quitting smoking will summon up your future spouse. The mathematical account of correlation has been pretty much fixed in place since the work of Galton and Pearson a century ago. Putting the idea of causation on a firm mathematical footing has been much more elusive.*

There's something slippery about our understanding of correlation and causation. Your intuition tends to grasp it quite firmly in some circumstances but lose its grip in others. When we say that HDL is correlated with a lower risk of heart attack, we're making a factual statement: "If you've got a higher level of HDL cholesterol, you're less likely to have a heart attack." It's hard not to think that the HDL is *doing* something— that the molecules in question are literally causing your cardiovascular health to improve, say, by "scrubbing" lipidic cruft off your arterial walls. If that were so—if the mere presence of a lot of HDL were working to your benefit—then it would be reasonable to expect any HDL-increasing intervention to reduce your risk of heart attack.

But it might be that HDL and heart attack are correlated for a different reason; say, that some other factor, one we haven't measured, tends both to increase HDL and decrease the risk of cardiovascular events. If that's the case, an HDL-increasing drug might or might not prevent heart attack; if the drug affects HDL by way of the mystery factor, it'll prob-

---

* Though see the work of Judea Pearl at UCLA, whose work is at the center of the most notable contemporary attack on the problem of formalizing causality.

ably help your heart, but if it boosts HDL in some other way, all bets are off. That's the situation with Tim and Sara. Their financial success is correlated, but it's not because Tim's fund is causing Sara's to take off, or the reverse. It's because there's a mystery factor, the Honda stock, that affects both Tim and Sara. Clinical researchers call this the *surrogate endpoint problem*. It's time consuming and expensive to check whether a drug improves average life span, because in order to record someone's life span you have to wait for them to die. HDL level is the surrogate endpoint, the easy-to-check biomarker that's supposed to stand in for "long life with no heart attack." But the correlation between HDL and absence of heart attack might not indicate any causal link.

Teasing apart correlations that come from causal relationships from those that don't is a maddeningly hard problem, even in cases you might think of as obvious, like the relation between smoking and lung cancer. At the turn of the twentieth century, lung cancer was an extremely rare disease. But by 1947, the disease accounted for nearly a fifth of cancer deaths among British men, killing fifteen times as many people as it had a few decades earlier. At first, many researchers thought that lung cancer was simply being diagnosed more effectively than before, but it soon became clear that the increase in cases was too big and too fast to be accounted for by any such effect. Lung cancer really was on the rise. But no one was sure what to blame. Maybe it was smoke from factories, maybe increased levels of car exhaust, or maybe some substance not even thought of as a pollutant. Or maybe it was cigarette smoking, whose popularity had exploded during the same period.

By the early 1950s, large studies in England and America had shown a powerful association between cigarette smoking and lung cancer. Among nonsmokers, lung cancer was still a rare disease, but for smokers, the risk was spectacularly higher. A famous paper of Doll and Hill from 1950 found that among 649 male lung cancer patients in twenty London hospitals, only two were nonsmokers. That's not as impressive as it sounds by modern standards; in midcentury London, smoking was an extremely popular habit, and nonsmokers were much rarer than they are now. Even so, in a population of 649 male patients admitted for complaints other than lung cancer, twenty-seven were nonsmokers, a lot more than two. What's more, the association got stronger as smoking got heavier. Of the lung cancer patients, 168 went through more than twenty-five cigarettes a day, while only eighty-four men hospitalized for some other condition smoked that much.

Doll and Hill's data showed that lung cancer and smoking were *correlated*; their relation was not one of strict determination (some heavy smokers don't get lung cancer, while some nonsmokers do), but neither were the two phenomena independent. Their relation lay in that fuzzy, intermediate zone that Galton and Pearson had been the first to map.

The mere assertion of correlation is very different from an explanation. Doll and Hill's study doesn't show that smoking causes cancer; as they write, "The association would occur if carcinoma of the lung caused people to smoke or if both attributes were end-effects of a common cause." That lung cancer causes smoking, as they point out, is not very reasonable; a tumor can't go back in time and give someone a pack-a-day habit. But the problem of the common cause is more troubling.

Our old friend R. A. Fisher, the founding hero of modern statistics, was a vigorous skeptic of the tobacco-cancer link on exactly those grounds. Fisher was the natural intellectual heir to Galton and Pearson; in fact, he succeeded Pearson in 1933 as the Galton Chair of Eugenics at University College, London. (In deference to modern sensibilities, the position is now called the Galton Chair of Genetics.)

Fisher felt it was premature even to rule out the cancer-causes-smoking theory:

> Is it possible then, that lung cancer—that is to say, the pre-cancerous condition which must exist and is known to exist for years in those who are going to show overt lung cancer—is one of the causes of smoking cigarettes? I don't think it can be excluded. I don't think we know enough to say that it is such a cause. But the pre-cancerous condition is one involving a certain amount of slight chronic inflammation. The causes of smoking cigarettes may be studied among your friends, to some extent, and I think you will agree that a slight cause of irritation—a slight disappointment, an unexpected delay, some sort of a mild rebuff, a frustration—are commonly accompa-

nied by pulling out a cigarette and getting a little compensation for life's minor ills in that way. And so, anyone suffering from a chronic inflammation in part of the body (something that does not give rise to conscious pain) is not unlikely to be associated with smoking more frequently, or smoking rather than not smoking. It is the kind of comfort that might be a real solace to anyone in the fifteen years of approaching lung cancer. And to take the poor chap's cigarettes away from him would be rather like taking away his white stick from a blind man. It would make an already unhappy person a little more unhappy than he need be.

One sees here both a brilliant and rigorous statistician's demand that all possibilities receive fair consideration, and a lifelong smoker's affection for his habit. (Some have also seen the influence of Fisher's work as a consultant to the Tobacco Manufacturer's Standing Committee, a British industry group; in my view, Fisher's reluctance to assert a causal relationship was consistent with his general statistical approach.) Fisher's suggestion that the men in Doll and Hill's sample might have been driven to smoke by precancerous inflammation never caught on, but his argument for a common cause gained more traction. Fisher, true to his academic title, was a devoted eugenicist, who believed that genetic differences determined a healthy portion of our fate and that the better sort of people were in grave danger, in these evolutionarily forgiving times, of being outbred by their natural inferiors. From Fisher's point of view, it was perfectly natural to imagine that a common genetic factor, as yet unmeasured, was behind both lung cancer and propensity to smoke cigarettes. That might seem rather speculative. But remember, at the time, the generation of lung cancer by smoking rested on equally mysterious grounds. No chemical component of tobacco had yet been shown to produce tumors in the lab.

There's an elegant way to test for genetic influence on smoking, by studying twins. Say two twin siblings "match" if either both are smokers or both are not. You might expect matching to be fairly common, since twins typically grow up in the same home, with the same parents, and in the same cultural conditions, and that's indeed what you see. But identical twins and fraternal twins are subject to these commonalities to exactly the same degree; so if identical twins are more likely to match than fraternal twins, it's evidence that heritable factors exert some influence on smoking. Fisher presented some small-scale results to that effect, from unpublished studies, and more recent work has borne out his intuition; smoking appears to be subject to at least some heritable effects.

Which, of course, isn't to say that those same genes are what give you lung cancer down the road. We know a lot more now about cancer and how tobacco brings it about. That smoking gives you cancer is no longer in serious dispute. And yet it's hard not to be somewhat sympathetic to Fisher's let's-not-be-hasty approach. It's *good* to be suspicious of correlations. The epidemiologist Jan Vandenbroucke wrote of Fisher's articles on tobacco, "To my surprise, I found extremely well-written and cogent papers that might have become textbook classics for their impeccable logic and clear exposition of data and argument if only the authors had been on the right side."

Over the course of the 1950s, scientific opinion on the question of lung cancer and smoking steadily converged toward consensus. True, there was still no clear biological mechanism for the generation of tumors by tobacco smoke, and there was still no case for the association between smoking and cancer that didn't rest on observed correlations. But by 1959, so many such correlations had been seen, and so many possible confounding factors ruled out, that U.S. Surgeon General Leroy E. Burney was willing to assert, "The weight of evidence at present implicates smoking as the principal factor in the increased incidence of lung cancer." Even then, this stance was not uncontroversial. John Talbott, the editor of the *Journal of the American Medical Association*, fired back just weeks later in a *JAMA* editorial: "A number of authorities who have examined the same evidence cited by Dr. Burney do not agree with his conclusions. Neither the proponents nor the opponents of the smoking theory have sufficient evidence to warrant the assumption of an all-or-none authoritative position. Until definitive studies are forthcoming, the physician can fulfill his responsibility by watching the situation closely, keeping courant of the facts, and advising his patients on the basis of his appraisal of those facts." Talbott, like Fisher before him, was accusing Burney and those who agreed with him of being, scientifically speaking, out in front of their skis.

Just how fierce the dispute remained, even within the scientific establishment, is made clear by the remarkable work of historian of medicine Jon Harkness. His exhaustive archival research has shown that the statement signed by the surgeon general was in fact written by a large group of scientists at the Public Health Service, with Burney himself having little direct involvement. As for Talbott's response, that too, was ghostwritten—by a rival group of PHS researchers! What looked like a tussle between government officialdom and the medical establishment was in fact a scientific in-fight projected onto a public screen.

We know how this story ends. Burney's successor as surgeon general, Luther Terry, convened a blue-ribbon commission on smoking and health in the early 1960s, and in January 1964, to nationwide press coverage, announced their findings in terms that made Burney look timid:

> In view of the continuing and mounting evidence from many sources, it is the judgment of the Committee that cigarette smoking contributes substantially to mortality from certain specific diseases and to the overall death rate. . . . **Cigarette smoking is a health hazard of sufficient importance in the United States to warrant appropriate remedial action** [boldface from the original report].

What had changed? By 1964, the association between smoking and cancer had appeared consistently across study after study. Heavier smokers suffered more cancer than lighter smokers, and cancer was most likely at the point of contact between tobacco and human tissue; cigarette smokers got more lung cancer, pipe smokers more lip cancer. Ex-smokers were less prone to cancer than smokers who kept up the habit. All these factors combined to lead the surgeon general's committee to the conclusion that smoking was not just correlated with lung cancer, but *caused* lung cancer, and that efforts to reduce tobacco consumption would be likely to lengthen American lives.

IT'S NOT ALWAYS WRONG TO BE WRONG

In an alternate universe, one where later research on tobacco came out differently, we might have found that Fisher's odd-sounding theory was right after all, and smoking was a consequence of cancer instead of the other way around. It wouldn't be the biggest reversal medical science has ever suffered, by a long shot. And what then? The surgeon general would have issued a press release saying, "Sorry, everyone can go back to smoking now." In the interim, tobacco companies would have lost a lot of money, and millions of smokers would have forgone billions of pleasurable cigarettes. All because the surgeon general had declared as a fact what was only a strongly supported hypothesis.

But what was the alternative? Imagine what you'd have to do in order to really know, with something like absolute assurance, that smoking causes lung cancer. You'd have to collect a large population of teenagers, select half of them at random, and force that half to spend the next fifty years smoking cigarettes on a regular schedule, while the other half would be required to abstain. Jerry Cornfield, an early pioneer of smoking research, called such an experiment "possible to conceive but impossible to conduct." Even if such an experiment were logistically possible, it would violate every ethical norm in existence about research on human subjects.

Makers of public policy don't have the luxury of uncertainty that scientists do. They have to form their best guesses and make decisions on the basis thereof. When the system works—as it unquestionably did, in the case of tobacco—the scientist and the policy maker work in concert, the scientist reckoning how uncertain we ought to be and the policy maker deciding how to act under the uncertainty thus specified.

Sometimes this leads to mistakes. We've already encountered the case of hormone replacement therapy, which was long thought to protect postmenopausal women against heart disease, based on observed correlations. Current recommendations, based on randomized experiments performed later, are more or less the opposite.

In 1976 and again in 2009, the U.S. government embarked on massive and expensive vaccination campaigns against the swine flu, having received warnings from epidemiologists each time that the currently prevailing strain was particularly likely to go catastrophically pandemic. In fact, both flus, while severe, fell well short of disastrous.

It's easy to criticize the policy makers in these scenarios for letting their decision making get ahead of the science. But it's not that simple. *It's not always wrong to be wrong.*

How can this be so? A quick expected value computation, like the ones in part III, helps unpack the seemingly paradoxical slogan. Suppose we're considering making a health recommendation—say, that people should stop eating eggplant because eggplant induces a small risk of sudden catastrophic heart failure. This conclusion is based on a series of studies that found eggplant eaters slightly more likely than non–eggplant eaters to keel over dead without warning. But there's no prospect of doing a large-scale randomized controlled trial where we force eggplants on some people and deny them to others. We have to make do with the information we have, which represents a correlation only. For all we know, there's a common genetic basis for eggplantophilia and cardiac arrest. There's no way to be sure.

Perhaps we are 75% sure that our conclusion is correct and that a campaign against eggplant would save a thousand American lives per year. But there's also a 25% chance our conclusion is wrong; and if it's wrong, we've induced many people to give up what might be a favorite vegetable, leading them to eat a less healthy diet overall, and causing, let's say, two hundred excess deaths annually.[*]

As always, we obtain the expected value by multiplying the result of each possible outcome by the corresponding probability, and then adding everything up. In this case, we find that

$$75\% \times 1000 + 25\% \times (-200) = 750 - 50 = 700.$$

So our recommendation has an expected value of seven hundred lives saved per year. Over the loud and well-financed complaints of the Eggplant Council, and despite our very real uncertainty, we go public.

Remember: the expected value doesn't represent what we literally expect to happen, but rather what we might expect to happen *on average* were the same decision to be repeated again and again. A public health decision isn't like flipping a coin; it's something you can do only once. On the other hand, eggplants are not the only environmental danger we may be called upon to assess. Maybe it will come to our attention next that cauliflower is associated with arthritis, or vibrating toothbrushes

* All numbers in this example made up with no regard for plausibility.

with autism. If, in each case, an intervention has an expected value of seven hundred lives a year, we should make them all, and on average we will expect to save seven hundred lives each time. In any individual case, we might end up doing more harm than good, but overall we're going to save a lot of lives. Like the lottery players on roll-down day, we risk losing on any given instance, but are almost assured to come out ahead in the long run.

And if we held ourselves to a stricter evidentiary standard, declining to issue any of these recommendations because we weren't *sure* we were right? Then the lives we would have saved would be lost instead.

It would be great if we could assign precise, objective probabilities to real-life health conundrums, but of course we can't. This is another way that the interaction of a drug with the human body differs from a coin you can flip or a lottery ticket you can scratch. We're stuck with the messy, vague probabilities that reflect our degree of belief in various hypotheses, the probabilities that R. A. Fisher loudly denied were probabilities at all. So we don't and can't know the exact expected value of launching a campaign against eggplant or vibrating toothbrushes, or tobacco. But often we can say with confidence that the expected value is positive. Again, that doesn't mean the campaign is sure to have good effects, only that the sum total of *all* similar campaigns, over time, is likely to do more good than harm. The very nature of uncertainty is that we don't know which of our choices will help, like attacking tobacco, and which will hurt, like recommending hormone replacement therapy. But one thing's for certain: refraining from making recommendations at all, on the grounds that they might be wrong, is a losing strategy. It's a lot like George Stigler's advice about missing planes. If you never give advice until you're sure it's right, you're not giving enough advice.

BERKSON'S FALLACY, OR: WHY ARE HANDSOME MEN SUCH JERKS?

That correlations can arise from unseen common causes is confusing enough, but that's not the end of the story. Correlations can also come from common *effects*. This phenomenon is known as *Berkson's fallacy,* after the medical statistician Joseph Berkson, who back in chapter 8 explained how blind reliance on p-values could lead you to conclude that a small group of people including an albino consisted of nonhumans.

Berkson himself was, like Fisher, a vigorous skeptic about the link between tobacco and cancer. Berkson, an MD, represented the old school of epidemiology, deeply suspicious of any claim whose support was more statistical than medical. Such claims, he felt, represented a trespass by naive theorists onto ground that rightfully belonged to the medical profession. "Cancer is a biologic, not a statistical problem," he wrote in 1958. "Statistics can soundly play an ancillary role in its elucidation. But if biologists permit statisticians to become arbiters of biologic questions, scientific disaster is inevitable."

Berkson was especially troubled by the fact that tobacco use was found to be correlated not only with lung cancer but with dozens of other diseases, afflicting every system of the human body. For Berkson, the idea that tobacco could be so thoroughgoingly poisonous was inherently implausible: "It is as though, in investigating a drug that previously had been indicated to relieve the common cold, the drug was found not only to ameliorate coryza, but to cure pneumonia, cancer, and many other diseases. A scientist would say, 'There must be something wrong with this method of investigation.'"

Berkson, like Fisher, was more apt to believe the "constitutional hypothesis," that some preexisting difference between nonsmokers and smokers accounted for the relative healthiness of the abstainers:

> If 85 to 95 per cent of a population are smokers, then the small minority who are not smokers would appear, on the face of it, to be of some special type of constitution. It is not implausible that they should be on the average relatively longevous, and this implies that death rates generally in this segment of the population will be relatively low. After all, the small group of persons who successfully resist the incessantly applied blandishments and reflex conditioning of the cigaret advertisers are a hardy lot, and, if they can withstand these assaults, they should have relatively little difficulty in fending off tuberculosis or even cancer!

Berkson also objected to the original study of Doll and Hill, which was conducted among patients in British hospitals. What Berkson had observed in 1938 was that selecting patients in this way can create the appearance of associations that aren't really there.

Suppose, for example, you want to know whether high blood pressure is a risk factor for diabetes. You might take a survey of the patients in your hospital, with the goal of finding out whether high blood pressure was more common among the nondiabetics or the diabetics. And you find, to your surprise, that high blood pressure is *less* common among the patients with diabetes. You might thus be inclined to conclude that high blood pressure was protective against diabetes, or at least against diabetic symptoms so severe as to require hospitalization. But before you start advising diabetic patients to ramp up their consumption of salty snacks, consider this table.

1,000 total population
300 people with high blood pressure
400 people with diabetes
120 people with both high blood pressure and diabetes

Suppose there are a thousand people in our town, of whom 30% have high blood pressure and 40% have diabetes. (We like salty snacks *and* sweet snacks in our town.) And let's suppose, furthermore, that there's no relation between the two conditions; so 30% of the 400 diabetics, or 120 people in all, suffer from high blood pressure as well.

If all the sick people in town wind up in the hospital, then your hospital population is going to consist of

180 people with high blood pressure but no diabetes
280 people with diabetes but no high blood pressure
120 people with both high blood pressure and diabetes

Of the 400 total diabetics in the hospital, 120, or 30%, have high blood pressure. But of the 180 nondiabetics in the hospital, 100% have high blood pressure! It would be nuts to conclude from this that high blood pressure keeps you from having diabetes. The two conditions are negatively correlated, but that's not because one causes the absence of the other. It's also not because there's a hidden factor that both raises your blood pressure and helps regulate your insulin. It's because the two conditions have a common *effect*—namely, they put you in the hospital.

To put it in words: if you're in the hospital, you're there for a reason. If you're not diabetic, that makes it more likely the reason is high blood pressure. So what looks at first like a causal relationship between high blood pressure and diabetes is really just a statistical phantom.

The effect can work the other way, too. In real life, having two diseases is more likely to land you in the hospital than having one. Maybe the 120 hypertensive diabetics in our town *all* end up in the hospital, but 90% of the relatively healthy folks with only one thing wrong with them stay home. What's more, there are other reasons to be in the hospital; for instance, on the first snowy day of the year, a lot of people try to clean out their snowblower with their hand and get their finger chopped off. So the hospital population might look like

> 10 people with no diabetes or high blood pressure but a severed finger
> 18 people with high blood pressure but no diabetes
> 28 people with diabetes but no high blood pressure
> 120 people with both high blood pressure and diabetes

Now, when you do your hospital study, you find that 120 out of 148 diabetics, or 81%, have high blood pressure. But only 18 of the 28 non-diabetics, or 64%, have high blood pressure. That makes it seem that high blood pressure makes you *more* likely to have diabetes. But again, it's an illusion; all we're measuring is the fact that the set of people who end up in the hospital is anything but a random sample of the population.

Berkson's fallacy makes sense outside the medical domain; in fact, it even makes sense outside the realm of features that can be precisely quantified. You may have noticed that, among the men* in your dating pool, the handsome ones tend not to be nice, and the nice ones tend not

---

* Or "people of your preferred gender, if any," obviously.

to be handsome. Is that because having a symmetrical face makes you cruel? Or because being nice to people makes you ugly? Well, it could be. But it doesn't have to be. I present below the Great Square of Men:

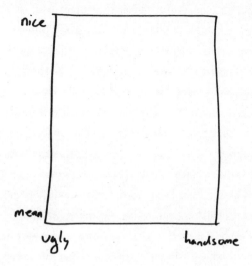

and I take as a working hypothesis that men are in fact equidistributed all over this square; in particular, there are nice handsome ones, nice ugly ones, mean handsome ones, and mean ugly ones, in roughly equal numbers.

But niceness and handsomeness have a common effect; they put these men in the group that you notice. Be honest—the mean uglies are the ones you never even consider. So inside the Great Square is a Smaller Triangle of Acceptable Men:

And now the source of the phenomenon is clear. The handsomest men in your triangle run the gamut of personalities, from kindest to cruelest. On average, they're about as nice as the average person in the whole population, which, let's face it, is not that nice. And by the same token, the nicest men are only averagely handsome. The ugly guys you like, though—they make up a tiny corner of the triangle, and they are pretty darn nice—they have to be, or they wouldn't be visible to you at all. The negative correlation between looks and personality in your dating pool is absolutely real. But if you try to improve your boyfriend's complexion by training him to act mean, you've fallen victim to Berkson's fallacy.

Literary snobbery works the same way. You know how popular novels are terrible? It's not because the masses don't appreciate quality. It's because there's a Great Square of Novels, and the only novels you ever hear about are the ones in the Acceptable Triangle, which are either popular or good. If you force yourself to read unpopular novels chosen essentially at random—I've been on a literary prize jury, so I've actually done this—you find that most of them, just like the popular ones, are pretty bad.

The Great Square is too simple by far, of course. There are many dimensions, not just two, along which you can rate your love interests or your weekly reading. So the Great Square is better described as a kind of Great Hypercube. And that's just for your own personal preferences! If you try to understand what happens in the whole population, you need to grapple with the fact that different people define attractiveness differently; they may differ about what weights to place on various criteria, or they may simply have incompatible preferences. The process of aggregating opinions, preferences, and desires from many different people presents yet another set of difficulties. Which means it's an opportunity to do more math. We turn to it now.

PART V

. . . . .

Existence

Includes: Derek Jeter's moral status, how to decide three-way elections, the Hilbert program, using the whole cow, why Americans are not stupid, "every two kumquats are joined by a frog," cruel and unusual punishment, "just as the work was completed the foundation gave way," the Marquis de Condorcet, the second incompleteness theorem, the wisdom of slime molds

SEVENTEEN

THERE IS NO SUCH THING AS PUBLIC OPINION

You're a good citizen of the United States of America, or some other more or less liberal democracy. Or maybe you're even an elected official. You think the government should, when possible, respect the people's will. So you want to know: What do the people want?

Sometimes you can poll the hell out of the people and it's still tough to be sure. For example: do Americans want small government? Well, sure we do—we say so constantly. In a January 2011 CBS News poll, 77% of respondents said cutting spending was the best way to handle the federal budget deficit, against only 9% who preferred raising taxes. That result isn't just a product of the current austerity vogue—year in, year out, the American people would rather cut government programs than pay more taxes.

But *which* government programs? That's where things get sticky. It turns out the things the U.S. government spends money on are things people kind of like. A Pew Research poll from February 2011 asked Americans about thirteen categories of government spending: in eleven of those categories, deficit or no deficit, more people wanted to increase spending than dial it down. Only foreign aid and unemployment insurance—which, combined, accounted for under 5% of 2010 spending—got the ax. That, too, agrees with years of data; the average American is always eager to slash foreign aid, occasionally tolerant of cuts to welfare or defense, and pretty gung ho for increased spending on every single other program our taxes fund.

Oh, yeah, and we want small government.

At the state level, the inconsistency is just as bad. Respondents to the Pew poll overwhelmingly favored a combination of cutting programs and raising taxes to balance state budgets. Next question: What about cutting funding for education, health care, transportation, or pensions? Or raising sales taxes, state income tax, or taxes on business? Not a single option drew majority support.

"The most plausible reading of this data is that the public wants a free lunch," economist Bryan Caplan wrote. "They hope to spend less on government without touching any of its main functions." Nobel Prize–winning economist Paul Krugman: "People want spending cut, but are opposed to cuts in anything except foreign aid. . . . The conclusion is inescapable: Republicans have a mandate to repeal the laws of arithmetic." The summary of a February 2011 Harris poll on the budget describes the self-negating public attitude toward the budget more colorfully: "Many people seem to want to cut down the forest but to keep the trees." It's an unflattering portrait of the American public. Either we are babies, unable to grasp that budget cuts will inevitably reduce funding to programs we support; or we are mulish, irrational children, who understand the math but refuse to accept it.

How are you supposed to know what the public wants when the public makes no sense?

RATIONAL PEOPLE, IRRATIONAL COUNTRIES

Let me stick up for the American people on this one, with the help of a word problem.

Suppose a third of the electorate thinks we should address the deficit by raising taxes without cutting spending; another third thinks we should cut defense spending; and the rest think we should drastically cut Medicare benefits.

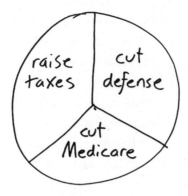

Two out of three people want to cut spending; so in a poll that asks "Should we cut spending or raise taxes?" the cutters are going to win by a massive 67–33 margin.

So what to cut? If you ask, "Should we cut the defense budget?" you'll get a resounding no: two-thirds of voters—the tax raisers joined by the Medicare cutters—want defense to keep its budget. And "Should we cut Medicare?" loses by the same amount.

That's the familiar self-contradicting position we see in polls: We want to cut! But we also want each program to keep all its funding! How did we get to this impasse? Not because the voters are stupid or delusional. *Each voter has a perfectly rational, coherent political stance.* But in the aggregate, their position is nonsensical.

When you dig past the front-line numbers of the budget polls, you see that the word problem isn't so far from the truth. Only 47% of Americans believed balancing the budget would require cutting programs that helped people like them. Just 38% agreed that there were worthwhile programs that would need to be cut. In other words: the infantile "average American," who wants to cut spending but demands to keep every single program, doesn't exist. The average American thinks there are plenty of non-worthwhile federal programs that are wasting our money and is ready and willing to put them on the chopping block to make ends meet. The problem is, there's no consensus on which programs are the worthless ones. In large part, that's because most Americans think the programs that benefit them personally are the ones that must, at all costs, be preserved. (I didn't say we weren't selfish, I just said we weren't stupid!)

The "majority rules" system is simple and elegant and feels fair, but it's at its best when deciding between just two options. Any more than two, and contradictions start to seep into the majority's preferences. As I write this, Americans are sharply divided over President Obama's signature domestic policy accomplishment, the Affordable Care Act. In an October 2010 poll of likely voters, 52% of respondents said they opposed the law, while only 41% supported it. Bad news for Obama? Not once you break down the numbers. Outright repeal of health care reform was favored by 37%, with another 10% saying the law should be weakened; but 15% preferred to leave it as is, and 36% said the ACA should be expanded to change the current health care system *more* than it currently does. That suggests that many of the law's opponents are to Obama's left, not his right. There are (at least) three choices here: leave the health care law alone, kill it, or make it stronger. And each of the three choices is opposed by most Americans.*

The incoherence of the majority creates plentiful opportunities to mislead. Here's how Fox News might report the poll results above:

Majority of Americans oppose Obamacare!

And this is how it might look on MSNBC:

Majority of Americans want to preserve or strengthen Obamacare!

These two headlines tell very different stories about public opinion. Annoyingly enough, both are true.

But both are incomplete. The poll watcher who aspires not to be wrong has to test each of the poll's options, to see whether it might break down into different-colored pieces. Fifty-six percent of the population disapproves of President Obama's policy in the Middle East? That impressive figure might include people from both the no-blood-for-oil left and the nuke-'em-all right, with a few Pat Buchananists and devoted libertarians in the mix. By itself, it tells us just about nothing about what the people really want.

Elections might seem an easier case. A pollster presents you with a simple binary choice, the same one you'll face at the ballot box: candidate 1, or candidate 2?

---

* Added in press: A CNN/ORC poll in May 2013 found that 43% favored the ACA, while 35% said it was too liberal and 16% said it wasn't liberal enough.

But sometimes there are more than two. In the 1992 presidential election, Bill Clinton drew 43% of the popular vote, ahead of George H. W. Bush with 38% and H. Ross Perot at 19%. To put it another way: a majority of voters (57%) thought Bill Clinton shouldn't be president. And a majority of voters (62%) thought George Bush shouldn't be president. And a really big majority of voters (81%) thought Ross Perot shouldn't be president. Not all those majorities can be satisfied at once; one of the majorities won't get to rule.

That doesn't seem like such a terrible problem—you can always award the presidency to the candidate with the highest vote tally, which, apart from Electoral College issues, is what the American electoral system does.

But suppose the 19% of voters who went with Perot broke down into 13% who thought Bush was the second-best choice and Clinton the worst of the bunch,* and 6% who thought Clinton was the better of the two major-party candidates. Then if you asked voters directly whether they preferred to have Bush or Clinton as president, 51%, a majority, would pick Bush. In that case, do you still think the public wants Clinton in the White House? Or is Bush, who most people preferred to Clinton, the people's choice? Why should the electorate's feelings about H. Ross Perot affect whether Bush or Clinton gets to be president?

I think the right answer is that there are no answers. Public opinion doesn't exist. More precisely, it exists sometimes, concerning matters about which there's a clear majority view. Safe to say it's the public's opinion that terrorism is bad and *The Big Bang Theory* is a great show. But cutting the deficit is a different story. The majority preferences don't meld into a definitive stance.

If there's no such thing as the public opinion, what's an elected official to do? The simplest answer: when there's no coherent message from the people, do whatever you want. As we've seen, simple logic demands that you'll sometimes be acting contrary to the will of the majority. If you're a mediocre politician, this is where you point out that the polling data contradicts itself. If you're a good politician, this is where you say, "I was elected to lead—not to watch the polls."

---

* People argue to this day about whether Perot took more votes from Bush or from Clinton, or whether the Perot voters would have just sat it out rather than vote for either of the major-party candidates.

And if you're a master politician, you figure out ways to turn the incoherence of public opinion to your advantage. In that February 2011 Pew poll, only 31% of respondents supported decreasing spending on transportation, and another 31% supported cutting funding for schools; but only 41% supported a tax hike on local businesses to pay for it all. In other words, each of the main options for cutting the state's deficit was opposed by a majority of voters. Which choice should the governor pick to minimize the political cost? The answer: don't pick one, pick two. The speech goes like this:

"I pledge not to raise taxes a single cent. I will give municipalities the tools they need to deliver top-quality public services at less cost to the taxpayers."

Now each locality, supplied with less revenue by the state, has to decide on its own between the remaining two options: cut roads or cut schools. See the genius here? The governor has specifically excluded raising taxes, the most popular of the three options, yet his firm stand has majority support: 59% of voters agree with the governor that taxes shouldn't rise. Pity the mayor or county executive who has to swing the axe. That poor sap has no choice but to execute a policy most voters won't like, and suffers the consequence while the governor sits pretty. In the budget game, as in so many others, playing first can be a big advantage.

VILLAINS OFTEN DESERVE WHIPPING, AND PERHAPS HAVING THEIR EARS CUT OFF

Is it wrong to execute mentally retarded prisoners? That sounds like an abstract ethical question, but it was a critical issue in a major Supreme Court case. More precisely, the question wasn't "Is it wrong to execute mentally retarded prisoners?" but "Do Americans believe it's wrong to execute mentally retarded prisoners?" That's a question about public opinion, not ethics—and as we've already seen, all but the very simplest questions about public opinion are lousy with paradox and confusion.

This one is not among the very simplest.

The justices encountered this question in the 2002 case *Atkins v. Virginia*. Daryl Renard Atkins and a confederate, William Jones, had robbed a man at gunpoint, kidnapped him, and then killed him. Each man testified that the other had been the triggerman, but the jury believed Jones, and Atkins was convicted of capital murder and sentenced to die.

Neither the quality of the evidence nor the severity of the crime was in dispute. The question before the court was not what Atkins had done, but what he was. Atkins's counsel argued before the Virginia Supreme Court that Atkins was mildly mentally retarded, with a measured IQ of 59, and as such could not be held sufficiently morally responsible to warrant the death penalty. The state supreme court rejected this argument, citing the U.S. Supreme Court's 1989 ruling in *Penry v. Lynaugh* that capital punishment of mentally retarded prisoners doesn't violate the Constitution.

This conclusion wasn't reached without great controversy among the Virginia justices. The constitutional questions involved were difficult enough that the U.S. Supreme Court agreed to revisit the case, and with it *Penry*. This time, the high court came down on the opposite side. In a 6–3 decision, they ruled that it would be unconstitutional to execute Atkins or any other mentally retarded criminal.

At first glance, this seems weird. The Constitution didn't change in any relevant way between 1989 and 2012; how could the document first license a punishment and then, twenty-three years later, forbid it? The key lies in the wording of the Eighth Amendment, which prohibits the state from imposing "cruel and unusual punishment." The question of what, precisely, constitutes cruelty and unusualness has been the subject of energetic legal dispute. The meaning of the words is hard to pin down; does "cruel" mean what the Founders would have considered cruel, or what we do? Does "unusual" mean unusual then, or unusual now? The makers of the Constitution were not unaware of this essential ambiguity. When the House of Representatives debated adoption of the Bill of Rights in August 1789, Samuel Livermore of New Hampshire argued that the vagueness of the language would allow softhearted future generations to outlaw necessary punishments:

> The clause seems to express a great deal of humanity, on which account I have no objection to it; but as it seems to have no meaning in

it, I do not think it necessary. What is meant by the term excessive bail? Who are to be the judges? What is understood by excessive fines? It lies with the court to determine. No cruel and unusual punishment is to be inflicted; it is sometimes necessary to hang a man, villains often deserve whipping, and perhaps having their ears cut off; but are we in future to be prevented from inflicting these punishments because they are cruel?

Livermore's nightmare came true; we do not now cut people's ears off, even if they were totally asking for it, and what's more, we hold that the Constitution forbids us from doing so. Eighth Amendment jurisprudence is now governed by the principle of "evolving standards of decency," first articulated by the Court in *Trop v. Dulles* (1958), which holds that contemporary American norms, not the prevailing standards of August 1789, provide the standard of what is cruel and what unusual.

That's where public opinion comes in. In *Penry*, Justice Sandra Day O'Connor's opinion held that opinion polls showing overwhelming public opposition to execution of mentally deficient criminals were not to be considered in the computation of "standards of decency." To be considered by the court, public opinion would need to be codified by state lawmakers into legislation, which represented "the clearest and most reliable objective evidence of contemporary values." In 1989, only two states, Georgia and Maryland, had made special provisions to prohibit execution of the mentally retarded. By 2002, the situation had changed, with such executions outlawed in many states; even the state legislature of Texas had passed such a law, though it was blocked from enactment by the governor's veto. The majority of the court felt the wave of legislation to be sufficient proof that standards of decency had evolved away from allowing Daryl Atkins to be put to death.

Justice Antonin Scalia was not on board. In the first place, he only grudgingly concedes that the Eighth Amendment can forbid punishments (like cutting off a criminal's ears, known in the penological context as "cropping") that were constitutionally permitted in the Framers' time.*

---

* On May 15, 1805, Massachusetts outlawed cropping, along with branding, whipping, and the pillory, as punishments for counterfeiting money; if those punishments had been understood to be forbidden by the Eighth Amendment at the time, the state law would not have been necessary

But even granting this point, Scalia writes, state legislatures have not demonstrated a national consensus against execution of the mentally retarded, as the precedent of *Penry* requires:

> The Court pays lip service to these precedents as it miraculously extracts a "national consensus" forbidding execution of the mentally retarded . . . from the fact that 18 States—less than half (47%) of the 38 States that permit capital punishment (for whom the issue exists)—have very recently enacted legislation barring execution of the mentally retarded. . . . That bare number of States alone—18— should be enough to convince any reasonable person that no "national consensus" exists. How is it possible that agreement among 47% of the death penalty jurisdictions amounts to "consensus"?

The majority's ruling does the math differently. By their reckoning, there are thirty states that prohibit execution of the mentally retarded: the eighteen mentioned by Scalia and the twelve that prohibit capital punishment entirely. That makes thirty out of fifty, a substantial majority.

Which fraction is correct? Akhil and Vikram Amar, brothers and constitutional law professors, explain why the majority has the better of it on mathematical grounds. Imagine, they ask, a scenario in which fortyseven state legislatures have outlawed capital punishment, but two of the three nonconforming states allow execution of mentally retarded convicts. In this case, it's hard to deny that the national standard of decency excludes the death penalty in general, and the death penalty for the mentally retarded even more so. To conclude otherwise concedes an awful lot of moral authority to the three states out of step with the national mood. The right fraction to consider here is 48 out of 50, not 1 out of 3.

In real life, though, there is plainly no national consensus against the death penalty itself. This confers a certain appeal to Scalia's argument.

(*A Historical Account of Massachusetts Currency*, by Joseph Barlow Felt, p. 214). Scalia's concession, by the way, doesn't reflect his current thinking: in a 2013 interview with *New York* magazine, he said he now believes the Constitution is A-OK with flogging, and presumably he feels the same way about cropping.

It's the twelve states that forbid the death penalty* that are out of step with general national opinion in favor of capital punishment; if they don't think executions should be allowed at all, how can they be said to have an opinion about which executions are permissible?

Scalia's mistake is the same one that constantly trips up attempts to make sense of public opinion; the inconsistency of aggregate judgments. Break it down like this. How many states believed in 2002 that capital punishment was morally unacceptable? On the evidence of legislation, only twelve. In other words, the majority of states, thirty-eight out of fifty, hold capital punishment to be morally acceptable.

Now, how many states think that executing a mentally retarded criminal is worse, legally speaking, than executing anyone else? Certainly the twenty states that are okay with both practices can't be counted among this number. Neither can the twelve states where capital punishment is categorically forbidden. There are only eighteen states that draw the relevant legal distinction; more than when *Penry* was decided, but still a small minority.

The majority of states, thirty-two out of fifty, hold capital punishment for mentally retarded criminals in the same legal standing as capital punishment generally.†

Putting those statements together seems like a matter of simple logic: if the majority thinks capital punishment in general is fine, and if the majority thinks capital punishment for mentally retarded criminals is no worse than capital punishment in general, then the majority must approve of capital punishment for mentally retarded criminals.

But this is wrong. As we've seen, "the majority" isn't a unified entity that follows logical rules. Remember, the majority of voters didn't want George H. W. Bush to be re-elected in 1992, and the majority of voters didn't want Bill Clinton to take over Bush's job; but, much as H. Ross Perot might have wished it, it doesn't follow that the majority wanted neither Bush nor Clinton in the Oval Office.

---

* Since 2002, the number has risen to seventeen.
† This is not precisely Scalia's computation; Scalia didn't go so far as to assert that the no-death-penalty states thought execution of mentally retarded criminals no worse than execution in general. Rather, his argument amounts to the claim that we have no information about their opinions in this matter, so we shouldn't count these states in our tally.

The Amar brothers' argument is more persuasive. If you want to know how many states think executing the mentally retarded is morally impermissible, you simply ask how many states outlaw the practice— and that number is thirty, not eighteen.

Which isn't to say Scalia's overall conclusion is wrong and the majority opinion correct; that's a legal question, not a mathematical one. And fairness compels me to point out that Scalia lands some mathematical blows as well. Justice Stevens's majority opinion, for instance, remarks that execution of mentally retarded prisoners is rare even in states that don't specifically prohibit the practice, suggesting a public resistance to such executions beyond that which state legislatures have made official. In only five states, Stevens writes, was such an execution carried out in the thirteen years between *Penry* and *Atkins.*

Just over six hundred people in all were executed in those years. Stevens offers a figure of 1% for the prevalence of mental retardation in the U.S. population. So if mentally retarded prisoners were executed at exactly the same rate as the general population, you'd expect about six or seven members of that population to have been put to death. Viewed in this light, as Scalia points out, the evidence shows no particular disinclination toward executing the mentally retarded. No Greek Orthodox bishop has ever been executed in Texas, but can you doubt Texas would kill a bishop if the necessity arose?

Scalia's real concern in *Atkins* is not so much the precise question before the court, which both sides agree affects a tiny segment of capital cases. Rather, he is worried about what he calls the "incremental abolition" of capital punishment by judicial decree. He quotes his own earlier opinion in *Harmelin v. Michigan:* "The Eighth Amendment is not a ratchet, whereby a temporary consensus on leniency for a particular crime fixes a permanent constitutional maximum, disabling the States from giving effect to altered beliefs and responding to changed social conditions."

Scalia is right to be troubled by a system in which the whims of one generation of Americans end up constitutionally binding our descendants. But it's clear his objection is more than legal; his concern is an America that loses the habit of punishment through enforced disuse, an America that is not only legally barred from killing mentally retarded murderers but that, by virtue of the court's lenient ratchet, has forgotten that it wants to. Scalia—much like Samuel Livermore two hundred years earlier—foresees and deplores a world in which the populace loses by inches its ability to impose effective punishments on wrongdoers. I can't manage to share their worry. The immense ingenuity of the human species in devising ways to punish people rivals our abilities in art, philosophy, and science. Punishment is a renewable resource; there is no danger we'll run out.

FLORIDA 2000, THE SLIME MOLD, AND HOW TO CHOOSE A WINGMAN

The slime mold *Physarum polycephalum* is a charming little organism. It spends much of its life as a tiny single cell, roughly related to the amoeba. But, under the right condition, thousands of these organisms coalesce into a unified collective called a plasmodium; in this form, the slime mold is bright yellow and big enough to be visible to the naked eye. In the wild it lives on rotting plants. In the laboratory it really likes oats.

You wouldn't think there'd be much to say about the psychology of the plasmodial slime mold, which has no brain or anything that could be called a nervous system, let alone feelings or thoughts. But a slime mold, like every living creature, makes decisions. And the interesting thing about the slime mold is that it makes *pretty good* decisions. In the slime mold's limited world, these decisions more or less come down to "move toward things I like" (oats) and "move away from things I don't like" (bright light). Somehow, the slime mold's decentralized thought process is able to get this job done very effectively. As in, you can train a slime mold to run through a maze. (This takes a long time and a lot of oats.) Biologists hope that by understanding how the slime mold navigates its world, they can open a window into the evolutionary dawn of cognition.

And even here, in the most primitive kind of decision-making imaginable, we encounter some puzzling phenomena. Tanya Latty and Madeleine Beekman of the University of Sydney were studying the way slime molds handled tough choices. A tough choice for a slime mold looks something like this: On one side of the petri dish is three grams of oats.

On the other side is five grams of oats, but with an ultraviolet light trained on it. You put a slime mold in the center of the dish. What does it do?

Under those conditions, they found, the slime mold chooses each option about half the time; the extra food just about balances out the unpleasantness of the UV light. If you were a classical economist of the kind Daniel Ellsberg worked with at RAND, you'd say that the smaller pile of oats in the dark and the bigger pile under the light have the same amount of utility for the slime mold, which is therefore ambivalent between them.

Replace the five grams with ten grams, though, and the balance is broken; the slime mold goes for the new double-size pile every time, light or no light. Experiments like this teach us about the slime mold's priorities and how it makes decisions when those priorities conflict. And they make the slime mold look like a pretty reasonable character.

But then something strange happened. The experimenters tried putting the slime mold in a petri dish with *three* options: the three grams of oats in the dark (3-dark), the five grams of oats in the light (5-light), and a single gram of oats in the dark (1-dark). You might predict that the slime mold would almost never go for 1-dark; the 3-dark pile has more oats in it and is just as dark, so it's clearly superior. And indeed, the slime mold just about never picks 1-dark.

You might also guess that, since the slime mold found 3-dark and 5-light equally attractive before, it would continue to do so in the new context. In the economist's terms, the presence of the new option shouldn't change the fact that 3-dark and 5-light have equal utility. But no: when 1-dark is available, the slime mold actually changes its preferences, choosing 3-dark more than three times as often as it does 5-light!

What's going on?

Here's a hint: the small, dark pile of oats is playing the role of H. Ross Perot in this scenario.

The mathematical buzzword in play here is "independence of irrelevant alternatives." That's a rule that says, whether you're a slime mold, a human being, or a democratic nation, if you have a choice between two options A and B, the presence of a third option C shouldn't affect which of A and B you like better. If you're deciding whether you'd rather have a Prius or a Hummer, it doesn't matter whether you also have the option of a Ford Pinto. You *know* you're not going to choose the Pinto. So what relevance could it have?

Or, to keep it closer to politics: in place of an auto dealership, put the state of Florida. In place of the Prius, put Al Gore. In place of the Hummer, put George W. Bush. And in place of the Ford Pinto, put Ralph Nader. In the 2000 presidential election, George Bush got 48.85% of Florida's votes and Al Gore got 48.84%. The Pinto got 1.6%.

So here's the thing about Florida in 2000. Ralph Nader was not going to win Florida's electoral votes. You know that, I know that, and every voter in the state of Florida knew that. What the voters of the state of Florida were being asked was not actually

"Should Gore, Bush, or Nader get Florida's electoral votes?"

but

"Should Gore or Bush get Florida's electoral votes?"

It's safe to say that virtually every Nader voter thought Al Gore would be a better president than George Bush.* Which is to say that a solid 51% majority of Florida voters preferred Gore over Bush. And yet the presence of Ralph Nader, the irrelevant alternative, means that Bush takes the election.

I'm not saying the election should have been decided differently. But what's true is that votes produce paradoxical outcomes, in which majorities don't always get their way and irrelevant alternatives control the outcome. Bill Clinton was the beneficiary in 1992, George W. Bush in 2000, but the mathematical principle is the same: it's hard to make sense of "what the voters really want."

But the way we settle elections in America isn't the only way. That might seem weird at first—what choice, other than the candidate who got the most votes, could possibly be fair?

Here's how a mathematician would think about this problem. In

---

* Yes, I, too, know that one guy who thought both Gore and Bush were tools of the capitalist overlords and it didn't make a difference who won. I am not talking about that guy.

fact, here's the way one mathematician—Jean-Charles de Borda, an eighteenth-century Frenchman distinguished for his work in ballistics—*did* think about the problem. An election is a machine. I like to think of it as a big cast-iron meat grinder. What goes into the machine is the preferences of the individual voters. The sausagey goop that comes out, once you turn the crank, is what we call the popular will.

What bothers us about Al Gore's loss in Florida? It's that more people preferred Gore to Bush than the reverse. Why doesn't our voting system know that? Because the people who voted for Nader had no way to express their preference for Gore over Bush. We're leaving some relevant data out of our computation.

A mathematician would say, "You shouldn't leave out information that might be relevant to the problem you're trying to solve!"

A sausage maker would put it, "If you're grinding meat, use the whole cow!"

And both would agree that you ought to find a way to take into account people's full set of preferences—not just which candidate they like the most. Suppose the Florida ballot had allowed voters to list all three candidates in their preferred order. The results might have looked something like this:

Bush, Gore, Nader	49%
Gore, Nader, Bush	25%
Gore, Bush, Nader	24%
Nader, Gore, Bush*	2%

The first group represents Republicans and the second group liberal Democrats. The third group is conservative Democrats for whom Nader is a little too much. The fourth group is, you know, people who voted for Nader.

How to make use of this extra information? Borda suggested a simple

* And surely there were some people who liked Nader best and preferred Bush to Gore, or who liked Bush best and preferred Nader to Gore, but my imagination is not strong enough to understand what sort of people these could possibly be, so I'm going to assume their numbers are too small to materially affect the computation.

and elegant rule. You can give each candidate points according to their placement: if there are three candidates, give 2 for a first-place vote, 1 for second, 0 for third. In this scenario, Bush gets 2 points from 49% of the voters and 1 point from 24% more, for a score of

$$2 \times 0.49 + 1 \times 0.24 = 1.22.$$

Gore gets 2 points from 49% of the voters and 1 point from another 51%, or a score of 1.49. And Nader gets 2 points from the 2% who like him best, and another point from the liberal 25%, coming in last at 0.29.

So Gore comes in first, Bush second, Nader third. And that jibes with the fact that 51% of the voters prefer Gore to Bush, 98% prefer Gore to Nader, and 73% prefer Bush to Nader. All three majorities get their way!

But what if the numbers were slightly shifted? Say you move 2% of the voters from "Gore, Nader, Bush" to "Bush, Gore, Nader." Then the tally looks like this:

Bush, Gore, Nader	51%
Gore, Nader, Bush	23%
Gore, Bush, Nader	24%
Nader, Gore, Bush	2%

Now a majority of Floridians like Bush better than Gore. In fact, an absolute majority of Floridians have Bush as their first choice. But Gore still wins the Borda count by a long way, 1.47 to 1.26. What puts Gore over the top? It's the presence of Ralph "Irrelevant Alternative" Nader, the same guy who spoiled Gore's bid in the actual 2000 election. Nader's presence on the ballot pushes Bush down to third place on many ballots, costing him points; while Gore enjoys the privilege of never being picked last, because the people who hate him hate Nader even more.

Which brings us back to the slime mold. Remember, the slime mold doesn't have a brain to coordinate its decision making, just thousands of nuclei enclosed in the plasmodium, each pushing the collective in one direction or another. Somehow the slime mold has to aggregate the information available to it into a decision.

If the slime mold were deciding purely on food quantity, it would rank 5-light first, 3-dark second, and 1-dark third. If it used only darkness, it would rank 3-dark and 1-dark tied for first, with 5-light third.

Those rankings are incompatible. So how does the slime mold decide to prefer 3-dark? What Latty and Beekman speculate is that the slime mold uses some form of *democracy* to choose between these two options, via something like the Borda count. Let's say 50% of the slime mold nuclei care about food and 50% care about light. Then the Borda count looks like this:

5-light, 3-dark, 1-dark	50%
1-dark and 3-dark tied, 5-light	50%

5-light gets 2 points from the half of the slime mold that cares about food, and 0 from the half of the slime mold that cares about light, for a point total of

$$2 \times (0.5) + 0 \times (0.5) = 1.$$

In a tie for first, we give both contestants 1.5 points; so 3-dark gets 1.5 points from half the slime mold and 1 from the other half, ending up with 1.25. And the inferior option 1-dark gets nothing from the food-loving half of the slime mold, which ranks it last, and 1.5 from the light-hating half, which has it tied for first, for a total of 0.75. So 3-dark comes in first, 5-light second, and 1-dark last, in exact conformity with the experimental result.

What if the 1-dark option weren't there? Then half the slime mold would rate 5-light above 3-dark, and the other half would rate 3-dark above 5-light; you get a tie, which is exactly what happened in the first experiment, where the slime mold chose between the dark three-gram pile of oats and the bright five-gram pile.

In other words: the slime mold likes the small, unlit pile of oats about as much as it likes the big, brightly lit one. But if you introduce a *really* small unlit pile of oats, the small dark pile looks better by comparison; so much so that the slime mold decides to choose it over the big bright pile almost all the time.

This phenomenon is called the "asymmetric domination effect," and slime molds are not the only creatures subject to it. Biologists have found jays, honeybees, and hummingbirds acting in the same seemingly irrational way.

Not to mention humans! Here we need to replace oats with romantic partners. Psychologists Constantine Sedikides, Dan Ariely, and Nils Olsen offered undergraduate research subjects the following task:

> You will be presented with several hypothetical persons. Think of these persons as prospective dating partners. You will be asked to choose *the one* person you would ask out for a date. Please assume that all prospective dating partners are: (1) University of North Carolina (or Duke University) students, (2) of the same ethnicity or race as you are, and (3) of approximately the same age as you are. The prospective dating partners will be described in terms of several attributes. A percentage point will accompany each attribute. This percentage point reflects the relative position of the prospective dating partner on that trait or characteristic, compared to UNC (DU) students who are of the same gender, race, and age as the prospective partner is.

Adam is in the 81st percentile of attractiveness, the 51st percentile of dependability, and the 65th percentile of intelligence, while Bill is the 61st percentile of attractiveness, 51st of dependability, and 87th of intelligence. The college students, like the slime mold before them, faced a tough choice. And just like the slime mold, they went 50-50, half the group preferring each potential date.

But things changed when Chris came into the picture. He was in the 81st percentile of attractiveness and 51st percentile of dependability, just like Adam, but in only the 54th percentile of intelligence. Chris was the irrelevant alternative; an option that was plainly worse than one of the choices already on offer. You can guess what happened. The presence of a slightly dumber version of Adam made the real Adam look better; given the choice between dating Adam, Bill, and Chris, almost two-thirds of the women chose Adam.

So if you're a single guy looking for love, and you're deciding which friend to bring out on the town with you, choose the one who's pretty much exactly like you—only slightly less desirable.

Where does irrationality come from? We've seen already that the apparent irrationality of popular opinion can arise from the collective behavior of perfectly rational individual people. But individual people, as we know from experience, are *not* perfectly rational. The story of the slime mold suggests that the paradoxes and incoherencies of our everyday behavior might themselves be explainable in a more systematic way. Maybe individual people seem irrational because they aren't really individuals! Each one of us is a little nation-state, doing our best to settle disputes and broker compromises between the squabbling voices that drive us. The results don't always make sense. But they somehow allow us, like the slime molds, to shamble along without making too many terrible mistakes. Democracy is a mess—but it kind of works.

USING THE WHOLE COW, IN AUSTRALIA AND VERMONT

Let me tell you how they do it in Australia.

The ballot down under looks a lot like Borda's. You don't just mark your ballot with the candidate you like best; you rank *all* the candidates, from your favorite to the one you hate the most.

The easiest way to explain what happens next is to see what Florida 2000 would have looked like under the Australian system.

Start by counting the first-place votes, and eliminate the candidate who got the fewest. In this case, that's Nader. Toss him! Now we're down to Bush vs. Gore.

But just because we threw Nader out doesn't mean we have to throw out the ballots of the people who voted for him. (Use the whole cow!) The next step—the "instant runoff"—is the really ingenious one. Cross Nader's name off every ballot and count the votes again, as if Nader had never existed. Now Gore has 51% of the first-place votes: the 49% he had from the first round, plus the votes that used to go to Nader. Bush still has the 49% he started with. He has fewer first-place votes, so he's eliminated. And Gore is the victor.

What about our slightly modified version of Florida 2000, where we moved 2% from "Gore, Nader, Bush" to "Bush, Gore, Nader"? In that situation, Gore still won the Borda count. By Aussie rules, it's a different story. Nader still gets knocked off in the first round; but now, since 51% of the ballots place Bush higher than Gore, Bush takes the prize.

The appeal of instant-runoff voting (or "preferential voting," as they call it in Australia) is obvious. People who like Ralph Nader can vote for him without worrying that they're throwing the race to the person they like least. For that matter, Ralph Nader can *run* without worrying about throwing the race to the person he likes least.*

Instant-runoff voting (IRV) has been around for more than 150 years. They use it not only in Australia but in Ireland and Papua New Guinea. When John Stuart Mill, who always had a soft spot for math, heard about the idea, he said it was "among the very greatest improvements yet made in the theory and practice of government."[†]

And yet—

Let's take a look at what happened in the 2009 mayoral race in Burlington, Vermont, one of the only U.S. municipalities to use the instant-runoff system.[‡] Get ready—a lot of numbers are about to come flying at your face.

The three main candidates were Kurt Wright, the Republican; Andy Montroll, the Democrat; and the incumbent, Bob Kiss, from the left-wing Progressive Party. (There were other minor candidates in the race, but in the interest of brevity I'm going to ignore their votes.) Here's the ballot count:

Montroll, Kiss, Wright	1332
Montroll, Wright, Kiss	767
Montroll	455
Kiss, Montroll, Wright	2043
Kiss, Wright, Montroll	371

---

* I'll concede it's not clear Ralph Nader actually worries about this.
† To be precise, Mill was actually talking about the closely related "single transferable vote" system.
‡ But not any more—in a narrowly decided referendum, Burlington voters repealed instant-runoff voting in 2010.

Kiss	568
Wright, Montroll, Kiss	1513
Wright, Kiss, Montroll	495
Wright	1289

(Not everyone was on board with the avant-garde voting system, as you can see: some people just marked their first choice.)

Wright, the Republican, gets 3297 first-place votes in all; Kiss gets 2982; and Montroll gets 2554. If you've ever been to Burlington, you probably feel safe in saying that a Republican mayor was not the people's will. But in the traditional American system, Wright would have won this election, thanks to vote splitting between the two more liberal candidates.

What actually happened was entirely different. Montroll, the Democrat, had the fewest first-place votes, so he was eliminated. In the next round, Kiss and Wright each kept the first-place votes they already had, but the 1332 ballots that used to say "Montroll, Kiss, Wright" now just said "Kiss, Wright," and they counted for Kiss. Similarly, the 767 "Montroll, Wright, Kiss" votes counted for Wright. Final vote: Kiss 4314, Wright 4064, and Kiss is reelected.

Sounds good, right? But wait a minute. Adding up the numbers a different way, you can check that 4067 voters liked Montroll better than Kiss, while only 3477 liked Kiss better than Montroll. And 4597 voters preferred Montroll to Wright, but only 3668 preferred Wright to Montroll.

In other words, a majority of voters liked the centrist candidate Montroll better than Kiss, and a majority of voters liked Montroll better than Wright. That's a pretty solid case for Montroll as the rightful winner— and yet Montroll was tossed in the first round. Here you see one of IRV's weaknesses. A centrist candidate who's liked pretty well by everyone, but is nobody's first choice, has a very hard time winning.

To sum up:

Traditional American voting method: Wright wins
Instant-runoff method: Kiss wins
Head-to-head matchups: Montroll wins

Confused yet? It gets even worse. Suppose those 495 voters who wrote "Wright, Kiss, Montroll" had decided to vote for Kiss instead, leaving the other two candidates off their ballot. And let's say 300 of the Wright-only voters switch to Kiss too. Now Wright has lost 795 of his first-place votes, setting him back to 2502; so he, not Montroll, gets eliminated in the first round. The election then goes down to Montroll vs. Kiss, and Montroll wins, 4067–3777.

See what just happened? We gave Kiss more votes—and instead of winning, he lost!

It's okay to be a little dizzy at this point.

But hold on to this to steady yourself: at least we have some reasonable sense of who *should* have won this election. It's Montroll, the Democrat, the guy who beats both Wright and Kiss head to head. Maybe we should toss all these Borda counts and runoffs and just elect the candidate who's preferred by the majority.

Do you get the feeling I'm setting you up for a fall?

THE RABID SHEEP WRESTLES WITH PARADOX

Let's make things a little simpler in Burlington. Suppose there were just three kinds of ballots:

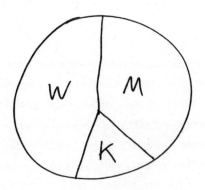

Montroll, Kiss, Wright	1332
Kiss, Wright, Montroll	371
Wright, Montroll, Kiss	1513

A majority of voters—everybody in the pie slices marked K and W—prefers Wright to Montroll. And another majority, the M and K slices, prefers Kiss to Wright. If most people like Kiss better than Wright, and most people like Wright better than Montroll, doesn't that mean Kiss should win again? There's just one problem: people like Montroll better than Kiss by a resounding 2845 to 371. There's a bizarre vote triangle: Kiss beats Wright, Wright beats Montroll, Montroll beats Kiss. *Every* candidate would lose a one-on-one race to one of the other two candidates. So how can anyone at all rightfully take office?

Vexing circles like this are called *Condorcet paradoxes*, after the French Enlightenment philosopher who first discovered them in the late eighteenth century. Marie-Jean-Antoine-Nicolas de Caritat, Marquis de Condorcet, was a leading liberal thinker in the run-up to the French Revolution, eventually becoming president of the Legislative Assembly. He was an unlikely politician—shy and prone to exhaustion, with a speaking style so quiet and hurried that his proposals often went unheard in the raucous revolutionary chamber. On the other hand, he became quickly exasperated with people whose intellectual standards didn't match his own. This combination of timidity and temper led his mentor Jacques Turgot to nickname him *"le mouton enragé,"* or "the rabid sheep."

The political virtue Condorcet did possess was a passionate, never-wavering belief in reason, and especially mathematics, as an organizing principle of human affairs. His allegiance to reason was standard stuff for the Enlightenment thinkers, but his further belief that the social and moral world could be analyzed by equations and formulas was novel. He was the first social scientist in the modern sense. (Condorcet's term was "social mathematics.") Condorcet, born into the aristocracy, quickly came to the view that universal laws of thought should take precedence over the whims of kings. He agreed with Rousseau's claim that the "general will" of the people should hold sway on governments but was not, like Rousseau, content to accept this claim as a self-evident principle. For Condorcet, the rule of the majority needed a mathematical justification, and he found one in the theory of probability.

Condorcet lays out his theory in his 1785 treatise *Essay on the Application of Analysis to the Probability of Majority Decisions*. A simple version: suppose a seven-person jury has to decide a defendant's guilt. Four

say the defendant is guilty, and only three believe he's innocent. Let's say
each of these citizens has a 51% chance of holding the correct view. In
that case, you might expect a 4–3 majority in the correct direction to be
more likely than a 4–3 majority favoring the incorrect choice.

It's a little like the World Series. If the Phillies and the Tigers are facing off, and we agree that the Phillies are a bit better than the Tigers—
say, they have a 51% chance of winning each game—then the Phillies are
more likely to win the Series 4–3 than to lose by the same margin. If the
World Series were best of fifteen instead of best of seven, Philadelphia's
advantage would be even greater.

Condorcet's so-called "jury theorem" shows that a sufficiently large
jury is very likely to arrive at the right outcome, as long as the jurors have
some individual bias toward correctness, no matter how small.* If the
majority of people believe something, Condorcet said, that must be
taken as strong evidence that it is correct. We are mathematically justified in trusting a sufficiently large majority—even when it contradicts
our own preexisting beliefs. "I must act not by what I think reasonable,"
Condorcet wrote, "but by what all who, like me, have abstracted from
their own opinion must regard as conforming to reason and truth." The
role of the jury is much like the role of the audience on *Who Wants to Be
a Millionaire?* When we have the chance to query a collective, Condorcet
thought, even a collective of unknown and unqualified peers, we ought
to value their majority opinion above our own.

Condorcet's wonkish approach made him a favorite of American
statesmen of a scientific bent, like Thomas Jefferson (with whom he
shared a fervent interest in standardizing units of measure). John Adams,
by contrast, had no use for Condorcet; in the margins of Condorcet's
books he assessed the author as a "quack" and a "mathematical charlatan." Adams viewed Condorcet as a hopelessly wild-eyed theorist whose
ideas could never work in practice, and as a bad influence on the similarly inclined Jefferson. Indeed, Condorcet's mathematically inspired Gi-
rondin Constitution, with its intricate election rules, was never adopted,
in France or anywhere else. On the positive side, Condorcet's practice of

---

* Of course, there are lots of assumptions in place here, most notably that the jurors' judgments
are arrived at independently from each other—surely not quite right in a context where the jurors
confer before voting.

following ideas to their logical conclusions led him to insist, almost alone among his peers, that the much-discussed Rights of Man belonged to women, too.

In 1770, the twenty-seven-year-old Condorcet and his mathematical mentor, Jean le Rond d'Alembert, a coeditor of the *Encylopédie*, made an extended visit to Voltaire's house at Ferney on the Swiss border. The mathophile Voltaire, then in his seventies and in faltering health, quickly adopted Condorcet as a favorite, seeing in the young up-and-comer his best hope of passing rationalistic Enlightenment principles to the next generation of French thinkers. It might have helped that Condorcet wrote a formal *éloge* (memorial appreciation) for the Royal Academy about Voltaire's old friend La Condamine, who had made Voltaire rich with his lottery scheme. Voltaire and Condorcet quickly struck up a vigorous correspondence, Condorcet keeping the older man abreast of the latest political developments in Paris.

Some friction between the two arose from another of Condorcet's *éloges*, the one for Blaise Pascal. Condorcet rightly praised Pascal as a great scientist. Without the development of probability theory, launched by Pascal and Fermat, Condorcet could not have done his own scientific work. Condorcet, like Voltaire, rejected the reasoning of Pascal's wager, but for a different reason. Voltaire found the idea of treating metaphysical matters like a dice game to be offensively unserious. Condorcet, like R. A. Fisher after him, had a more mathematical objection: he didn't accept the use of probabilistic language to talk about questions like God's existence, which weren't literally governed by chance. But Pascal's determination to view human thought and behavior through a mathematical lens was naturally appealing to the budding "social mathematician."

Voltaire, by contrast, viewed Pascal's work as fundamentally driven by religious fanaticism he had no use for, and rejected Pascal's suggestion that mathematics could speak to matters beyond the observable world as not only wrong but dangerous. Voltaire described Condorcet's *éloge* as "so beautiful that it was frightening . . . if he [Pascal] was such a great man, then the rest of us are total idiots for not being able to think like him. Condorcet will do us great harm if he publishes this book as it was sent to me." One sees here a legitimate intellectual difference, but also a mentor's jealous annoyance at his protégé's flirtation with a philosophical adversary. You can almost hear Voltaire saying, "Who's it gonna be, kid, him or me?" Condorcet managed never to make that choice (though he did bow to Voltaire and tone down his praise of Pascal in later editions). He split the difference, combining Pascal's devotion to the broad application of mathematical principles with Voltaire's sunny faith in reason, secularism, and progress.

When it came to voting, Condorcet was every inch the mathematician. A typical person might look at the results of Florida 2000 and say, "Huh, weird: a more left-wing candidate ended up swinging the election to the Republican." Or they might look at Burlington 2009 and say, "Huh, weird: the centrist guy who most people basically liked got thrown out in the first round." For a mathematician, that "Huh, weird" feeling comes as an intellectual challenge. Can you say in some precise way what *makes* it weird? Can you formalize what it would mean for a voting system *not* to be weird?

Condorcet thought he could. He wrote down an *axiom*—that is, a statement he took to be so self-evident as to require no justification. Here it is:

If the majority of voters prefer candidate A to candidate B,
then candidate B cannot be the people's choice.

Condorcet wrote admiringly of Borda's work, but considered the Borda count unsatisfactory for the same reason that the slime mold is considered irrational by the classical economist; in Borda's system, as with majority voting, the addition of a third alternative can flip the election from candidate A to candidate B. That violates Condorcet's axiom: if A would win a two-person race against B, then B can't be the winner of a three-person race that includes A.

Condorcet intended to build a mathematical theory of voting from his axiom, just as Euclid had built an entire theory of geometry on his five axioms about the behavior of points, lines, and circles:

- There is a line joining any two points.
- Any line segment can be extended to a line segment of any desired length.

- For every line segment L, there is a circle that has L as a radius.
- All right angles are congruent to each other.
- If P is a point and L is a line not passing through P, there is exactly one line through P parallel to L.

Imagine what would happen if someone constructed a complicated geometric argument showing that Euclid's axioms led, inexorably, to a contradiction. Does that seem completely impossible? Be warned—geometry harbors many mysteries. In 1924, Stefan Banach and Alfred Tarski showed how to take a sphere apart into six pieces, move the pieces around, and reassemble them into two spheres, each the same size as the first. How can it be? Because some natural set of axioms that our experience might lead us to believe about three-dimensional bodies, their volumes, and their motions simply can't all be true, however intuitively correct they may seem. Of course, the Banach-Tarski pieces are shapes of infinitely complex intricacy, not things that can be realized in the crude physical world. So the obvious business model of buying a platinum sphere, breaking it into Banach-Tarski pieces, putting the pieces together into two new spheres, and repeating until you have a wagonload of precious metal is not going to work.

If there were a contradiction in Euclid's axioms, geometers would freak out, and rightly so—because it would mean that one or more of the axioms they relied on was not, in fact, correct. We could even put it more pungently—if there's a contradiction in Euclid's axioms, then points, lines, and circles, as Euclid understood them, *do not exist.*

That's the disgusting situation that faced Condorcet when he discovered his paradox. In the pie chart above, Condorcet's axiom says Montroll cannot be elected, because he loses the head-to-head matchup to Wright. The same goes for Wright, who loses to Kiss, and for Kiss, who loses to Montroll. There is no such thing as the people's choice. It just doesn't exist.

Condorcet's paradox presented a grave challenge to his logically grounded worldview. If there is an objectively correct ranking of candidates, it can hardly be the case that Kiss is better than Wright, who is better than Montroll, who is better than Kiss. Condorcet was forced to concede that in the presence of such examples, his axiom had to be weakened: the majority could sometimes be wrong. But the problem remained of piercing the fog of contradiction to divine the people's actual will—for Condorcet never really doubted there was such a thing.

EIGHTEEN

"OUT OF NOTHING I HAVE CREATED A STRANGE NEW UNIVERSE"

C ondorcet thought that questions like "Who is the best leader?" had something like a *right answer*, and that citizens were something like scientific instruments for investigating those questions, subject to some inaccuracies of measurement, to be sure, but on average quite accurate. For him, democracy and majority rule were ways not to be wrong, via math.

We don't talk about democracy that way now. For most people, nowadays, the appeal of democratic choice is that it's *fair*; we speak in the language of rights and believe on moral grounds that people should be able to choose their own rulers, whether these choices are wise or not.

This is not just an argument about politics—it's a fundamental question that applies to every field of mental endeavor. Are we trying to figure out what's *true*, or are we trying to figure out what conclusions are licensed by our rules and procedures? Hopefully the two notions frequently agree; but all the difficulty, and thus all the conceptually interesting stuff, happens at the points where they diverge.

You might think it's obvious that figuring out what's true is always our proper business. But that's not always the case in criminal law, where the difference presents itself quite starkly in the form of defendants who committed the crime but who cannot be convicted (say, because evidence was obtained improperly) or who are innocent of the crime but are convicted anyway. What's justice here—to punish the guilty and free the innocent, or to follow criminal procedure wherever it leads us? In experimental science, we've already seen the dispute with R. A. Fisher on one side and Jerzy Neyman and Egon Pearson on the other. Are we, as Fisher thought, trying to figure out which hypotheses we should actually believe are true? Or are we to follow the Neyman-Pearson philosophy, under which we resist thinking about the truth of hypotheses at all and merely ask: Which hypotheses are we to certify as correct, whether they're really true or not, according to our chosen rules of inference?

Even in math, supposedly the land of certainty, we face these problems. And not in some arcane precinct of contemporary research, but in plain old classical geometry. The subject is founded on Euclid's axioms, which we wrote down in the last chapter. That fifth one—

If P is a point and L is a line not passing through P, there is exactly one line through P parallel to L.

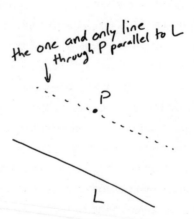

—is a bit funny, isn't it? It's somehow a bit more complicated, a bit less obvious, than the rest. That's how geometers saw it for centuries, at any rate.* Euclid himself is thought to have disliked it, proving the first twenty-eight propositions in the *Elements* using only the first four axioms.

An inelegant axiom is like a stain in the corner of the floor; it doesn't get in your way, per se, but it's *maddening*, and one spends an inordinate amount of time scrubbing and scouring and trying to make the surface nice and clean. In the mathematical context, this amounted to trying to show that the fifth axiom, the so-called parallel postulate, followed from all the others. If that were so, the axiom could be removed from Euclid's list, leaving it shiny and pristine.

After two thousand years of scrubbing, the stain was still there.

In 1820, the Hungarian noble Farkas Bolyai, who had given years of his life to the problem without success, warned his son János against following the same path:

> You must not attempt this approach to parallels. I know this way to the very end. I have traversed this bottomless night, which extinguished all light and joy in my life. I entreat you, leave the science of parallels alone. . . . I was ready to become a martyr who would remove the flaw from geometry and return it purified to mankind. I accomplished monstrous, enormous labors; my creations are far better than those of others and yet I have not achieved complete satisfaction. . . . I turned back when I saw that no man can reach the bottom of this night. I turned back unconsoled, pitying myself and all mankind. Learn from my example. . . .

Sons don't always take advice from their fathers, and mathematicians don't always quit easily. The younger Bolyai kept working on the parallels, and by 1823 he had formed the outline of a solution to the ancient problem. He wrote back to his father, saying:

> I have discovered such wonderful things that I was amazed, and it would be an everlasting piece of bad fortune if they were lost. When

---

* The version of the fifth postulate I've written here is actually not Euclid's original, but a logically equivalent version, originally stated by Proclus in the fifth century CE and made popular by John Playfair in 1795. Euclid's version is a bit longer.

you, my dear Father, see them, you will understand; at present I can say nothing except this: that *out of nothing I have created a strange new universe.*

János Bolyai's insight was to come at the problem from behind. Rather than try to prove the parallel postulate from the other axioms, he allowed his mind the freedom to wonder: What if the parallel axiom were false? Does a contradiction follow? And he found that the answer was no—that there was *another* geometry, not Euclid's but something else, in which the first four axioms were correct but the parallel postulate was not. Thus, there can be no proof of the parallel postulate from the other axioms; such a proof would rule out the possibility of Bolyai's geometry. But there it was.

Sometimes, a mathematical development is "in the air"—for reasons only poorly understood, the community is ready for a certain advance to come, and it comes from several sources at once. Just as Bolyai was constructing his non-Euclidean geometry in Austria-Hungary, Nikolai Lobachevskii* was doing the same in Russia. And the great Carl Friedrich Gauss, an old friend of the senior Bolyai, had formulated many of the same ideas in work that had not yet seen print. (When informed of Bolyai's paper, Gauss responded, somewhat ungraciously, "To praise it would amount to praising myself.")

To describe the so-called hyperbolic geometry of Bolyai, Lobachevskii, and Gauss would take a little more space than we have here. But as Bernhard Riemann observed a few decades later, there is a simpler non-Euclidean geometry, one that's not a crazy new universe at all: it is the geometry of the sphere.

Let's revisit the first four axioms:

• There is a Line joining any two Points.
• Any Line segment can be extended to a Line segment of any desired length.

---

* The eponym for Tom Lehrer's song "Lobachevsky," surely the greatest comic musical number of all time about mathematical publishing.

- For every Line segment L, there is a Circle which has L as a radius.
- All Right Angles are congruent to each other.

You might notice I've made a small typographical change, capitalizing the geometric terms *point, line, circle,* and *right angle.* I've done this, not to give the axioms an old-timey look on the page, but to emphasize that, from a strictly logical point of view, it doesn't matter what "points" and "lines" are called; they could be called "frogs" and "kumquats" and the structure of logical deduction from the axioms should be just the same. It's just like Gino Fano's seven-point plane, where the "lines" don't look like the lines we learned in school, but it doesn't matter—the point is that they act like lines so far as the rules of geometry are concerned. It would be *better,* in a way, to call points frogs and lines kumquats, because the point is to free ourselves from preconceptions about what the words *Point* and *Line* really mean.

Here is what they mean in Riemann's spherical geometry. A Point is a *pair* of points on the sphere which are *antipodal,* or diametrically opposite each other. A Line is a "great circle"—that is, a circle on the sphere's surface, and a Line segment is a segment of such a circle. A Circle is a circle, now allowed to be of any size.

With these definitions, Euclid's first four axioms are true! Given any two Points—that is, any two *pairs* of antipodal points on the sphere—there is a Line—that is, a great circle—that joins them.* What's more (though this is not one of the axioms) any two Lines intersect in a single Point.

You might complain about the second axiom; how can we say that a Line segment can be extended to *any* length, when it can never be longer than the length of a Line itself, which is the circumference of the sphere? This is a reasonable objection, but comes down to a question of interpretation. Riemann interpreted the axiom to mean that lines were *boundless,* not that they were of infinite extent. Those two notions are subtly

---

* This is not supposed to be immediately obvious, but it's not hard to convince yourself it's true—I highly recommend getting out a tennis ball and a Sharpie and checking for yourself!

different. Riemann's Lines, which are circles, have finite length but are
boundless; one can travel along them forever without having to stop.

But the fifth axiom is a different story. Suppose we have a Point P,
and a Line L not containing P. Is there exactly one Line through P that is
parallel to L? No, for a very simple reason: in spherical geometry, *there
are no such things* as parallel lines! Any two great circles on a sphere must
intersect.

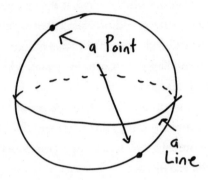

ONE-PARAGRAPH PROOF: Any great circle C cuts the sphere's
surface into two equal parts, each one of which has the same area; call
this area A. Now suppose another great circle, C', is parallel to C. Since
it doesn't intersect with C, it must be entirely enclosed on one side or the
other of C, on one of those two area-A half-spheres. But this means that
the area enclosed by C' is *less* than A, impossible, since every great circle
encloses area exactly A.

So the parallel postulate fails, in spectacular fashion. (In Bolyai's geometry, the situation is just the opposite: there are too many parallel lines,
not two few, and in fact there are *infinitely many* lines through P parallel
to L. As you can imagine, this geometry is a bit harder to visualize.)

If that strange condition, where no two lines are ever parallel, sounds
familiar, it's because we've been here before. It's just the same phenomenon we saw in the projective plane, which Brunelleschi and his fellow
painters used to develop the theory of perspective.* There, too, every
pair of lines met. And this is no coincidence—one can prove that Rie-

---

* The painters didn't develop, or need, a formal geometric theory of the projective plane, but
they understood how it translated into brushstrokes on the canvas, which was enough for their
purposes.

mann's geometry of Points and Lines on a sphere is the *same* as that of the projective plane.

When interpreted as statements about Points and Lines on a sphere, the first four axioms are true but the fifth is not. If the fifth axiom were a logical consequence of the first four axioms, then the existence of the sphere would present a contradiction; the fifth axiom would be both true (by virtue of the truth of the first four axioms) and not (by virtue of what we know about spheres). By the good old reductio ad absurdum, this means that spheres do not exist. But spheres do exist. So the fifth axiom cannot be proved from the first four, QED.

This might seem like a lot of work to get a stain off the floor. But the motivation for proving statements of this kind is not just an obsessive attention to aesthetics (though I can't deny those feelings play a role). Here's the thing; once you understand that the first four axioms apply to many different geometries, then any theorem Euclid proves from only those axioms must be true, not only in Euclid's geometry, but in *all* the geometries where those axioms hold. It's a kind of mathematical force multiplier; from one proof, you get many theorems.

And these theorems are not just about abstract geometries made up to prove a point. Post-Einstein, we understand that non-Euclidean geometry is not just a game; like it or not, it's the way space-time actually looks.

This is a story told in mathematics again and again: we develop a method that works for one problem, and if it is a *good* method, one that really contains a new idea, we typically find that the same proof works in many different contexts, which may be as different from the original as a sphere is from a plane, or more so. At the moment, the young Italian mathematician Olivia Caramello is making a splash with claims that theories governing many different fields of mathematics are closely related beneath the skin—if you like technical terms, they are "classified by the same Grothendieck topos"—and, that, as a result, theorems proved in one field of mathematics can be carried over for free into theorems in another area, which on the surface appear totally different. It's too early to say whether Caramello has truly "created a strange new universe," as Bolyai did—but her work is very much in keeping with the long tradition in mathematics of which Bolyai was a part.

The tradition is called "formalism." It's what G. H. Hardy was talking about when he remarked, admiringly, that mathematicians of the nineteenth century finally began to ask what things like

$$1 - 1 + 1 - 1 + \ldots$$

should be *defined* to be, rather than what they *were*. In this way they avoided the "unnecessary perplexities" that had dogged the mathematicians of earlier times. In the purest version of this view, mathematics becomes a kind of game played with symbols and words. A statement is a theorem precisely if it follows by logical steps from the axioms. But what the axioms and theorems refer to, what they *mean*, is up for grabs. What is a Point, or a Line, or a frog, or a kumquat? It can be anything that behaves the way the axioms demand, and the meaning we should choose is whichever one suits our present needs. A purely formal geometry is a geometry you can in principle do without ever having seen or imagined a point or a line; it is a geometry in which it's irrelevant what points and lines, understood in the usual way, are actually like.

Hardy would certainly have recognized Condorcet's anguish as perplexity of the most unnecessary kind. He would have advised Condorcet not to ask who the best candidate really was, or even who the public really intended to install in office, but rather which candidate we should *define* to be the public choice. And this formalist take on democracy is more or less general in the free world today. In the contested 2000 presidential election in Florida, thousands of Palm Beach County voters who believed they were voting for Al Gore in fact recorded votes for the paleoconservative Reform Party candidate Pat Buchanan instead, thanks to the confusingly designed "butterfly ballot." Had Gore received those votes instead, he would have won the state, and the presidency.

But Gore doesn't get those votes; he never even seriously argued for them. Our electoral system is formalist: what counts is the mark made on the ballot, not whatever feature of the voter's mind we may take it to indicate. Condorcet would have cared about the voter's intent; we, at least officially, do not. Condorcet would have cared, too, about the Floridians who voted for Ralph Nader. Presuming, as seems safe, that most of those people preferred Gore to Bush, we see that Gore is the candidate who Condorcet's axiom declares the victor: a majority preferred him to Bush, and an even greater majority preferred him to Nader. But those preferences aren't relevant to the system we have. We define the public will to be that mark that appears most frequently on the pieces of paper collected at the voting booth.

Even that number, of course, is open to argument: How do we count a partially punched ballot, the so-called hanging chad? What to do with votes mailed from overseas military bases, some of which couldn't be certified as having been cast on or before Election Day? And to what extent were Florida counties to recount the ballots in an attempt to get as precise a reckoning as possible of the actual votes?

It was this last question that made its way to the Supreme Court, where the matter was finally decided. Gore's team had asked for a recount in selected counties, and the Florida Supreme Court had agreed, but the U.S. Supreme Court said no, fixing the total in place with Bush holding a 537-vote lead, and granting him the election. More counting would presumably have resulted in a more accurate enumeration of the votes; but this, the court said, is not the overriding goal of an election. Recounting some counties but not others, they said, would be unfair to the voters whose ballots were not revisited. The proper business of the state is not to count the votes as accurately as possible—to know what actually happened—but to obey the formal protocol that tells us, in Hardy's terms, who the winner should be defined to be.

More generally, formalism in the law manifests itself as an adherence to procedure and the words of the statutes, even when—or especially when—they cut against what common sense prescribes. Justice Antonin Scalia, the fiercest advocate of legal formalism there is, puts it very directly: "Long live formalism. It is what makes a government a government of laws and not of men."

In Scalia's view, when judges try to understand what the law intends—its spirit—they're inevitably bamboozled by their own prejudices and desires. Better to stick to the words of the Constitution and the statutes, treating them as axioms from which judgments can be derived by something like logical deduction.

In questions of criminal justice, Scalia is equally devoted to formalism: the truth is, by definition, whatever a properly convened trial process determines it to be. Scalia makes this stance strikingly clear in his dissenting opinion in the 2009 case *In Re Troy Anthony Davis*, where he argued that a convicted murderer shouldn't be granted a new evidentiary hearing, despite the fact that seven of the nine witnesses against him had recanted their testimony:

> This Court has *never* held that the Constitution forbids the execution of a convicted defendant who has had a full and fair trial but is later able to convince a habeas court that he is "actually" innocent.

(Emphasis on "never" and scare quotes around "actually" both Scalia's.)

As far as the court is concerned, Scalia says, what matters is the verdict arrived at by the jury. Davis was a murderer whether he killed anyone or not.

Chief Justice John Roberts isn't a fervent advocate of formalism like Scalia, but he's broadly in sympathy with Scalia's philosophy. In his confirmation hearing in 2005, he famously described his job in baseball terms:

> Judges and justices are servants of the law, not the other way around. Judges are like umpires. Umpires don't make the rules; they apply them. The role of an umpire and a judge is critical. They make sure everybody plays by the rules. But it is a limited role. Nobody ever went to a ball game to see the umpire.

Roberts, knowingly or not, was echoing Bill Klem, the "Old Arbitrator," an umpire in the National League for almost forty years, who said, "The best-umpired game is the game in which the fans cannot recall the umpires who worked it."

But the umpire's role is not as limited as Roberts and Klem make it sound, because baseball is a formalist sport. To see this, you need look no further than game 1 of the 1996 American League Championship Series, in which the Baltimore Orioles faced the New York Yankees in the Bronx. Baltimore was leading in the bottom of the eighth when Yankee shortstop Derek Jeter launched a long fly ball to right field off Baltimore reliever Armando Benitez; well hit, but playable for center fielder Tony Tarasco, who settled under the ball and prepared to make the catch. That's when twelve-year-old Yankee fan Jeffrey Maier, sitting in the front row of the bleachers, reached over the fence and pulled the ball into the stands.

Jeter knew it wasn't a home run. Tarasco and Benitez knew it wasn't a home run. Fifty-six thousand Yankee fans knew it wasn't a home run. The only person in Yankee Stadium who didn't see Maier reach over the fence was the only one who mattered, umpire Rich Garcia. Garcia called the ball a homer. Jeter didn't try to correct the umpire's call, much less refuse to jog around the bases and collect his game-tying run. No one would have expected that of him. That's because baseball is a formalist sport. What a thing *is* is what an umpire declares it to be, and nothing else. Or, as Klem put it, in what must be the bluntest assertion of an ontological stance ever made by a professional sports official: "It ain't nothin' till I call it."

This is changing, just a bit. Since 2008, umpires have been allowed to consult video replay when they're unsure of what actually took place on the field. This is good for getting calls right instead of wrong, but many longtime baseball fans feel it's somehow foreign to the spirit of the sport. I'm one of them. I'll bet John Roberts is too.

Not everybody shares Scalia's view of the law (note that his opinion in *Davis* was in the minority). As we saw in *Atkins v. Virginia*, the words of the Constitution, like "cruel and unusual," leave a remarkable amount of space for interpretation. If even the great Euclid left some ambiguity in his axioms, how can we expect any different from the framers? Legal realists, like judge and University of Chicago professor Richard Posner, argue that Supreme Court jurisprudence is *never* the exercise in formal rule following that Scalia says it is:

> Most of the cases the Supreme Court agrees to decide are toss-ups, in the sense that they cannot be decided by conventional legal reasoning, with its heavy reliance on constitutional and statutory language and previous decisions. If they could be decided by those essentially

semantic methods, they would be resolved uncontroversially at the level of a state supreme court or federal court of appeals and never get reviewed by the Supreme Court.

In this view, the hard questions about law, the ones that make it all the way to the Supremes, are left indeterminate by the axioms. The justices are thus in the same position Pascal was when he found he couldn't reason his way to any conclusion about God's existence. And yet, as Pascal wrote, we don't have the choice not to play the game. The court must decide, whether it can do so by conventional legal reasoning or not. Sometimes it takes Pascal's route: if reason does not determine the judgment, make the judgment that seems to have the best consequences. According to Posner, this is the path the justices finally adopted in *Bush v. Gore*, with Scalia on board. The decision they arrived at, Posner says, was not really supported by the Constitution or judicial precedent; it was a decision made pragmatically, in order to close off the possibility of many more months of electoral chaos.

THE SPECTER OF CONTRADICTION

Formalism has an austere elegance. It appeals to people like G. H. Hardy, Antonin Scalia, and me, who relish that feeling of a nice rigid theory shut tight against contradiction. But it's not easy to hold to principles like this consistently, and it's not clear it's even wise. Even Justice Scalia has occasionally conceded that when the literal words of the law seem to require an absurd judgment, the literal words have to be set aside in favor of a reasonable guess as to what Congress must have meant. In just the same way, no scientist really wants to be bound strictly by the rules of significance, no matter what they say their principles are. When you run two experiments, one testing a clinical treatment that seems theoretically promising and the other testing whether dead salmon respond emotionally to romantic photos, and both experiments succeed with p-values of .03, you don't really want to treat the two hypotheses the same. You want to approach absurd conclusions with an extra coat of skepticism, rules be damned.

Formalism's greatest champion in mathematics was David Hilbert, the German mathematician whose list of twenty-three problems, delivered in Paris at the 1900 International Congress of Mathematics, set the course for much of twentieth-century math. Hilbert is so revered that any work that touches even tangentially on one of his problems takes on a little extra shine, even a hundred years later. I once met a historian of German culture in Columbus, Ohio, who told me that Hilbert's predilection for wearing sandals with socks is the reason that fashion choice is still noticeably popular among mathematicians today. I could find no evidence this was actually true, but it suits me to believe it, and it gives a correct impression of the length of Hilbert's shadow.

Many of Hilbert's problems fell quickly; others, like number 18, concerning the densest possible packing of spheres, were settled only recently, as we saw in chapter 12. Some are still open, and hotly pursued. Solving number 8, the Riemann Hypothesis, will get you a million-dollar prize from the Clay Foundation. At least once, the great Hilbert got it wrong: in his tenth problem, he asked for an algorithm that would take any equation and tell you whether it had a solution in which all the variables took whole-number values. In a series of papers in the 1960s and '70s, Martin Davis, Yuri Matijasevic, Hilary Putnam and Julia Robinson showed that no such algorithm existed. (Number theorists everywhere breathed a sigh of relief—it might have been a bit dispiriting had it transpired that a formal algorithm could autosolve the problems we've spent years on.)

Hilbert's second problem was different from the others, because it was not so much a mathematical question as a question about mathematics itself. He began with a full-throated endorsement of the formalist approach to mathematics:

> When we are engaged in investigating the foundations of a science, we must set up a system of axioms which contains an exact and complete description of the relations subsisting between the elementary ideas of that science. The axioms so set up are at the same time the definitions of those elementary ideas; and no statement within the realm of the science whose foundation we are testing is held to be correct unless it can be derived from those axioms by means of a finite number of logical steps.

By the time of the Paris lecture, Hilbert had already revisited Euclid's axioms and rewritten them to remove any trace of ambiguity; at the same time, he had rigorously squeezed out any appeal to geometric intuition. His version of the axioms really does make just as much sense if "point" and "line" are replaced by "frog" and "kumquat." Hilbert himself famously remarked, "One must be able to say at all times— instead of points, straight lines, and planes—tables, chairs, and beer mugs." One early fan of Hilbert's new geometry was the young Abraham Wald, who, while still a student at Vienna, had shown how some of Hilbert's axioms could be derived from the others, and were thus expendable.*

Hilbert was not content to stop with geometry. His dream was to create a purely formal mathematics, where to say a statement was true was precisely to say it obeyed the rules laid down at the beginning of the game, no more, no less. It was a mathematics Antonin Scalia would have liked. The axioms Hilbert had in mind for arithmetic, first formulated by the Italian mathematician Guiseppe Peano, hardly seem the sort of thing about which there could be any interesting questions or controversy. They say things like "Zero is a number," "If x equals y and y equals z, then x equals z," and "If the number directly following x is the same as the number directly following y, then x and y are the same." They're the truths we hold to be self-evident.

What's remarkable about these Peano axioms is that from these bare beginnings one can generate a great deal of mathematics. The axioms themselves seem to refer only to whole numbers, but Peano himself showed that, starting from his axioms and proceeding purely by definition and logical deduction, one could define the rational numbers and prove their basic properties.† The mathematics of the nineteenth century had been plagued by confusion and crises as widely accepted definitions in analysis and geometry were found to be logically flawed. Hilbert saw

---

* Some historians trace the current hypermathematization of economics back this far, saying that the habit of axioms passes from Hilbert to economics through Wald and the other young mathematicians in 1930s Vienna, who combined a Hilbertian style with strong applied interests: see E. Roy Weintraub's *How Economics Became a Mathematical Science,* where this idea is fully worked out.

† It's probably not a coincidence that Peano was yet another devotee of artificial languages constructed on rational principles: he created his own such language, Latino Sine Flexione, in which he wrote some of his later mathematical works.

formalism as a way of starting over clean, building on a foundation so basic as to be completely incontrovertible.

But a specter was haunting Hilbert's program—the specter of contradiction. Here's the nightmare scenario. The community of mathematicians, working together in concert, rebuilds the entire apparatus of number theory, geometry, and calculus, starting from the bedrock axioms and laying on new theorems, brick by brick, each layer glued to the last by the rules of deduction. And then, one day, a mathematician in Amsterdam proves that a certain mathematical assertion is the case, while another mathematician in Kyoto proves that it is not.

Now what? Starting from assertions one cannot possibly doubt, one has arrived at a contradiction. Reductio ad absurdum. Do you conclude that the axioms were wrong? Or that there's something wrong with the structure of logical deduction itself? And what do you do with the decades of work based on those axioms?*

Thus, the second problem among those Hilbert presented to the mathematicians gathered in Paris:

> But above all I wish to designate the following as the most important among the numerous questions which can be asked with regard to the axioms: To prove that they are not contradictory, that is, that a definite number of logical steps based upon them can never lead to contradictory results.

One is tempted simply to assert that such a terrible outcome can't happen. How could it? The axioms are obviously true. But it was no less obvious to the ancient Greeks that a geometric magnitude must be a ratio of two whole numbers; that's how their notion of measurement worked, until the whole framework got mugged by the Pythagorean Theorem and the stubbornly irrational square root of 2. Mathematics has a nasty habit of showing that, sometimes, what's obviously true is absolutely wrong. Consider the case of Gottlob Frege, the German logician who, like Hilbert, was laboring to shore up the logical underpinnings

---

* Ted Chiang's 1991 short story "Division by Zero" contemplates the psychological consequences suffered by a mathematician unfortunate enough to uncover such an inconsistency.

of mathematics. Frege's focus was not number theory, but the theory of sets. He, too, started from a sequence of axioms, which seemed so obvious as to hardly need stating. In Frege's set theory, a set was nothing other than a collection of objects, called elements. Often we use curly brackets {} to denote the sets whose elements are thereby enclosed; so that {1,2,pig} is the set whose elements are the number 1, the number 2, and a pig.

When some of those elements enjoy a certain property and others don't, there's a set that is the collection of all those elements with the specified property. To make it a little more down to earth: there is a set of pigs, and among those, the ones that are yellow form a set, the set of yellow pigs. Hard to find much to take issue with here. But these definitions are really, really general. A set can be a collection of pigs, or real numbers, or ideas, possible universes, or other sets. And it's that last one that causes all the problems. Is there a set of all sets? Sure. A set of all infinite sets? Why not? In fact, both of these sets have a curious property: *they are elements of themselves.* The set of infinite sets, for example, is certainly itself an infinite set; its elements include sets like

{the integers}
{the integers, and also a pig}
{the integers, and also the Eiffel Tower}

and so on and so on. Clearly there's no end.

We might call such a set ouroboric, after the mythical snake so hungry it chows down on its own tail and consumes itself. So the set of infinite sets is ouroboric, but {1,2,pig} is not, because none of its elements is the set {1,2,pig} itself; all its elements are either numbers or farm animals, but not sets.

Now here comes the punch line. Let **NO** be the set of all non-ouroboric sets. **NO** seems like a weird thing to think about, but if Frege's definition allows it into the world of sets, so must we.

Is **NO** ouroboric or not? That is, is **NO** an element of **NO**? By definition, if **NO** is ouroboric, then **NO** cannot be in **NO**, which consists only of non-ouroboric sets. But to say **NO** is not an element of **NO** is precisely to say **NO** is non-ouroboric; it does not contain itself.

But wait a minute—if **NO** is non-ouroboric, then it *is* an element of **NO**, which is the set of all non-ouroboric sets. Now **NO** is an element of **NO** after all, which is to say that **NO** is ouroboric.

If **NO** is ouroboric, it isn't, and if it isn't, it is.

This, more or less, was the content of a letter the young Bertrand Russell wrote to Frege in June of 1902. Russell had met Peano in Paris at the International Congress—whether he attended Hilbert's talk isn't known, but he was certainly on board with the program of reducing all of mathematics to a pristine sequence of deductions from basic axioms.* Russell's letter starts out like a fan letter to the older logician: "I find myself in full accord with you on all main points, especially in your rejection of any psychological element in logic and in the value you attach to a conceptual notation for the foundations of mathematics and of formal logic, which, incidentally, can hardly be distinguished."

But then: "I have encountered a difficulty only on one point."

And Russell explains the quandary of **NO**, now known as Russell's paradox.

Russell closes the letter by expressing regret that Frege had not yet published the second volume of his *Grundgesetze* ("Foundations"). In fact, the book was finished and already in press when Frege received Russell's letter. Despite the respectful tone ("I have encountered a difficulty," not "Hi, I've just borked your life's work"), Frege understood at once what Russell's paradox meant for his version of set theory. It was too late to change the book, but he hurriedly appended a postscript recording Russell's devastating insight. Frege's explanation is perhaps the saddest sentence ever written in a technical work of mathematics: *"Einem wissenschaftlichen Schriftsteller kann kaum etwas Unerwünschteres begegnen, als dass ihm nach Vollendung einer Arbeit eine der Grundlagen seines Baues erschüttert wird."* Or: "A scientist can hardly encounter anything more undesirable than, just as a work is completed, to have its foundation give way."

---

* If we're to be precise, Russell was not a formalist, like Hilbert, who declared that the axioms were just strings of symbols with no defined meaning; rather, he was a "logicist," whose view was that the axioms were actually true statements about logical facts. Both groups shared a vigorous interest in figuring out which statements could be deduced from the axioms. The extent to which you care about this distinction is a good measure of whether you would enjoy going to graduate school in analytic philosophy.

Hilbert and the other formalists didn't want to leave open the possibility of a contradiction embedded like a time bomb in the axioms; he wanted a mathematical framework in which consistency was guaranteed. It wasn't that Hilbert really thought there was *likely* to be a contradiction hidden in arithmetic. Like most mathematicians, and even most normal people, he believed that the standard rules of arithmetic were true statements about the whole numbers, so they couldn't really contradict one another. But this was not satisfying—it relied on the presupposition that the set of whole numbers *actually existed*. This was a sticking point for many. Georg Cantor, a few decades earlier, had for the first time put the notion of the infinite on some kind of firm mathematical footing. But his work had not been digested easily or accepted universally, and there was a substantial group of mathematicians who felt that any proof relying on the existence of infinite sets ought to be considered suspect. That there was such a thing as the number 7, all were willing to accept. That there was such a thing as the set of *all* numbers was the question at issue. Hilbert knew very well what Russell had done to Frege and was keenly aware of the dangers posed by casual reasoning about infinite sets. "A careful reader," he wrote in 1926, "will find that the literature of mathematics is glutted with inanities and absurdities which have had their source in the infinite." (The tone here would not be out of place in one of Antonin Scalia's sweatier dissents.) Hilbert sought a *finitary* proof of consistency, one that did not make reference to any infinite sets, one that a rational mind couldn't help but wholly believe.

But Hilbert was to be disappointed. In 1931, Kurt Gödel proved in his famous second incompleteness theorem that there could be no finitary proof of the consistency of arithmetic. He had killed Hilbert's program with a single stroke.

So should you be worried that all of mathematics might collapse tomorrow afternoon? For what it's worth, I'm not. I *do* believe in infinite sets, and I find the proofs of consistency that use infinite sets to be convincing enough to let me sleep at night.

Most mathematicians are like me, but there are some dissenters. Edward Nelson, a logician at Princeton, circulated a proof of the *inconsistency* of arithmetic in 2011. (Fortunately for us, Terry Tao found a mistake in Nelson's argument within a few days.) Vladimir Voevodsky, a Fields Medalist now at the Institute for Advanced Study in Princeton, made a splash in 2010 when he said that he, too, saw no reason to feel sure that arithmetic is consistent. He and a large international group of collaborators have their own proposal for a new foundation of mathematics. Hilbert had started out with geometry, but had quickly come to see the consistency of arithmetic as the more fundamental problem. Voevodsky's group, by contrast, argues that geometry is the fundamental thing after all—not any geometry that would have been familiar to Euclid, but a more modern kind, called *homotopy theory*. Will these foundations be immune to skepticism and contradiction? Ask me in twenty years. These things take time.

Hilbert's style of mathematics survived the death of his formalist program. Even before Gödel's work, Hilbert had made it clear he didn't really intend for mathematics to be *created* in a fundamentally formalist way. That would be too difficult! Even if geometry can be recast as an exercise in manipulating meaningless strings of symbols, no human being can generate geometric ideas without drawing pictures, without imagining figures, without thinking of the objects of geometry as *real things*. My philosopher friends typically find this point of view, usually called Platonism, fairly disreputable; how can a fifteen-dimensional hypercube be a real thing? I can only reply that they seem as real to me as, say, mountains. After all, I can *define* a fifteen-dimensional hypercube. Can you do the same for the mountain?

But we are Hilbert's children; when we have beers with the philosophers on the weekend, and the philosophers hassle us about the status of the objects we study,* we retreat into our formalist redoubt, protesting: sure, we use our geometric intuition to figure out what's going on, but the way we finally *know* that what we say is true is that there's a formal proof behind the picture. In the famous formulation of Philip Davis and Reuben Hersh, "The typical working mathematician is a Platonist on weekdays and a formalist on Sundays."

Hilbert didn't want to destroy Platonism; he wanted to make the

---

* They really do this!

world safe for Platonism, by placing subjects like geometry on a formal foundation so unshakable that we could feel as morally sturdy the whole week as we do on Sunday.

GENIUS IS A THING THAT HAPPENS

I have made much of Hilbert's role, as is right, but there's a risk that by paying so much attention to the names at the top of the marquee I'll give a misimpression of mathematics as an enterprise in which a few solitary geniuses, marked at birth, blaze a path for the rest of humankind to trot along. It's easy to tell the story that way. In some cases, like that of Srinivasa Ramanujan, it's not so far off. Ramanujan was a prodigy from southern India who, from childhood, produced astonishingly original mathematical ideas, which he described as divine revelations from the goddess Namagiri. He worked for years completely in isolation from the main body of mathematics, with access to only a few books to acquaint him with the contemporary state of the subject. By 1913, when he finally made contact with the greater world of number theory, he had filled a series of notebooks with something like four thousand theorems, many of which are still the subject of active investigation today. (The goddess provided Ramanujan with theorem statements, but no proofs—those are for us, Ramanujan's successors, to fill in.)

But Ramanujan is an outlier, whose story is so often told precisely because it's so uncharacteristic. Hilbert started out a very good but not exceptional student, by no means the brightest young mathematician in Königsberg; that was Hermann Minkowski, two years younger. Minkowski went on to a distinguished mathematical career himself, but he was no Hilbert.

One of the most painful parts of teaching mathematics is seeing students damaged by the cult of the genius. The genius cult tells students it's not worth doing mathematics unless you're the *best* at mathematics, because those special few are the only ones whose contributions matter. We don't treat any other subject that way! I've never heard a student say, "I like *Hamlet*, but I don't really belong in AP English—that kid who sits in the front row knows *all* the plays, and he started reading Shakespeare when he was nine!" Athletes don't quit their sport just because one of their teammates outshines them. And yet I see promising young mathematicians quit every year, even though they love mathematics, because someone in their range of vision was "ahead" of them.

We lose a lot of math majors this way. Thus, we lose a lot of future mathematicians; but that's not the whole of the problem. I think we need more math majors who *don't* become mathematicians. More math major doctors, more math major high school teachers, more math major CEOs, more math major senators. But we won't get there until we dump the stereotype that math is only worthwhile for kid geniuses.

The cult of the genius also tends to undervalue hard work. When I was starting out, I thought "hardworking" was a kind of veiled insult—something to say about a student when you can't honestly say they're smart. But the ability to work hard—to keep one's whole attention and energy focused on a problem, systematically turning it over and over and pushing at everything that looks like a crack, despite the lack of outward signs of progress—is not a skill everybody has. Psychologists nowadays call it "grit," and it's impossible to do math without it. It's easy to lose sight of the importance of work, because mathematical inspiration, when it finally does come, can feel effortless and instant. I remember the first theorem I ever proved; I was in college, working on my senior thesis, and I was completely stuck. One night I was at an editorial meeting of the campus literary magazine, drinking red wine and participating fitfully in the discussion of a somewhat boring short story, when all at once something turned over in my mind and I understood how to get past the block. No details, but it didn't matter; there was no doubt in my mind that the thing was done.

That's the way mathematical creation often presents itself. Here's the French mathematician Henri Poincaré's famous account of a geometric breakthrough he made in 1881:

> Having reached Coutances, we entered an omnibus to go some place or other. At the moment when I put my foot on the step the idea came to me, without anything in my former thoughts seeming to have

paved the way for it, that the transformations I had used to define the Fuchsian functions were identical with those of non-Euclidean geometry. I did not verify the idea; I should not have had time, as, upon taking my seat in the omnibus, I went on with a conversation already commenced, but I felt a perfect certainty. On my return to Caen, for conscience's sake I verified the result at my leisure.[*]

But it didn't *really* happen in the space of a footstep, Poincaré explains. That moment of inspiration is the product of weeks of work, both conscious and unconscious, which somehow prepare the mind to make the necessary connection of ideas. Sitting around waiting for inspiration leads to failure, no matter how much of a whiz kid you are.

It can be hard for me to make this case, because I was one of those prodigious kids myself. I knew I was going to be a mathematician when I was six years old. I took courses way above my grade level and won a neckful of medals in math contests. And I was pretty sure, when I went off to college, that the competitors I knew from Math Olympiad were the great mathematicians of my generation. It didn't exactly turn out that way. That group of young stars produced many excellent mathematicians, like Terry Tao, the Fields Medal–winning harmonic analyst. But most of the mathematicians I work with now weren't ace mathletes at thirteen; they developed their abilities and talents on a different timescale. Should they have given up in middle school?

What you learn after a long time in math—and I think the lesson applies much more broadly—is that there's *always* somebody ahead of you, whether they're right there in class with you or not. People just starting out look to people with good theorems, people with some good theorems look to people with lots of good theorems, people with lots of good theorems look to people with Fields Medals, people with Fields Medals look to the "inner circle" Medalists, and those people can always look toward the dead. Nobody ever looks in the mirror and says, "Let's face it, I'm smarter than Gauss." And yet, in the last hundred

---

[*] From Poincaré's essay "Mathematical Creation," highly recommended reading if you care about mathematical creativity, or for that matter any kind of creativity.

years, the joined effort of all these dummies-compared-to-Gauss has produced the greatest flowering of mathematical knowledge the world has ever seen.

Mathematics, mostly, is a communal enterprise, each advance the product of a huge network of minds working toward a common purpose, even if we accord special honor to the person who places the last stone in the arch. Mark Twain is good on this: "It takes a thousand men to invent a telegraph, or a steam engine, or a phonograph, or a telephone or any other important thing—and the last man gets the credit and we forget the others."

It's something like football. There are moments, of course, when one player seizes control of the game totally, and these are moments we remember and honor and recount for a long time afterward. But they're not the normal mode of football, and they're not the way most games are won. When the quarterback completes a dazzling touchdown pass to a streaking wide receiver, you are seeing the work of many people in concert: not only the quarterback and the receiver, but the offensive linemen who prevented the defense from breaking through just long enough to allow the quarterback to set and throw, that prevention in turn enabled by the running back who pretended to take a handoff in order to distract the attention of the defenders for a critical moment; and then, too, there's the offensive coordinator who called the play, and his many clipboarded assistants, and the training staff who keep the players in condition to run and throw . . . One doesn't call all those people geniuses. But they create the conditions under which genius can take place.

Terry Tao writes:

The popular image of the lone (and possibly slightly mad) genius— who ignores the literature and other conventional wisdom and manages by some inexplicable inspiration (enhanced, perhaps, with a liberal dash of suffering) to come up with a breathtakingly original solution to a problem that confounded all the experts—is a charming and romantic image, but also a wildly inaccurate one, at least in the world of modern mathematics. We do have spectacular, deep and remarkable results and insights in this subject, of course, but they are the hard-won and cumulative achievement of years, decades, or even centuries of steady work and progress of many good and great mathematicians; the advance from one stage of understanding to the next can be highly non-trivial, and sometimes rather unexpected, but still builds upon the foundation of earlier work rather than starting totally anew. . . . Actually, I find the reality of mathematical research today—in which progress is obtained naturally and cumulatively as a consequence of hard work, directed by intuition, literature, and a bit of luck—to be far more satisfying than the romantic image that I had as a student of mathematics being advanced primarily by the mystic inspirations of some rare breed of "geniuses."

It's not wrong to say Hilbert was a genius. But it's more right to say that what Hilbert accomplished was genius. Genius is a thing that happens, not a kind of person.

POLITICAL LOGIC

Political logic is not a formal system in the sense that Hilbert and the mathematical logicians meant, but mathematicians with a formalist outlook couldn't help but approach politics with the same kind of methodological sympathies. They were encouraged in this by Hilbert himself, who in his 1918 lecture "Axiomatic Thought" advocated that the other sciences adopt the axiomatic approach that had been so successful in mathematics.

For example, Gödel, whose theorem ruled out the possibility of definitively banishing contradiction from arithmetic, was also worried about the Constitution, which he was studying in preparation for his 1948 U.S. citizenship test. In his view, the document contained a contradiction that could allow a Fascist dictatorship to take over the country in a perfectly constitutional manner. Gödel's friends Albert Einstein and Oskar Morgenstern begged him to avoid this matter in his exam, but, as Morgenstern recalls it, the conversation ended up going like this:

The examiner: Now, Mr. Gödel, where do you come from?

Gödel: Where I come from? Austria.

The examiner: What kind of government did you have in Austria?

Gödel: It was a republic, but the constitution was such that it finally was changed into a dictatorship.

The examiner: Oh! This is very bad. This could not happen in this country.

Gödel: Oh, yes, I can prove it.

Fortunately, the examiner hurriedly changed the subject and Gödel's citizenship was duly granted. As to the nature of the contradiction Gödel found in the Constitution, it seems to have been lost to mathematical history. Perhaps for the best!

Hilbert's commitment to logical principle and deduction often led him, like Condorcet, to adopt a surprisingly modern outlook in non-mathematical matters.* At some political cost to himself, he refused to sign the 1914 Declaration to the Cultural World, which defended the kaiser's war in Europe with a long list of denials, each one starting "It is not true": "It is not true that Germany violated the neutrality of Belgium," and so on. Many of the greatest German scientists, like Felix Klein, Wilhelm Roentgen, and Max Planck, signed the declaration. Hilbert said, quite simply, that he was unable to verify to his exacting standards that the assertions in question were not true.

A year later, when the faculty at Göttingen balked at offering a position to the great algebraist Emmy Noether, arguing that students could not possibly be asked to learn mathematics from a woman, Hilbert responded: "I do not see how the sex of the candidate is an argument against her admission. We are a university, not a bathhouse."

But reasoned analysis of politics has its limits. As an old man in the 1930s, Hilbert seemed quite unable to grasp what was happening to his

* Though: Amir Alexander, in his book *Infinitesimal* (New York: FSG, 2014) argues that in the 17th century, it was the pure formalist position, represented by classical Euclidean geometry, that was allied with rigid hierarchies and Jesuitical orthodoxy, while the more intuitive and less rigorous pre-Newtonian theory of infinitesimals was tied to a more forward-looking and democratic ideology.

home country as the Nazis consolidated their power. His first PhD student, Otto Blumenthal, visited Hilbert in Göttingen in 1938 to celebrate his seventy-sixth birthday. Blumenthal was a Christian but came from a Jewish family, and for that reason had been removed from his academic position at Aachen. (It was the same year that Abraham Wald, in German-occupied Austria, left for the United States.)

Constance Reid, in her biography of Hilbert, recounts the conversation at the birthday party:

> "What subjects are you lecturing on this semester?" Hilbert asked.
> "I do not lecture anymore," Blumenthal gently reminded him.
> "What do you mean, you do not lecture?"
> "I am not allowed to lecture anymore."
> "But that is completely impossible! This cannot be done. Nobody has the right to dismiss a professor unless he has committed a crime. Why do you not apply for justice?"

THE PROGRESS OF THE HUMAN MIND

Condorcet, too, held fast to his formalist ideas about politics even when they didn't conform well to reality. The existence of Condorcet cycles meant that any voting system that obeyed his basic, seemingly inarguable axiom—when the majority prefers A to B, B cannot be the winner—can fall prey to self-contradiction. Condorcet spent much of the last decade of his life grappling with the problem of the cycles, developing more and more intricate voting systems intended to evade the problem of collective inconsistency. He never succeeded. In 1785 he wrote, rather forlornly, "We cannot usually avoid being presented with decisions of this kind, which we might call equivocal, except by requiring a large plurality or allowing only very enlightened men to vote. . . . If we cannot find voters who are sufficiently enlightened, we must avoid making a bad choice by accepting as candidates only those men in whose competence we can trust."

But the problem wasn't the voters; it was the math. Condorcet, we now understand, was doomed to failure from the start. Kenneth Arrow, in his 1951 PhD thesis, proved that even a much weaker set of axioms than Condorcet's, a set of requirements that seem as hard to doubt as Peano's rules of arithmetic, leads to paradoxes.* It was a work of great elegance, which helped earn Arrow a Nobel Prize in economics in 1972, but it surely would have disappointed Condorcet, just as Gödel's Theorem had disappointed Hilbert.

Or maybe not—Condorcet was a tough man to disappoint. When the Revolution gathered speed, his mild-mannered brand of republicanism was quickly crowded out by the more radical Jacobins; Condorcet was first politically marginalized, then forced into hiding to avoid the guillotine. And yet Condorcet's belief in the inexorability of progress guided by reason and math didn't desert him. Sequestered in a Paris safe house, knowing he might not have much time left, he wrote his *Sketch for a Historical Picture of the Progress of the Human Mind*, laying out his vision of the future. It is an astonishingly optimistic document, describing a world from which the errors of royalism, sex prejudice, hunger, and old age would be eliminated in turn by the force of science. This passage is typical:

May it not be expected that the human race will be meliorated by new discoveries in the sciences and the arts, and, as an unavoidable consequence, in the means of individual and general prosperity; by farther progress in the principles of conduct, and in moral practice; and lastly, by the real improvement of our faculties, moral, intellectual and physical, which may be the result either of the improvement of the instruments which increase the power and direct the exercise of those faculties, or of the improvement of our natural organization itself?

---

* One voting system to which Arrow's Theorem *doesn't* apply is "approval voting," in which you don't have to declare all your preferences; you just vote for as many of the people on the ballot as you want, and the candidate who gets the most votes wins. Most mathematicians I know consider approval voting or its variants to be superior to both plurality voting and IRV; it has been used to elect popes, secretaries-general of the United Nations, and officials of the American Mathematical Society, but never yet government officials in the United States.

Nowadays, the *Sketch* is best known indirectly; it inspired Thomas Malthus, who considered Condorcet's predictions hopelessly sunny, to write his much more famous, and much bleaker, account of humanity's future.

Shortly after the above passage was written, in March 1794 (or, in the rationalized revolutionary calendar, in Germinal of Year 2) Condorcet was captured and arrested. Two days later he was found dead—some say it was suicide, others that he was murdered.

Just as Hilbert's style of mathematics persisted despite the destruction of his formal program by Gödel, Condorcet's approach to politics survived his demise. We no longer hope to find voting systems that satisfy his axiom. But we have committed ourselves to Condorcet's more fundamental belief, that a quantitative "social mathematics"—what we now call "social science"—ought to have a part in determining the proper conduct of government. These were "the instruments which increase the power and direct the exercise of [our] faculties" that Condorcet wrote about with such vigor in the *Sketch*.

Condorcet's idea is so thoroughly intertwined with the modern way of doing political business that we hardly see it as a choice. But it is a choice. I think it's the right one.

HOW TO BE RIGHT

B etween my sophomore and junior years of college, I had a summer job working for a researcher in public health. The researcher—it will be clear in a minute why I don't use his name—wanted to hire a math major because he wanted to know how many people were going to have tuberculosis in the year 2050. That was my summer job, to figure that out. The researcher gave me a big folder of papers about tuberculosis: how transmissible it was under various circumstances, the typical course of infection and the length of the maximally contagious period, survival curves and medication compliance rates and breakdowns of all of the above by age, race, sex, and HIV status. Big folder. Lots of papers. And I got to work, doing what math majors do: I made a model of TB prevalence, using the data the researcher had given me to estimate how the TB infection rates in different population groups would change and interact over time, decade by decade, until 2050, when the simulation terminated.

And what I learned was this: I did not have a clue how many people were going to have tuberculosis in the year 2050. Each one of those empirical studies had some uncertainty built in; they thought the transmission rate was 20%, but maybe it was 13%, or maybe it was 25%, though they were pretty sure it wasn't 60% or 0%. Each one of these little local uncertainties percolated through the simulation, and the uncertainties about different parameters of the model fed back into each other, and by 2050, the noise had engulfed the signal. I could make the simulation come out any which way. Maybe there was going to be no such thing as tuberculosis in 2050, or maybe most of the world's population would be infected. I had no principled way to choose.

This was not what the researcher wanted to hear. It was not what he was paying me for. He was paying me for a number, and he patiently repeated his request for one. I know there's uncertainty, he said, that's how medical research is, I get that, just *give me your best guess*. It didn't matter how much I protested that any single guess would be worse than no guess at all. He insisted. And he was my boss, and eventually I gave in. I have no doubt he told many people, afterward, that X million people were going to have tuberculosis in the year 2050. And I'll bet if anyone asked him how he knew this, he would say, I hired a guy who did the math.

THE CRITIC WHO COUNTS

That story might make it seem I'm recommending the coward's way of not being wrong: namely, never saying anything at all, responding to every difficult question with shrugs and equivocation: *Well, it certainly could be like this, but on the other hand, you see, it very well could be like that . . .*

People like that, the quibblers and the naysayers and the maybesayers, don't make things happen. When one wants to denounce those people, it's customary to quote Theodore Roosevelt, from his speech "Citizenship in a Republic," delivered in Paris in 1910, shortly after the end of his presidency:

> It is not the critic who counts; not the man who points out how the strong man stumbles, or where the doer of deeds could have done them better. The credit belongs to the man who is actually in the arena, whose face is marred by dust and sweat and blood; who strives valiantly; who errs, who comes short again and again, because

there is no effort without error and shortcoming; but who does actually strive to do the deeds; who knows great enthusiasms, the great
devotions; who spends himself in a worthy cause; who at the best
knows in the end the triumph of high achievement, and who at the
worst, if he fails, at least fails while daring greatly, so that his place
shall never be with those cold and timid souls who neither know victory nor defeat.

That's the part people always quote, but the whole speech is fantasti-
cally interesting, longer and more substantive than anything a U.S. president would deliver nowadays. You can find issues there we've discussed
elsewhere in this book, as where Roosevelt touches on the diminishing
utility of money—

> The truth is that, after a certain measure of tangible material success
> or reward has been achieved, the question of increasing it becomes
> of constantly less importance compared to the other things that can
> be done in life.

—and the "Less like Sweden" fallacy that if a thing is good, more of
it must be better, and vice versa:

> It is just as foolish to refuse all progress because people demanding
> it desire at some points to go to absurd extremes, as it would be to
> go to these absurd extremes simply because some of the measures
> advocated by the extremists were wise.

But the main theme, to which Roosevelt returns throughout the
speech, is that the survival of civilization depends on the triumph of
the bold, commonsensical, and virile against the soft, intellectual, and
infertile.* He was speaking at the Sorbonne, the temple of French academia, the same place where David Hilbert had presented his twenty-

* Roosevelt's view that analytic "book-learning" stands in opposition to virility is expressed more
directly by Shakespeare, who in the opening scene of *Othello* has Iago derisively call his rival Cas-
sio "a great arithmetician . . . That never set a squadron in a field / Nor the division of a battle
knows / more than a spinster." This is the point in the play where every mathematician in the
audience figures out Iago is the bad guy.

three problems just ten years before. A statue of Blaise Pascal looked on from the balcony. Hilbert had urged the mathematicians in his audience into ever-deeper flights of abstraction from geometric intuition and the physical world. Roosevelt's goal was just the opposite: he pays lip service to the accomplishments of the French academics, but makes it clear their book learning is of only secondary importance in the production of national greatness: "I speak in a great university which represents the flower of the highest intellectual development; I pay all homage to intellect and to elaborate and specialized training of the intellect; and yet I know I shall have the assent of all of you present when I add that more important still are the commonplace, every-day qualities and virtues."

And yet—when Roosevelt says, "The closet philosopher, the refined and cultured individual who from his library tells how men ought to be governed under ideal conditions, is of no use in actual governmental work," I think of Condorcet, who spent his time in the library doing just that, and who contributed more to the French state than most of his time's more practical men. And when Roosevelt sneers at the cold and timid souls who sit on the sidelines and second-guess the warriors, I come back to Abraham Wald, who as far as I know went his whole life without lifting a weapon in anger, but who nonetheless played a serious part in the American war effort, precisely by counseling the doers of deeds how to do them better. He was unsweaty, undusty, and unbloody, but he was *right*. He was a critic who counted.

FOR THIS IS ACTION

Against Roosevelt I place John Ashbery, whose poem "Soonest Mended" is the greatest summation I know of the way uncertainty and revelation can mingle, without dissolving together, in the human mind. It's a more complex and accurate portrait of life's enterprise than Roosevelt's hard-charging man's man, sore and broken but never doubting his direction. Ashbery's sad-comic vision of citizenship might almost be a reply to Roosevelt's "Citizenship in a Republic":

And you see, both of us were right, though nothing
Has somehow come to nothing; the avatars
Of our conforming to the rules and living
Around the home have made—well, in a sense, "good citizens" of us,
Brushing the teeth and all that, and learning to accept
The charity of the hard moments as they are doled out,
For this is action, this not being sure, this careless
Preparing, sowing the seeds crooked in the furrow,
Making ready to forget, and always coming back
To the mooring of starting out, that day so long ago.

For this is action, this not being sure! It is a sentence I often repeat to myself like a mantra. Theodore Roosevelt would surely have denied that "not being sure" was a kind of action. He would have called it cowardly fence sitting. The Housemartins—the greatest Marxist pop band ever to pick up guitars—took Roosevelt's side in their 1986 song "Sitting on a Fence," a withering portrait of a wishy-washy political moderate:

Sitting on a fence is a man who swings from poll to poll
Sitting on a fence is a man who sees both sides of both sides. . . .
But the real problem with this man
Is he says he can't when he can . . .

But Roosevelt and the Housemartins are wrong, and Ashbery is right. For him, not being sure is the move of a strong person, not a weakling: it is, elsewhere in the poem, "a kind of fence-sitting / Raised to the level of an esthetic ideal."

And math is part of it. People usually think of mathematics as the realm of certainty and absolute truth. In some ways that's right. We traffic in necessary facts: $2 + 3 = 5$ and all that.

But mathematics is also a means by which we can reason about the uncertain, taming if not altogether domesticating it. It's been that way since the time of Pascal, who started by helping gamblers understand the whims of chance and ended up figuring the betting odds on the most

cosmic uncertainty of all.* Math gives us a way of being unsure in a principled way: not just throwing up our hands and saying "huh," but rather
making a firm assertion: "I'm not sure, this is why I'm not sure, and this
is roughly how not-sure I am." Or even more: "I'm unsure, and you
should be too."

A MAN WHO SWINGS FROM POLL TO POLL

The paladin of principled uncertainty in our time is Nate Silver, the
online-poker-player-turned-baseball-statistics-maven-turned-political-
analyst whose *New York Times* columns about the 2012 presidential election drew more public attention to the methods of probability theory
than they have ever before enjoyed. I think of Silver as a kind of Kurt
Cobain of probability. Both were devoted to cultural practices that had
previously been confined to a small, inward-looking cadre of true believ-
ers (for Silver, quantitative forecasting of sports and politics, for Cobain,
punk rock). And both proved that if you carried their practice out in
public, with an approachable style but without compromising the source
material, you could make it massively popular.

What made Silver so good? In large part, it's that he was willing to
talk about uncertainty, willing to treat uncertainty not as a sign of weakness but as a real thing in the world, a thing that can be studied with
scientific rigor and employed to good effect. If it's September 2012 and
you ask a bunch of political pundits, "Who's going to be elected president in November?" a bunch of them are going to say, "Obama is," and
a somewhat smaller bunch are going to say, "Romney is," and the point
is that all of those people are wrong, because the right answer is the
kind of answer that Silver, almost alone in the broad-reach media, was
willing to give: "Either one might win, but Obama is substantially more
likely to win."

* Ashbery starts the second and final section of "Soonest Mended" with the lines "These then
were some hazards of the course / Yet though we knew the course *was* hazards and nothing else":
Ashbery, who lived in France for a decade, certainly means the English word's sense of danger to
be closely followed by the echo of the French *hasard*, which means "chance," fitting the poem's
overall atmosphere of rigorous uncertainty. Pascal would have called the gambling games he discussed with Fermat *jeux de hasard*, and the word's ultimate origin is the Arabic word for dice.

Traditional political types greeted this response with the same disrespect I got from my tuberculosis boss. They wanted an *answer*. They didn't understand that Silver was giving them one.

Josh Jordan, in the *National Review*, wrote: "On September 30, leading into the debates, Silver gave Obama an 85 percent chance and predicted an Electoral College count of 320–218. Today, the margins have narrowed—but Silver still gives Obama a 67 percent chance and an Electoral College lead of 288–250, which has led many to wonder if he has observed the same movement to Romney over the past three weeks as everyone else has."

Had Silver noticed the movement to Romney? Clearly, yes. He gave Romney a 15% chance of winning at the end of September, and a 33% chance on October 22—more than twice as much. But Jordan didn't notice that Silver had noticed, because Silver was still estimating—correctly—that Obama had a better chance of winning than Romney did. For traditional political reporters like Jordan, that meant his answer hadn't changed.

Or take Dylan Byers in *Politico*: "So should Mitt Romney win on Nov. 6, it's difficult to see how people can continue to put faith in the predictions of someone who has never given that candidate anything higher than a 41 percent chance of winning (way back on June 2) and—one week from the election—gives him a one-in-four chance, even as the polls have him almost neck-and-neck with the incumbent. . . . For all the confidence Silver puts in his predictions, he often gives the impression of hedging."

If you care at all about math, this is the kind of thing that makes you want to stab yourself in the hand with a fork. What Silver offers isn't hedging; it's honesty. When the weather report says there's a 40% chance of rain, and it rains, do you lose faith in its predictions? No—you recognize that the weather is inherently uncertain, and that a definitive statement of whether it will or won't rain tomorrow is usually the wrong thing to offer.*

---

* There are other, more sophisticated reasons to be skeptical about Silver's approach, though these weren't dominant among the Washington press corps. For instance, one could follow R. A. Fisher's line and say that the language of probability is inappropriate for one-off events, and applies only to things like coin flips that can in principle be repeated again and again.

Of course, Obama did win in the end, and with a comfortable margin, leaving Silver's critics looking a little foolish.

The irony is that if the critics had wanted to catch Silver in a mistaken prediction, they missed a great chance. They could have asked him, "How many states are you going to get wrong?" As far as I know, nobody ever asked Silver this question, but it's easy to figure out how he would have answered it. On October 26, Silver estimated that Obama had a 69% chance of winning New Hampshire. If you forced him to predict then and there, he'd call it for Obama. So you could say that Silver estimated his chance of being wrong about New Hampshire to be 0.31. Put in other words, the *expected number of wrong answers* he would give about New Hampshire was 0.31. Remember—the expected value isn't the value you expect, but rather a probabilistic compromise among the possible outcomes—in this case, he's either going to give zero wrong answers about New Hampshire (an outcome with probability 0.69) or one (an outcome with probability 0.31), which gives an expected value of

$$(0.69) \times 0 + (0.31) \times 1 = 0.31$$

via the method we set up in chapter 11.

Silver was more certain about North Carolina, giving Obama only a 19% chance of winning. But that still means he estimated a 19% probability that his Romney call would end up wrong; that is, he gave himself another 0.19 expected wrong answers. Here's a list of the states Silver considered potentially competitive on October 26:

State	Obama win probability	Expected wrong answers
OR	99%	.01
NM	97%	.03
MN	97%	.03
MI	98%	.02
PA	94%	.06
WI	86%	.14
NV	78%	.22
OH	75%	.25

State	Obama win probability	Expected wrong answers
NH	69%	.31
IA	68%	.32
CO	57%	.43
VA	54%	.46
FL	35%	.35
NC	19%	.19
MO	2%	.02
AZ	3%	.03
MT	2%	.02

Since expected value is additive, Silver's best guess at the number of competitive states he'd pick wrong is just the sum of the contributions of each of these states, which comes to 2.83. In other words, he'd probably have said, if anyone had asked him, "On average I'm likely to get about three states wrong."

Actually, he got all fifty right.*

Even the most seasoned political pundit might have trouble pulling off an attack on Silver for being more accurate than he said he would be. The twistiness this incites in the mind is healthy; follow it! When you reason correctly, as Silver does, you find that you always think you're right, but you don't think you're always right. As the philosopher W. O. V. Quine put it, "To believe something is to believe that it is true; therefore a reasonable person believes each of his beliefs to be true; yet experience has taught him to expect that some of his beliefs, he knows not which, will turn out to be false. A reasonable person believes, in short, that each of his beliefs is true and that some of them are false."

Formally, this is very similar to the apparent contradictions in American public opinion we unraveled in chapter 17. The American people think each government program is worthy of continued funding, but

* To be precise, it was his *final* prediction that got all the states right; on October 26, he had everything correct except Florida, where polls swung from leaning Romney to just about even in the last two weeks of the campaign.

that doesn't mean they think *all* government programs are worthy of continued funding.

Silver bypassed the sclerotic conventions of political reporting and told the public a truer story. Instead of saying who was going to win, or who had the "momentum," he reported what he thought the chances were. Instead of saying how many electoral votes Obama was going to win, he presented a probability distribution: say, Obama had a 67% chance of getting the 270 electoral votes he needed for reelection, a 44% chance of breaking 300, a 21% chance of getting to 330, and so on. Silver was being uncertain, *rigorously* uncertain, in public, and the public ate it up. I wouldn't have thought it was possible.

This is action, this not being sure!

AGAINST PRECISION

One criticism of Silver to which I'm somewhat sympathetic is that it's misleading to make statements like "Obama has a 73.1% chance of winning as of today." The decimal suggests a precision of measurement that's probably not there; you don't want to say that something meaningful has happened if his model gives 73.1% one day and 73.0% the next. This is a criticism of Silver's presentation, not his actual program, but it carried a lot of weight with political writers who felt readers were being bullied into acceptance by an impressively precise-looking number.

There's such a thing as being too precise. The models we use to score standardized tests could give SAT scores to several decimal places, if we let them, but we shouldn't—students are anxious enough as it is, without having to worry about their classmate nosing ahead of them by a hundredth of a point.

The fetish of perfect precision affects elections, not just in the fevered poll-watching period but after the election takes place. The Florida 2000 election, remember, rode on a difference of a few hundred votes between George W. Bush and Al Gore, a hundredth of a percent of the total votes cast. It was of critical importance, by our law and custom, to determine which candidate it was who could claim a few hundred more ballots than the other. But as a way of thinking about who Floridians wanted to be president, this is absurd; the imprecision caused by ballots spoiled, ballots lost, ballots miscounted, is much greater than the tiny difference in the final count. We don't know who got more votes in Florida. The difference between judges and mathematicians is that judges have to find a way to pretend we know, while mathematicians are free to tell the truth.

The journalist Charles Seife included in his book *Proofiness* a very funny and mildly depressing chronicle of the similarly close contest between Democrat Al Franken and Republican Norm Coleman to represent Minnesota in the U.S. Senate. It would be great to say that Franken took office because a cold analytical procedure showed exactly 312 more Minnesotans wanted to see him seated in the chamber. In reality, though, that number reflects the result of an extended legal tussle over questions like whether a ballot with a mark for Franken and a write-in for "Lizard People" was legally cast. Once you get down to this kind of issue, the question of who "really" got more votes doesn't even make sense. The signal is lost in the noise. And I tend to side with Seife, who argues that elections this close should be decided by coin flip.* Some will balk at the idea of choosing our leaders by chance. But that's actually the coin flip's most important benefit! Close elections are *already* determined by chance. Bad weather in the big city, a busted voting machine in an outlying town, a poorly designed ballot leading elderly Jews to vote for Pat Buchanan—any of these chance events can make the difference when the electorate is stuck at 50–50. Choosing by coin flip helps keep us from pretending that the people have spoken for the winning candidate in a closely divided race. Sometimes the people speak and they say, "I dunno."

You might think I'd be really into decimal places. The conjoined twin of the stereotype that mathematicians are always certain is the stereotype that we are always precise, determined to compute everything to as many decimal places as possible. It isn't so. We want to compute everything to as many decimal places as *necessary*. There is a young man in China named Lu Chao who learned and recited 67,890 digits of pi.

---

* Of course, if you wanted to set this up correctly, you'd have to modify the coin flip to give a greater chance of winning to the candidate who appears to be slightly ahead, etc., etc.

That's an impressive feat of memory. But is it interesting? No, because the digits of pi are not interesting. As far as anyone knows, they're as good as random. Pi *itself* is interesting, to be sure. But pi is not its digits; it is merely specified by its digits, in the same way the Eiffel Tower is specified by the longitude and latitude 48.8586° N, 2.2942° E. Add as many decimal places to those numbers as you want, and they still won't tell you what makes the Eiffel Tower the Eiffel Tower.

Precision isn't just about digits. Benjamin Franklin wrote cuttingly of a member of his Philadelphia set, Thomas Godfrey: "He knew little out of his way, and was not a pleasing companion; as, like most great mathematicians I have met with, he expected universal precision in everything said, or was for ever denying or distinguishing upon trifles, to the disturbance of all conversation."

This stings because it's only partially unfair. Mathematicians can be persnickety about logical niceties. We're the kind of people who think it's funny, when asked, "Do you want soup or salad with that?" to reply, "Yes."

THAT DOES NOT COMPUTE

And yet even mathematicians don't, except when cracking wise, try to make themselves beings of pure logic. It can be dangerous to do so! For example: If you're a purely deductive thinker, once you believe two contradictory facts you are logically obliged to believe that *every statement is false*. Here's how that goes. Suppose I believe both that Paris is the capital of France and that it's not. This seems to have nothing to do with whether the Portland Trail Blazers were NBA champions in 1982. But now watch this trick. Is it the case that Paris is the capital of France *and* the Trail Blazers won the NBA championship? It is not, because I know that Paris is not the capital of France.

If it's not true that Paris is the capital of France and the Trail Blazers were the champs, then either Paris *isn't* the capital of France or the Trail Blazers weren't NBA champs. But I know that Paris is the capital of France, which rules out the first possibility. So the Trail Blazers did not win the 1982 NBA championship.

It is not hard to check that an argument of exactly the same form, but standing on its head, proves that every statement is also true.

This sounds weird, but as a logical deduction it's irrefutable; drop one tiny contradiction anywhere into a formal system and the whole thing goes to hell. Philosophers of a mathematical bent call this brittleness in formal logic *ex falso quodlibet*, or, among friends, "the principle of explosion." (Remember what I said about how much math people love violent terminology?)

Ex falso quodlibet is how Captain James T. Kirk used to disable dictatorial AIs—feed them a paradox and their reasoning modules frazzle and halt. *That* (they plaintively remark, just before the power light goes out) *does not compute.* Bertrand Russell did to Gottlob Frege's set theory what Kirk did to uppity robots. His one sneaky paradox brought the whole edifice down.

But Kirk's trick doesn't work on human beings. We don't reason this way, not even those of us who do math for a living. We are tolerant of contradiction, to a point. As F. Scott Fitzgerald said, "The test of a first-rate intelligence is the ability to hold two opposed ideas in the mind at the same time, and still retain the ability to function."

Mathematicians use this ability as a basic tool of thought. It's essential for the reductio ad absurdum, which requires you to hold in your mind a proposition you believe to be false and reason as if you think it's true: suppose the square root of 2 *is* a rational number, even though I'm trying to prove it's not. . . . It is lucid dreaming of a very systematic kind. And we can do it without short-circuiting ourselves.

In fact, it's a common piece of folk advice—I know I heard it from my Ph.D. advisor, and presumably he from his, etc.—that when you're working hard on a theorem you should try to prove it by day and disprove it by night. (The precise frequency of the toggle isn't critical; it's said of the topologist R. H. Bing that his habit was to split each month between two weeks trying to prove the Poincaré Conjecture and two weeks trying to find a counterexample.')

Why work at such cross-purposes? There are two good reasons. The

* In the end he didn't succeed at either task; the Poincaré Conjecture was eventually proved by Grigori Perelman in 2003.

first is that you might, after all, be wrong; if the statement you think is true is really false, all your effort to prove it is doomed to be useless. Disproving by night is a kind of hedge against that gigantic waste.

But there's a deeper reason. If something is true and you try to disprove it, you will fail. We are trained to think of failure as bad, but it's not *all* bad. You can learn from failure. You try to disprove the statement one way, and you hit a wall. You try another way, and you hit another wall. Each night you try, each night you fail, each night a new wall, and if you are lucky, those walls start to come together into a structure, and that structure is the structure of the proof of the theorem. For if you have really understood *what's keeping you* from disproving the theorem, you very likely understand, in a way inaccessible to you before, why the theorem is true. This is what happened to Bolyai, who bucked his father's well-meaning advice and tried, like so many before him, to prove that the parallel postulate followed from Euclid's other axioms. Like all the others, he failed. But unlike the others, he was able to understand the shape of his failure. What was blocking all his attempts to prove that there was no geometry without the parallel postulate was the existence of just such a geometry! And with each failed attempt he learned more about the features of the thing he didn't think existed, getting to know it more and more intimately, until the moment when he realized it was really there.

Proving by day and disproving by night is not just for mathematics. I find it's a good habit to put pressure on all your beliefs, social, political, scientific, and philosophical. Believe whatever you believe by day; but at night, argue against the propositions you hold most dear. Don't cheat! To the greatest extent possible you have to think as though you believe what you don't believe. And if you can't talk yourself out of your existing beliefs, you'll know a lot more about *why* you believe what you believe. You'll have come a little closer to a proof.

This salutary mental exercise is not at all what F. Scott Fitzgerald was talking about, by the way. His endorsement of holding contradictory beliefs comes from "The Crack-Up," his 1936 essay about his own irreparable brokenness. The opposing ideas he has in mind there are "the sense of futility of effort and the sense of the necessity to struggle." Samuel

Beckett later put it more succinctly: "I can't go on, I'll go on." Fitzgerald's characterization of a "first-rate intelligence" is meant to deny his own intelligence that designation; as he saw it, the pressure of the contradiction had made him effectively cease to exist, like Frege's set theory or a computer downed by Kirkian paradox. (The Housemartins, elsewhere in "Sitting on a Fence," more or less summarize "The Crack-Up": "I lied to myself right from the start / and I just worked out that I'm falling apart.") Unmanned and undone by self-doubt, drowned in books and introspection, he had become exactly the kind of sad young literary man who made Theodore Roosevelt puke.

David Foster Wallace was interested in paradox too. In his characteristically mathematical style; he put a somewhat tamed version of Russell's paradox at the center of his first novel, *The Broom of the System*. It isn't too strong to say his writing was driven by his struggle with contradictions. He was in love with the technical and analytic, but he saw that the simple dicta of religion and self-help offered better weapons against drugs, despair, and killing solipsism. He knew it was supposed to be the writer's job to get inside other people's heads, but his chief subject was the predicament of being stuck fast inside one's own. Determined to record and neutralize the influence of his own preoccupations and prejudices, he knew this determination was itself among those preoccupations and subject to those prejudices. This is Phil 101 stuff, to be sure, but as any math student knows, the old problems you meet freshman year are some of the deepest you ever see. Wallace wrestled with the paradoxes just the way mathematicians do. You believe two things that seem in opposition. And so you go to work—step by step, clearing the brush, separating what you know from what you believe, holding the opposing hypotheses side by side in your mind and viewing each in the adversarial light of the other until the truth, or the nearest you can get to it, comes clear.

As for Beckett, he had a richer and more sympathetic view of contradiction, which is so ever-present in his work that it takes on every possible emotional color somewhere or other in the corpus. "I can't go on, I'll go on" is bleak; but Beckett also draws on the Pythagoreans' proof of the irrationality of the square root of 2, turning it into a joke between drunks:

"But betray me," said Neary, "and you go the way of Hippasos."

"The Akousmatic, I presume," said Wylie. "His retribution slips my mind."

"Drowned in a puddle," said Neary, "for having divulged the incommensurability of side and diagonal."

"So perish all babblers," said Wylie.

It's not clear how much higher math Beckett knew, but in his late prose piece *Worstward Ho*, he sums up the value of failure in mathematical creation more succinctly than any professor ever has:

Ever tried. Ever failed. No matter. Try again. Fail again. Fail better.

WHEN AM I GOING TO USE THIS?

The mathematicians we've encountered in this book are not just puncturers of unjustified certainties, not just critics who count. They found things and they built things. Galton uncovered the idea of regression to the mean; Condorcet built a new paradigm for social decision making; Bolyai created an entirely novel geometry, "a strange new universe"; Shannon and Hamming made a geometry of their own, a space where digital signals lived instead of circles and triangles; Wald got the armor on the right part of the plane.

Every mathematician creates new things, some big, some small. All mathematical writing is creative writing. And the entities we can create mathematically are subject to no physical limits; they can be finite or infinite, they can be realizable in our observable universe or not. This sometimes leads outsiders to think of mathematicians as voyagers in a psychedelic realm of dangerous mental fire, staring straight at visions that would drive lesser beings mad, sometimes indeed being driven mad themselves.

It's not like that, as we've seen. Mathematicians aren't crazy, and we aren't aliens, and we aren't mystics.

What's true is that the sensation of mathematical understanding—of suddenly knowing what's going on, with total certainty, *all the way to the bottom*—is a special thing, attainable in few if any other places in life. You feel you've reached into the universe's guts and put your hand on the wire. It's hard to describe to people who haven't experienced it.

We are not free to say whatever we like about the wild entities we make up. They require definition, and having been defined, they are no more psychedelic than trees and fish; they are what they are. To do mathematics is to be, at once, touched by fire and bound by reason. This is no contradiction. Logic forms a narrow channel through which intuition flows with vastly augmented force.

The lessons of mathematics are simple ones and there are no numbers in them: that there is structure in the world; that we can hope to understand some of it and not just gape at what our senses present to us; that our intuition is stronger with a formal exoskeleton than without one. And that mathematical certainty is one thing, the softer convictions we find attached to us in everyday life another, and we should keep track of the difference if we can.

Every time you observe that more of a good thing is not always better; or you remember that improbable things happen a lot, given enough chances, and resist the lure of the Baltimore stockbroker; or you make a decision based not just on the most likely future, but on the cloud of all possible futures, with attention to which ones are likely and which ones are not; or you let go of the idea that the beliefs of groups should be subject to the same rules as beliefs of individuals; or, simply, you find that cognitive sweet spot where you can let your intuition run wild on the network of tracks formal reasoning makes for it; without writing down an equation or drawing a graph, you are doing mathematics, the extension of common sense by other means. When are you going to use it? You've been using mathematics since you were born and you'll probably never stop. Use it well.

ACKNOWLEDGMENTS

It has been about eight years since I first had the idea of writing this book. That *How Not to Be Wrong* is now in your hands, and not just an idea, is testament to the wise guidance of my agent, Jay Mandel, who patiently asked me every year whether I was ready to take a try at writing something and, when I finally said "yes," helped me refine the concept from "I want to yell at people, at length, about how great math is" to something more like an actual book.

I'm very fortunate to have placed the book with The Penguin Press, which has a long tradition of helping academics speak to a wide audience while still allowing them to totally nerd out. I benefited tremendously from the insights of Colin Dickerman, who acquired the book and helped see it through to near-finished form, and Scott Moyers, who took over for the final push. Both of them were very understanding with a novice author as the project transformed itself into something quite different from the book I had originally proposed. I have also benefited greatly from the advice and assistance of Mally Anderson, Akif Saifi, Sarah Hutson, and Liz Calamari at The Penguin Press and Laura Stickney at Penguin UK.

I also owe thanks to the editors of *Slate*, especially Josh Levin, Jack Shafer, and David Plotz, who decided in 2001 that what *Slate* needed was a math column. They've been running my stuff ever since, helping me learn how to talk about math in a way that non-mathematicians can understand. Some parts of this book are adapted from my *Slate* pieces and have benefited from their editing. I'm also very grateful to my editors at other publications: at the *New York Times*, the *Washington Post*, the *Boston Globe*, and the *Wall Street Journal*. (The book also contains some repurposed bits and pieces from my articles in the *Post* and the *Globe*.) I'm especially thankful for Heidi Julavits at the *Believer* and Nicholas Thompson at *Wired*, who were the first to assign me long pieces and taught me critical lessons about how to keep a mathematical narrative moving for thousands of words at a stretch.

Elise Craig did an excellent job fact-checking portions of this book; if you find a mistake, it's in the other portions. Greg Villepique copy-edited the book, removing many errors of usage and fact. He is a tireless foe of unnecessary hyphens.

Barry Mazur, my PhD advisor, taught me much of what I know about number theory; what's more, he serves as a model for the deep connections between mathematics and other modes of thinking, expressing, and feeling.

For the Russell quote that opens the book I'm indebted to David Foster Wallace, who marked the quote as a potential epigraph in his working notes for *Everything and More*, his book about set theory, but didn't end up using it.

Much of *How Not to Be Wrong* was written while I was on sabbatical from my position at the University of Wisconsin–Madison; I thank the Wisconsin Alumni Research Foundation for enabling me to extend that leave to a full year with a Romnes Faculty Fellowship and my colleagues at Madison for supporting this idiosyncratic and not-exactly-academic project.

I also want to thank Barriques Coffee on Monroe Street in Madison, Wisconsin, where much of this book was produced.

The book itself has benefited from suggestions and close readings from many friends, colleagues, and strangers who answered my e-mail, including: Laura Balzano, Meredith Broussard, Tim Carmody, Tim Chow, Jenny Davidson, Jon Eckhardt, Steve Fienberg, Peli Grietzer, the Hieratic Conglomerate, Gil Kalai, Emmanuel Kowalski, David Krakauer, Lauren Kroiz, Tanya Latty, Marc Mangel, Arika Okrent, John Quiggin, Ben Recht, Michel Regenwetter, Ian Roulstone, Nissim Schlam-Salman, Gerald Selbee, Cosma Shalizi, Michelle Shih, Barry Simon, Brad Snyder, Elliott Sober, Miranda Spieler, Jason Steinberg, Hal Stern, Stephanie Tai, Bob Temple, Ravi Vakil, Robert Wardrop, Eric Wepsic, Leland Wilkinson, and Janet Wittes. Inevitably there are others; I apologize to anyone I have missed. I want to single out several readers who gave especially important feedback: Tom Scocca, who read the whole thing with a keen eye and an unsparing stance; Andrew Gelman and Stephen Stigler, who kept me honest about the history of statistics; Stephen Burt, who kept me honest about poetry; Henry Cohn, who carried out an amazing close reading on a big chunk of the book and fed me the quote about Winston Churchill and the projective plane; Lynda Barry, who told me it was okay to draw the pictures myself; and my parents, both applied statisticians, who read everything and told me when it was getting too abstract.

I thank my son and daughter for being patient through the many working weekends the book has made necessary, and my son in particular for drawing one of the pictures. And most of all, Tanya Schlam, both first and final reader of everything you've seen here, and the person whose support and love made it possible even to conceive this project. She has helped me understand, even more than mathematics has, how to be right.

NOTES

When Am I Going to Use This?

3: **Abraham Wald was born in 1902:** Biographical material about Abraham Wald is drawn from Oscar Morgenstern, "Abraham Wald, 1902–1950," *Econometrica* 19, no. 4 (Oct. 1951): 361–67.

4: **The Statistical Research Group (SRG):** Historical material about the SRG is largely drawn from W. Allen Wallis, "The Statistical Research Group, 1942–1945," *Journal of the American Statistical Association* 75, no. 370 (June 1980): 320–30.

4: **"When we made recommendations":** Ibid., 322.

5: **"the most extraordinary group":** Ibid., 322.

5: **the joke around SRG:** Ibid., 329.

5: **So here's the question:** I learned about Wald and the missing bullet holes from Howard Wainer's book *Uneducated Guesses: Using Evidence to Uncover Misguided Education Policies* (Princeton, NJ: Princeton University Press, 2011) where he applies Wald's insights to the similarly complicated and partial statistics obtained in education studies.

7: **through the wars in Korea and Vietnam:** Marc Mangel and Francisco J. Samaniego, "Abraham Wald's Work on Aircraft Survivability," *Journal of the American Statistical Association* 79, no. 386 (June 1984): 259–67.

8: **"all of the most abstract sort":** Jacob Wolfowitz, "Abraham Wald, 1902–1950," *Annals of Mathematical Statistics* 23, no. 1 (Mar. 1952): 1–13.

8: **A 2006 study by Savant Capital:** Amy L. Barrett and Brent R. Brodeski, "Survivor Bias and Improper Measurement: How the Mutual Fund Industry Inflates Actively Managed Fund Performance," www .savantcapital.com/uploadedFiles/Savant_CMS_Website/Press_Coverage/Press_Releases/Older_ releases/sbiasstudy[1].pdf (accessed Jan. 13, 2014).

9: **a comprehensive 2011 study in the *Review of Finance*:** Martin Rohleder, Hendrik Scholz, and Marco Wilkens, "Survivorship Bias and Mutual Fund Performance: Relevance, Significance, and Methodical Differences," *Review of Finance* 15 (2011): 441–74; see table. We have converted monthly excess return to annual excess return, so the numbers in the text don't match those in the table.

9: **Wald's actual report:** Abraham Wald, A Method of Estimating Plane Vulnerability Based on Damage of Survivors (Alexandria, VA: Center for Naval Analyses, repr., CRC 432, July 1980).

16: **books of its own:** For the Riemann Hypothesis, I like John Derbyshire's *Prime Obsession* and Marcus du Sautoy's *The Music of the Primes*. For Gödel's Theorem, there's of course Douglas Hofstadter's *Gödel, Escher, Bach*, which is, to be fair, only tangentially about the theorem as one mantra in its meditation on self-reference in art, music, and logic.

Chapter 1: Less like Sweden

21: **a blog entry with the provocative title:** Daniel J. Mitchell, "Why Is Obama Trying to Make America More Like Sweden when Swedes Are Trying to Be Less Like Sweden?" Cato Institute, Mar. 16, 2010, www.cato.org/blog/why-obama-trying-make-america-more-sweden-when-swedes-are-trying-be-less-sweden (accessed Jan. 13, 2014).

24:    ***Est modus in rebus***: Horace, *Satires* 1.1.106, trans. Basil Dufallo, in "Satis/Satura: Reconsidering
       the 'Programmatic Intent' of Horace's Satires 1.1," *Classical World* 93 (2000): 579–90.
26:    **the government does take in some amount of revenue:** Laffer was always very clear on the point
       that the Laffer curve was not his invention; Keynes had understood and written about the idea
       very clearly, and the basic idea goes back (at least) as far as the fourteenth-century historian Ibn
       Khaldun.
26:    **"Thomas Edison was a nut":** Jonathan Chait, "Prophet Motive," *New Republic*, Mar. 31, 1997.
27:    **"and he can talk about it for six months":** Hal R. Varian, "What Use Is Economic Theory?" (1989),
       http://people.ischool.berkeley.edu/~hal/Papers/theory.pdf (accessed Jan. 13, 2014).
27:    **"I came into the Big Money":** David Stockman, *The Triumph of Politics: How the Reagan Revolution
       Failed* (New York: Harper & Row, 1986), 10.
28:    **"Subsequent history failed to confirm Laffer's conjecture":** N. Gregory Mankiw, *Principles of Mi-
       croeconomics, vol. 1* (Amsterdam: Elsevier, 1998), 166.
29:    **his acid assessment of supply-side theory:** Martin Gardner, "The Laffer Curve," *The Night Is Large:
       Collected Essays, 1938-1995* (New York: St. Martin's, 1996), 127–39.
30:    **during congressional testimony:** In 1978, during consideration of the Kemp-Roth tax-cut bill.

Chapter 2: Straight Locally, Curved Globally

34:    **one of the very few ancient Greeks to wear pants:** Christoph Riedweg, *Pythagoras: His Life, Teach-
       ing, and Influence* (Ithaca, NY: Cornell University Press, 2005), 2.
40:    **"the ghosts of departed quantities":** George Berkeley, *The Analyst: A Discourse Addressed to an
       Infidel Mathematician* (1734), ed. David R. Wilkins, www.maths.tcd.ie/pub/HistMath/People/
       Berkeley/Analyst/Analyst.pdf (accessed Jan. 13, 2014).
42:    **Most people, if you press them:** David O. Tall and Rolph L. E. Schwarzenberger, "Conflicts in the
       Learning of Real Numbers and Limits," *Mathematics Teaching* 82 (1978): 44–49.
44:    ***2-adic numbers:*** In Cauchy's theory, a series converging to a limit x means that when you sum
       more and more terms, the total gets closer and closer to x. This requires that we have in mind an
       idea of what it means for two numbers to be "close" to each other. It turns out that the familiar
       notion of closeness is not the only one! In the 2-adic world, two numbers are said to be close to
       each other when their difference is a multiple of a large power of 2. When we say the series 1 + 2
       + 4 + 8 + 16 + . . . converges to –1, we are saying that the partial sums 1, 3, 7, 15, 31 . . . are getting
       closer and closer to –1. With the usual meaning of "close," that's not true; but using 2-adic close-
       ness, it's a different story. The numbers 31 and –1 differ by 32, which is 2^5, a pretty small 2-adic
       number. Sum a few more terms and you get 511, which differs from –1 by only 512, smaller
       (2-adically) still. Much of the math you know—calculus, logarithms and exponentials, geometry—
       has a 2-adic analogue (and indeed a p-adic analogue for any p), and the interaction between all
       these different notions of closeness is a crazy and glorious story of its own.
45:    **including the Italian mathematician/priest Guido Grandi:** Material on Grandi and his series is
       largely drawn from Morris Kline, "Euler and Infinite Series," *Mathematics Magazine* 56, no. 5
       (Nov. 1983): 307–14.
49:    **Cauchy was interested in the truth:** The story of Cauchy's calculus class is taken from *Duel at
       Dawn*, Amir Alexander's immensely interesting historical study of the interaction between math
       and culture at the beginning of the nineteenth century. See also Michael J. Barany, "Stuck in the
       Middle: Cauchy's Intermediate Value Theorem and the History of Analytic Rigor," *Notices of the
       American Mathematical Society* 60, no. 10 (Nov. 2013): 1334–38, for a somewhat contrary point of
       view concerning the modernness of Cauchy's approach.

Chapter 3: Everyone Is Obese

50:    **a paper in the journal *Obesity*:** Youfa Wang et al., "Will All Americans Become Overweight or
       Obese? Estimating the Progression and Cost of the US Obesity Epidemic," *Obesity* 16, no. 10
       (Oct. 2008): 2323–30.
50:    **"obesity apocalypse":** abcnews.go.com/Health/Fitness/story?id=5499878&page=1.
50:    **"We're Getting Fatter":** *Long Beach Press-Telegram*, Aug. 17, 2008.
51:    **We're not all going to be overweight:** My discussion of the Wang obesity study largely agrees with
       that in Carl Bialik's article "Obesity Study Looks Thin" (*Wall Street Journal*, Aug. 15, 2008), which
       I learned about after writing this chapter.
51:    **the North Carolina Career Resource Network:** The figures here are from www.soicc.state.nc.us/
       soicc/planning/c2c.htm, which has since been taken down.
61:    **had already begun to slow:** Katherine M. Flegal et al., "Prevalence of Obesity and Trends in the

Distribution of Body Mass Index Among US Adults, 1999–2010," *Journal of the American Medical Association* 307, no. 5 (Feb. 1, 2012), 491–97.

Chapter 4: How Much Is That in Dead Americans?

62: **"The Israeli military reports":** Daniel Byman, "Do Targeted Killings Work?" *Foreign Affairs* 85, no. 2 (Mar.–Apr. 2006), 95.

62: **"the equivalent, on a proportional basis":** "Expressing Solidarity with Israel in the Fight Against Terrorism," H. R. Res. 280, 107th Congress (2001).

62: **Newt Gingrich:** Some of the material in this chapter is adapted from my article "Proportionate Response," *Slate,* July 24, 2006.

62: **"Remember that when Israel loses eight people":** From *Meet the Press*, July 16, 2006, transcript at www.nbcnews.com/id/13839698/page/2/#.Uf_Gc2TEo9E (accessed Jan. 13, 2014).

62: **"When Israel killed 1,400 Palestinians in Gaza":** Ahmed Moor, "What Israel Wants from the Palestinians, It Takes," *Los Angeles Times*, Sept. 17, 2010.

62: **"Some 45,000 Nicaraguans":** Gerald Caplan, "We Must Give Nicaragua More Aid," *Toronto Star,* May 8, 1988.

63: **"equivalent to 27 million Americans":** David K. Shipler, "Robert McNamara and the Ghosts of Vietnam," *New York Times Magazine*, Aug. 10, 1997, pp. 30–35.

64: **they have the most people:** The brain cancer data is all from "State Cancer Profiles," National Cancer Institute, http://statecancerprofiles.cancer.gov/cgi-bin/deathrates/deathrates.pl?00&076&00&2&001&1&1&1 (accessed Jan. 13, 2014).

65: **much more or much less likely you'll get brain cancer:** The example of brain cancer rates owes much to a similar treatment of county-by-county kidney cancer statistics in Howard Wainer's book *Picturing the Uncertain World* (Princeton, NJ: Princeton University Press, 2009), which develops the idea much more thoroughly than I do here.

67: **10,000 times in all:** John E. Kerrich, "Random Remarks," *American Statistician* 15, no. 3 (June 1961), 16–20.

69: **Who wins this kind of contest?:** The scores for 1999 are taken from "A Report Card for the ABCs of Public Education Volume I: 1998-1999 Growth and Performance of Public Schools in North Carolina—25 Most Improved K-8 Schools," www.ncpublicschools.org/abc_results/results_99/99ABCsTop25.pdf (accessed Jan. 13, 2014).

69: **shooting percentage is as much a function:** Kirk Goldsberry, "Extra Points: A New Way to Understand the NBA's Best Scorers," *Grantland,* Oct. 9, 2013, www.grantland.com/story/_/id/9795591/kirk-goldsberry-introduces-new-way-understand-nba-best-scorers (accessed Jan. 13, 2014), suggests one way of going beyond shooting percentage to develop more informative measures of offensive performance.

70: **A study by Thomas Kane and Douglas Staiger:** Thomas J. Kane and Douglas O. Staiger, "The Promise and Pitfalls of Using Imprecise School Accountability Measures," *Journal of Economic Perspectives* 16, no. 4 (Fall 2002), 91–114.

71: **I'll spare you here:** But see Kenneth G. Manton et al., "Empirical Bayes Procedures for Stabilizing Maps of U.S. Cancer Mortality Rates," *Journal of the American Statistical Association* 84, no. 407 (Sept. 1989): 637–50; and Andrew Gelman and Phillip N. Price, "All Maps of Parameter Estimates Are Misleading," *Statistics in Medicine* 18, no. 23 (1999): 3221–34) if you want the no-holds-barred technical treatment.

72: **the *gendarme's hat*:** Stephen M. Stigler, *Statistics on the Table: The History of Statistical Concepts and Methods* (Cambridge, MA: Harvard University Press, 1999), 95.

73: **decipherable formulae:** See, e.g., Ian Hacking, *The Emergence of Probability: A Philosophical Study of Early Ideas About Probability, Induction, and Statistical Inference,* 2d ed. (Cambridge, UK: Cambridge University Press, 2006), ch. 18.

74: **King Leopold's war in the Congo:** White's figures here are taken from Matthew White, "30 Worst Atrocities of the 20th Century," http://users.erols.com/mwhite28/atrox.htm (accessed Jan. 13, 2014).

Chapter 5: More Pie than Plate

77: **A recent working paper by economists Michael Spence and Sandile Hlatshwayo:** A. Michael Spence and Sandile Hlatshwayo, "The Evolving Structure of the American Economy and the Employment Challenge," Council on Foreign Relations, Mar. 2011, www.cfr.org/industrial-policy/evolving-structure-american-economy-employment-challenge/p24366 (accessed Jan. 13, 2014).

77: **From *The Economist*:** "Move Over," *Economist*, July 7, 2012.

77:   **Bill Clinton's latest book:** William J. Clinton, *Back to Work: Why We Need Smart Government for a Strong Economy* (New York: Random House, 2011), 167.

78:   ***ficta*, or fake:** Jacqueline A. Stedall, *From Cardano's Great Art to Lagrange's Reflections: Filling a Gap in the History of Algebra* (Zurich: European Mathematical Society, 2011), 14.

80:   **"over 50 percent of U.S. job growth in June came from our state":** Milwaukee Journal Sentinel, PolitiFact,   www.politifact.com/wisconsin/statements/2011/jul/28/republican-party-wisconsin/ wisconsin-republican-party-says-more-than-half-nat (accessed Jan. 13, 2014).

80:   **"Something we are doing here must be working":** WTMJ News, Milwaukee, "Sensenbrenner, Voters Take Part in Contentious Town Hall Meeting over Federal Debt," www.todaystmj4.com/news/ local/126122793.html (accessed Jan. 13, 2014).

80:   **more than thirteen thousand in the same month:** All the job data here comes from the June 2011 Regional and State Employment and Unemployment (Monthly) News Release by the Bureau of Labor Statistics, July 22, 2011, www.bls.gov/news.release/archives/laus_07222011.htm.

80:   ***New York Times* op-ed:** Steven Rattner, "The Rich Get Even Richer," *New York Times*, Mar. 26, 2012, A27.

81:   **which they've helpfully put online:** elsa.berkeley.edu/~saez/TabFig2010.xls (accessed Jan. 13, 2014).

82:   **released a statement asserting:** Mitt Romney, "Women and the Obama Economy," Apr. 10, 2012, available at www.scribd.com/doc/88740691/Women-And-The-Obama-Economy-Infographic.

83:   **one month later, in February 2009:** The computations and argument here are drawn from Glenn Kessler, "Are Obama's Job Policies Hurting Women?" *Washington Post*, Apr. 10, 2012.

84:   **"true but false":** Ibid.

Chapter 6: The Baltimore Stockbroker and the Bible Code

89:   **"After Abraham was weaned":** Maimonides, *Laws of Idolatry* 1.2, from Isadore Twersky, *A Maimonides Reader* (New York: Behrman House, Inc., 1972), 73.

91:   **respite for Slovakia's Jews:** Yehuda Bauer, *Jews for Sale? Nazi-Jewish Negotiations, 1933–1945* (New Haven: Yale University Press, 1996), 74–90.

92:   **Witztum and his colleagues:** Doron Witztum, Eliyahu Rips, and Yoav Rosenberg, "Equidistant Letter Sequences in the Book of Genesis," *Statistical Science* 9, no. 3 (1994): 429–38.

93:   **"Our referees were baffled":** Robert E. Kass, "In This Issue," *Statistical Science* 9, no. 3 (1994): 305–6.

94:   **"they have disgraced mathematics":** Shlomo Sternberg, "Comments on *The Bible Code*," *Notices of the American Mathematical Society* 44, no. 8 (Sept. 1997): 938–39.

97:   **a practice called *incubation*:** Alan Palmiter and Ahmed Taha, "Star Creation: The Incubation of Mutual Funds," *Vanderbilt Law Review* 62 (2009): 1485–1534. Palmiter and Taha explicitly draw the analogy between the Baltimore stockbroker and fund incubation.

97:   **the same returns as the median fund:** Ibid., 1503.

98:   **a much less miraculous one in a thousand:** Leonard A. Stefanski, "The North Carolina Lottery Coincidence," *American Statistician* 62, no. 2 (2008): 130–34.

99:   **"*what is improbable is probable*":** Aristotle, *Rhetoric* 2.24, trans. W. Rhys Roberts, classics.mit .edu/Aristotle/rhetoric.mb.txt (accessed Jan. 14, 2014).

99:   **"the 'one chance in a million' will undoubtedly occur":** Ronald A. Fisher, *The Design of Experiments* (Edinburgh: Oliver & Boyd, 1935), 13–14.

100: **McKay and Bar-Natan found:** Brendan McKay and Dror Bar-Natan, "Equidistant Letter Sequences in Tolstoy's *War and Peace*," cs.anu.edu.au/~bdm/dilugim/WNP/main.pdf (accessed Jan. 14, 2014).

100: **"equally appalling":** Brendan McKay, Dror Bar-Natan, Maya Bar-Hillel, and Gil Kalai, "Solving the Bible Code Puzzle," *Statistical Science* 14, no. 2 (1999): 150–73, section 6.

100: **In a later paper:** Ibid.

101: **a full-page ad:** *New York Times*, Dec. 8, 2010, A27.

101: **Witztum, Rips, and Rosenberg insist:** See, e.g., Witztum's article "Of Science and Parody: A Complete Refutation of MBBK's Central Claim," www.torahcode.co.il/english/paro_hb.htm (accessed Jan. 14, 2014).

Chapter 7: Dead Fish Don't Read Minds

102: **"Neural correlates":** Craig M. Bennett et al., "Neural Correlates of Interspecies Perspective Taking in the Post-Mortem Atlantic Salmon: An Argument for Proper Multiple Comparisons Correction," *Journal of Serendipitous and Unexpected Results* 1 (2010): 1–5.

102: **"One mature Atlantic Salmon"**: Ibid., 2

103: **The joke, like all jokes:** Gershon Legman, *Rationale of the Dirty Joke: An Analysis of Sexual Humor* (New York: Grove, 1968; repr. Simon & Schuster, 2006).

104: **are said to understand it:** See, e.g., Stanislas Dehaene, *The Number Sense: How the Mind Creates Mathematics* (New York: Oxford University Press, 1997)

110: **in cryptic rhymed verse:** Richard W. Feldmann, "The Cardano-Tartaglia Dispute," *Mathematics Teacher* 54, no. 3 (1961): 160–63.

115: **and part-time mathematician:** Material on Arbuthnot drawn from chapter 18 of Ian Hacking, *The Emergence of Probability* (New York: Cambridge University Press, 1975) and chapter 6 of Stephen M. Stigler, *The History of Statistics* (Cambridge, MA: Harvard University Press/Belknap Press, 1986).

116: **to look like the one we have:** See Elliot Sober's *Evidence and Evolution: The Logic Behind the Science* (New York: Cambridge University Press, 2008) for a thorough discussion of the many, many strands, both classical and contemporary, of this "argument by design."

116: **"It can hardly be supposed":** Charles Darwin, *The Origin of Species*, 6th ed. (London: 1872), 421.

116: **"the backbone of psychological research":** Richard J. Gerrig and Philip George Zimbardo, *Psychology and Life* (Boston: Allyn & Bacon, 2002).

117: **"only in his underwear":** David Bakan. "The Test of Significance in Psychological Research," *Psychological Bulletin* 66, no. 6 (1966): 423–37.

118: **"New evidence has become available":** Quoted in Ann Furedi, "Social Consequences: The Public Health Implications of the 1995 'Pill Scare,'" *Human Reproduction Update* 5, no. 6 (1999): 621–26.

119: **"The government warned Thursday":** Edith M. Lederer, "Government Warns Some Birth Control Pills May Cause Blood Clots," Associated Press, Oct. 19, 1995.

119: **stopped taking their contraceptives:** Sally Hope, "Third Generation Oral Contraceptives: 12% of Women Stopped Taking Their Pill Immediately They Heard CSM's Warning," *BMJ: British Medical Journal* 312, no. 7030 (1996): 576.

119: **13,600 more abortions:** Furedi, "Social Consequences," 623.

119: **"possibly one":** Klim McPherson, "Third Generation Oral Contraception and Venous Thromboembolism," *BMJ: British Medical Journal* 312, no. 7023 (1996): 68.

120: **A study by sociologists at CUNY:** Julia Wrigley and Joanna Dreby, "Fatalities and the Organization of Child Care in the United States, 1985–2003," *American Sociological Review* 70, no. 5 (2005): 729–57.

120: **who died of sudden infant death syndrome:** All statistics about infant deaths are drawn from the Centers for Disease Control. Sherry L. Murphy, Jiaquan Xu, and Kenneth D. Kochanek, "Deaths: Final Data for 2010," www.cdc.gov/nchs/data/nvsr/nvsr61/nvsr61_04.pdf.

121: **a frustrated novelist:** Biographical material on Skinner is drawn from his autobiographical article "B. F. Skinner . . . An Autobiography" in Peter B. Dews, ed., *Festschrift for BF Skinner* (New York: Appleton-Century-Crofts, 1970), 1–22, and from his autobiography, *Particulars of My Life*, particularly pp. 262–63.

121: **"At night, he stops":** Skinner, "Autobiography," 6.

122: **"a one-act play about a quack":** Ibid., 8.

122: **"All that makes a writer":** Skinner, *Particulars*, 262.

122: **"A violent reaction . . . at fault":** Skinner, "Autobiography," 7.

122: **"Literature must be demolished":** Skinner, *Particulars*, 292.

122: **"No one has ever touched a soul":** John B. Watson, *Behaviorism* (Livingston, NJ: Transaction Publishers, 1998), 4.

123: **"You'll never get out":** Skinner, "Autobiography," 12.

123: **"emitting whole lines ready-made":** Skinner, "Autobiography," 6.

123: **intentions of the author:** Joshua Gang, "Behaviorism and the Beginnings of Close Reading," *ELH (English Literary History)* 78, no. 1 (2011): pp. 1–25.

123: **"Proof that there was a process":** B. F. Skinner, "The Alliteration in Shakespeare's Sonnets: A Study in Literary Behavior," *Psychological Record* 3 (1939): 186–92. I learned about Skinner's work on alliteration from Persi Diaconis and Frederick Mosteller's classic paper "Methods for Studying Coincidences," *Journal of the American Statistical Association* 84, no. 408 (1989), 853–61, essential reading if you want to go further into the ideas discussed in this chapter.

124: **"out of a hat":** Skinner, "Alliteration in Shakespeare's Sonnets," 191.

124: **As documented by literary historians:** See, e.g., Ulrich K. Goldsmith, "Words out of a Hat? Alliteration and Assonance in Shakespeare's Sonnets," *Journal of English and Germanic Philology* 49, no. 1 (1950), 33–48.

124: **"Many writers indulge":** Herbert D. Ward, "The Trick of Alliteration," *North American Review* 150, no. 398 (1890): 140–42.

125: **one of the most famous contemporary papers:** Thomas Gilovich, Robert Vallone, and Amos Tversky, "The Hot Hand in Basketball: On the Misperception of Random Sequences," *Cognitive Psychology* 17, no. 3 (1985): 295–314.

128: **Kevin Korb and Michael Stillwell:** Kevin B. Korb and Michael Stillwell, "The Story of the Hot Hand: Powerful Myth or Powerless Critique?" (paper presented at the International Conference on Cognitive Science, 2003), www.csse.monash.edu.au/~korb/iccs.pdf.

128: **slightly more likely to make the next one:** Gur Yaar and Shmuel Eisenmann, "The Hot (Invisible?) Hand: Can Time Sequence Patterns of Success/Failure in Sports Be Modeled as Repeated Random Independent Trials?" *PLoS One*, vol. 6, no. 10 (2011): e24532.

129: **vexed, murky, unsettled question:** In this connection, I really like the 2011 paper "Differentiating Skill and Luck in Financial Markets with Streaks," by Andrew Mauboussin and Samuel Arbesman—it's an especially impressive piece of work considering the first author was a high school senior when it was written! I don't think its conclusions are decisive, but I think it represents a very good way to think about these difficult problems. Available at papers.ssrn.com/sol3/papers.cfm?abstract_id=1664031.

129: **A 2009 study by John Huizinga and Sandy Weil:** Personal communication with Huizinga.

129: **even more intriguing results along these lines:** Yigal Attali, "Perceived Hotness Affects Behavior of Basketball Players and Coaches," *Psychological Science* 24, no. 7 (July 1, 2013): 1151–56.

Chapter 8: Reductio ad Unlikely

132: **there were only eighty-eight:** Allison Klein, "Homicides Decrease in Washington Region," *Washington Post*, Dec. 31, 2012.

133: **the study of the heavenly bodies:** David W. Hughes and Susan Cartwright, "John Michell, the Pleiades, and Odds of 496,000 to 1," *Journal of Astronomical History and Heritage* 10 (2007): 93–99.

134–35: **[figures]:** The two pictures of points strewn across the square were generated by Yuval Peres of Microsoft Research and are taken from his "Gaussian Analytic Functions," http://research.microsoft.com/en-us/um/people/peres/GAF/GAF.html.

136: **"The force with which such a conclusion":** Ronald A. Fisher, *Statistical Methods and Scientific Inference* (Edinburgh: Oliver & Boyd, 1959), 39.

136: **Joseph Berkson:** Joseph Berkson, "Tests of Significance Considered as Evidence," *Journal of the American Statistical Association* 37, no. 219 (1942), 325–35.

138: **In 2013, Yitang "Tom" Zhang:** The story of Zhang's work on the bounded gap conjecture is adapted from my article "The Beauty of Bounded Gaps," *Slate*, May 22, 2013. See Yitang Zhang, "Bounded Gaps Between Primes," *Annals of Mathematics*, forthcoming.

Chapter 9: The International Journal of Haruspicy

145: **a parable I learned from the statistician Cosma Shalizi:** You can read Shalizi's own version at his blog, at http://bactra.org/weblog/698.html.

147: **John Ioannidis:** John P. A. Ioannidis, "Why Most Published Research Findings Are False," *PLoS Medicine* 2, no. 8 (2005): e124, available at www.plosmedicine.org/article/info:doi/10.1371/journal.pmed.0020124.

149: **Low power is a special danger:** For an assessment of the perils of low-powered studies in neuroscience, see Katherine S. Button et al., "Power Failure: Why Small Sample Size Undermines the Reliability of Neuroscience," *Nature Reviews Neuroscience* 14 (2013): 365–76.

149: **A recent paper in *Psychological Science*:** Kristina M. Durante, Ashley Rae, and Vladas Griskevicius, "The Fluctuating Female Vote: Politics, Religion, and the Ovulatory Cycle," *Psychological Science* 24, no. 6 (2013): 1007–16. I am grateful to Andrew Gelman for conversations about the methodology of this paper and for his blog post about it (http://andrewgelman.com/2013/05/17/how-can-statisticians-help-psychologists-do-their-research-better), from which my analysis is largely drawn.

149: **as likely to push you in the *opposite* direction:** See Andrew Gelman and David Weakliem, "Of Beauty, Sex, and Power: Statistical Challenges in Estimating Small Effects," *American Scientist* 97 (2009): 310–16, for a worked-out example of this phenomenon, in the context of the question of whether good-looking people have more daughters than sons. (Nope.)

150: **But when Chabris's team tested those SNPs:** Christopher F. Chabris et al., "Most Reported Genetic Associations with General Intelligence Are Probably False Positives," *Psychological Science* 23, no. 11 (2012): 1314–23.

150: **Amgen set out to replicate:** C. Glenn Begley and Lee M. Ellis, "Drug Development: Raise Standards for Preclinical Cancer Research," *Nature* 483, no. 7391 (2012): 531–33.

153: **calls these practices "p-hacking":** Uri Simonsohn, Leif Nelson, and Joseph Simmons, "P-Curve: A Key to the File Drawer," *Journal of Experimental Psychology: General,* forthcoming. The curves sketched in this section are the "p-curves" described in this paper.

154: **political science to economics to psychology to sociology:** Some representative references: Alan Gerber and Neil Malhotra, "Do Statistical Reporting Standards Affect What Is Published? Publication Bias in Two Leading Political Science Journals," *Quarterly Journal of Political Science* 3, no. 3 (2008): 313–26; Alan S. Gerber and Neil Malhotra, "Publication Bias in Empirical Sociological Research: Do Arbitrary Significance Levels Distort Published Results?" *Sociological Methods & Research* 37, no. 1 (2008): 3–30; and E. J. Masicampo and Daniel R. Lalande, "A Peculiar Prevalence of P Values Just Below .05," *Quarterly Journal of Experimental Psychology* 65, no. 11 (2012): 2271–79.

156: **the U.S. Supreme Court ruled unanimously:** *Matrixx Initiatives, Inc. v. Siracusano,* 131 S. Ct. 1309, 563 U.S., 179 L. Ed. 2d 398 (2011).

156: **a paper by Robert Rector:** Robert Rector and Kirk A. Johnson, "Adolescent Virginity Pledges and Risky Sexual Behaviors," Heritage Foundation (2005), www.heritage.org/research/reports/2005/06/adolescent-virginity-pledges-and-risky-sexual-behaviors (accessed Jan. 14, 2014).

156: **"If a variable is not statistically significant":** Robert Rector, Kirk A. Johnson, and Patrick F. Fagan, "Understanding Differences in Black and White Child Poverty Rates," Heritage Center for Data Analysis report CDA01-04 (2001), p. 15 (n. 20), quoted in Jordan Ellenberg, "Sex and Signifance," *Slate,* July 5, 2005, http://thf_media.s3.amazonaws.com/2001/pdf/cda01-04.pdf (accessed Jan. 14, 2014).

159: **"the ordinary humanity of his fellows":** Michael Fitzgerald and Ioan James, *The Mind of the Mathematician* (Baltimore: Johns Hopkins University Press, 2007), 151, quoted in "The Widest Cleft in Statistics: How and Why Fisher Opposed Neyman and Pearson," by Francisco Louçã, Department of Economics of the School of Economics and Management, Lisbon, Working Paper 02/2008/DE/UECE, available at www.iseg.utl.pt/departamentos/economia/wp/wp022008deuece.pdf (accessed Jan. 14, 2014). Note that the Fitzgerald-James book seems intent on arguing that a large number of successful mathematicians through history had Asperger's syndrome, so their assessment of Fisher's social development should be read with that in mind.

160: **"I am a little sorry":** Letter to Hick of Oct. 8, 1951, in J. H. Bennett, ed., *Statistical Inference and Analysis: Selected Correspondence of R. A. Fisher* (Oxford: Clarendon Press, 1990), 144. Quoted in Louçã, "Widest Cleft."

160: **"A scientific fact should be regarded":** Ronald A. Fisher, "The Arrangement of Field Experiments," *Journal of the Ministry of Agriculture of Great Britain* 33 (1926): 503–13, quoted in Jerry Dallal's short article "Why p = 0.05?" (www.jerrydallal.com/LHSP/p05.htm)—a good introduction to Fisher's thought on this issue.

162: **"in fact no scientific worker":** Ronald A. Fisher, *Statistical Methods and Scientific Inference* (Edinburgh: Oliver & Boyd, 1956), 41–42, also quoted in Dallal, "Why p = 0.05?"

Chapter 10: Are You There, God? It's Me, Bayesian Inference

163: **the Guest Marketing Analytics team at Target:** Charles Duhigg, "How Companies Learn Your Secrets," *New York Times Magazine,* Feb. 16, 2012.

164: **the computation was reproduced on a Nokia 6300:** Peter Lynch and Owen Lynch, "Forecasts by PHONIAC," *Weather* 63, no. 11 (2008): 324–26.

164: **a typical five-day forecast:** Ian Roulstone and John Norbury, *Invisible in the Storm: The Role of Mathematics in Understanding Weather* (Princeton, NJ: Princeton University Press, 2013), 281.

164: **"One meteorologist remarked":** Edward N. Lorenz, "The Predictability of Hydrodynamic Flow," *Transactions of the New York Academy of Sciences,* series 2, vol. 25, no. 4 (1963): 409–32.

165: **Lorenz thought it was about two weeks:** Eugenia Kalnay, *Atmospheric Modeling, Data Assimilation, and Predictability* (Cambridge, UK: Cambridge University Press, 2003), 26.

165: **Netflix launched a $1 million competition:** Jordan Ellenberg, "This Psychologist Might Outsmart the Math Brains Competing for the Netflix Prize," *Wired,* Mar. 2008, pp. 114–22.

165: **which makes dud recommendations less of a big deal:** Xavier Amatriain and Justin Basilico, "Netflix Recommendations: Beyond the 5 Stars," techblog.netflix.com/2012/04/netflix-recommendations-beyond-5-stars.html (accessed Jan. 14, 2014).

171: **psychic powers were a hot topic:** A good contemporary account of the ESP craze can be found in Francis Wickware, "Dr. Rhine and ESP," *Life,* Apr. 15, 1940.

173: **were highly nonuniform:** Thomas L. Griffiths and Joshua B. Tenenbaum, "Randomness and Coincidences: Reconciling Intuition and Probability Theory," *Proceedings of the 23rd Annual Conference of the Cognitive Science Society,* 2001.

173: **17 is the most common choice:** Personal communication, Gary Lupyan.
173: **they most frequently pick 7:** Griffiths and Tenenbaum, "Randomness and Coincidences," fig. 2.
173: **Two graduate students at Columbia:** Bernd Beber and Alexandra Scacco, "The Devil Is in the Digits," *Washington Post*, June 20, 2009.
181: **"If the views of the last section of Mr. Keynes's book":** Ronald A. Fisher, "Mr. Keynes's Treatise on Probability," *Eugenics Review* 14, no.1 (1922): 46–50.
182: **"You know, the most amazing thing happened to me tonight":** Quoted by David Goodstein and Gerry Neugebauer in a special preface to the Feynman Lectures, reprinted in Richard Feynman, *Six Easy Pieces* (New York: Basic Books, 2011), xxi.
184: **The Cat in the Hat, the Cleanest Man in School, and the Creation of the Universe:** The discussion in this section owes a great deal to Elliott Sober's book *Evidence and Evolution* (New York: Cambridge University Press, 2008).
186: **fifteen editions in fifteen years:** Aileen Fyfe, "The Reception of William Paley's *Natural Theology* in the University of Cambridge," *British Journal for the History of Science* 30, no. 106 (1997): 324.
186: **"I could almost formerly have said it by heart":** Letter from Darwin to John Lubbock, Nov. 22, 1859, Darwin Correspondence Project, www.darwinproject.ac.uk/letter/entry-2532 (accessed Jan. 14, 2014).
189: **Nick Bostrom:** Nick Bostrom, "Are We Living in a Computer Simulation?" *Philosophical Quarterly* 53, no. 211 (2003): 243–55.
190: **a good argument that we're all sims:** Bostrom's argument in favor of SIMS has more to it than this one; it's controversial, but not immediately dismissible.

Chapter 11: What to Expect When You're Expecting to Win the Lottery

195: **seventeenth-century Genoa:** All information on the Genoese lottery from David R. Bellhouse, "The Genoese Lottery," *Statistical Science* 6, no. 2 (May 1991): 141–48.
196: **two new college buildings:** Stoughton Hall and Holworthy Hall.
196: **"That the chance of gain is naturally overvalued":** Adam Smith, *The Wealth of Nations* (New York: Wiley, 2010), bk. 1, ch. 10, p. 102.
199: **the "Million Act" of 1692:** The story of Halley and the mispriced annuity comes from chapter 13 of Ian Hacking, *The Emergence of Probability* (New York: Cambridge University Press, 1975).
200: **needed to be greater:** See Edwin W. Kopf, "The Early History of the Annuity," *Proceedings of the Casualty Actuarial Society* 13 (1926): 225–66.
201: **as many as 100 million tickets:** personal communication from Powerball PR department.
203: **three times in 2012:** "Jackpot History," www.lottostrategies.com/script/jackpot_history/draw_date/101 (accessed Jan. 14, 2014).
206: **pick numbers other players won't:** See John Haigh, "The Statistics of Lotteries," ch. 23 of Donald B. Hausch and William Thomas Ziemba, eds., *Handbook of Sports and Lottery Markets* (Amsterdam: Elsevier, 2008), for a survey of known results about which combinations lottery players prefer and which ones they shun.
209: **the exhaustive and, frankly, kind of thrilling account:** Letter from Gregory W. Sullivan, Inspector General of the Commonwealth of Massachusetts, to Steven Grossman, State Treasurer of Massachusetts, July 27, 2012. The Sullivan report is the source for the material here about high-volume betting in Cash WinFall, except where otherwise specified; it's available at www.mass.gov/ig/publications/reports-and-recommendations/2012/lottery-cash-winfall-letter-july-2012.pdf (accessed Jan. 14, 2014).
210: **They called their team Random Strategies:** I couldn't verify how early the name Random Strategies was actually chosen; it's possible the team didn't use this name when they were making their initial plays in 2005.
211: **an extra profit-making venture:** Phone interview, Gerald Selbee, Feb. 11, 2013.
214: **"squarely within the square":** Thanks to François Dorais for this translation.
214: **a provincial aristocrat from Burgundy:** The material on Buffon's early life is drawn from chapters 1 and 2 of Jacques Roger, *Buffon: A Life in Natural History, trans.* Sarah Lucille Bonnefoi (Ithaca, NY: Cornell University Press, 1997).
215: **"But if instead of throwing":** From the translation of Buffon's "Essay on Moral Arithmetic" by John D. Hey, Tibor M. Neugebauer, and Carmen M. Pasca, in Axel Ockenfels and Abdolkarim Sadrieh, *The Selten School of Behavioral Economics* (Berlin/Heidelberg: Springer-Verlag, 2010), 54.
222: **"Nothing seems to happen":** Pierre Deligne, "Quelques idées maîtresses de l'œuvre de A. Grothendieck," in *Matériaux pour l'histoire des mathématiques au XXe siècle: Actes du colloque à la mémoire de Jean Dieudonné, Nice, 1996 (Paris: Société Mathématique de France, 1998).* The original is "rien ne semble de passer et pourtant à la fin de l'exposé un théorème clairement non trivial est

là." Translation by Colin McCarty, from his article "The Rising Sea: Grothendieck on Simplicity and Generality," part 1, from *Episodes in the History of Modern Algebra (1800–1950)* (Providence: American Mathematical Society, 2007), 301–22.

223: **"The unknown thing to be known":** From Grothendeick's memoir *Récoltes et Semailles*, translated and quoted in McCarty, "Rising Sea," 302.

224: **Gerald Selbee told me:** Phone interview, Gerald Selbee, Feb. 11, 2013. All information about Selbee's role is taken from this interview.

227: **friend of *Boston Globe* reporter Andrea Estes:** E-mail from Andrea Estes, Feb. 5, 2013.

227: **the *Globe* ran a front-page story:** Andrea Estes and Scott Allen, "A Game with a Windfall for a Knowing Few," *Boston Globe*, July 31, 2011.

227: **In the early eighteenth century:** The story of Voltaire and the lottery is drawn from Haydn Mason, *Voltaire* (Baltimore: Johns Hopkins University Press, 1981), 22–23, and from Brendan Mackie's article "The Enlightenment Guide to Winning the Lottery," www.damninteresting.com/the-enlightenment-guide-to-winning-the-lottery (accessed Jan. 14, 2014).

230: **"As long as the Lottery announced to the public":** Letter from Gregory W. Sullivan to Steven Grossman.

231: **"It's a private lottery for skilled people":** Estes and Allen, "Game with a Windfall."

Chapter 12: Miss More Planes!

233: **"If you never miss the plane":** Or at least everyone says he used to say this. I couldn't find any evidence he ever put it in writing.

236: **"The Social Security Administration's inspector general on Monday said":** "Social Security Kept Paying Benefits to 1,546 Deceased," *Washington Wire* (blog), *Wall Street Journal*, June 24, 2013.

236: **Nicholas Beaudrot observed:** Nicholas Beaudrot, "The Social Security Administration Is Incredibly Well Run," www.donkeylicious.com/2013/06/the-social-security-administration-is.html.

239: **"For, to talk frankly with you about Geometry":** Letter from Pascal to Fermat, August 10, 1660.

241: **devoted a long essay:** The Voltaire here is all from the 25th of his "Philosophical Letters," which consists of remarks on the Pensées.

246: **a widely circulated blog post:** N. Gregory Mankiw, "My personal work incentives," Oct. 26, 2008, gregmankiw.blogspot.com/2008/10/blog-post.html. Mankiw returned to the same theme in his column "I Can Afford Higher Taxes, but They'll Make Me Work Less," *New York Times*, BU3, Oct. 10, 2010.

247: **Fran Lebowitz tells a story:** In the 2010 movie *Public Speaking*.

247: **"I dreamed about this problem some time":** Both quotes from Buffon's "Essays on Moral Arithmetic," 1777.

249: **After graduating third from his class at Harvard:** Biographical material on Ellsberg drawn from Tom Wells, *Wild Man: The Life and Times of Daniel Ellsberg* (New York: St. Martin's, 2001); and Daniel Ellsberg, *Secrets: A Memoir of Vietnam and the Pentagon Papers* (New York: Penguin, 2003).

250: **"There is the artist to study":** Daniel Ellsberg, "The Theory and Practice of Blackmail," RAND Corporation, July 1968, unpublished at the time, available at www.rand.org/content/dam/rand/pubs/papers/2005/P3883.pdf (accessed Jan. 14, 2014).

250: **now known as Ellsberg's paradox:** Daniel Ellsberg, "Risk, Ambiguity, and the Savage Axioms," *Quarterly Journal of Economics* 75, no.4 (1961): 643–69.

Chapter 13: Where the Train Tracks Meet

255: **it worked for Long-Term Capital Management:** LTCM itself didn't survive for long, but the principal actors walked away rich men and stuck around in the financial sector despite the LTCM debacle.

262: **The eye must generate light:** Otto-Joachim Gruesser and Michael Hagner, "On the History of Deformation Phosphenes and the Idea of Internal Light Generated in the Eye for the Purpose of Vision," *Documenta Ophthalmologica* 74, no. 1–2 (1990): 57–85.

265: **"There is an ending as far as I'm concerned":** David Foster Wallace, interviewed at *Word* e-zine, May 17, 1996, www.badgerinternet.com/~bobkat/jest11a.html (accessed Jan. 14, 2014).

267: **"A base del nostro studio":** Gino Fano, "Sui postulati fondamentali della geometria proiettiva," *Giornale di matematiche* 30.S 106 (1892).

267: **"As a basis for our study":** Translation adapted from that in C. H. Kimberling, "The Origins of Modern Axiomatics: Pasch to Peano," *American Mathematical Monthly* 79, no.2 (Feb. 1972): 133–36.

268: **If you do Cartesian geometry using the Boolean number system:** Highly abbreviated explanation:

Remember, the projective plane can be thought of as the set of lines through the origin in three-dimensonal space, and the lines in the projective plane are planes through the origin. A plane through the origin in 3-space has an equation of the form ax + by + cz = 0. So a plane through the origin in 3-space over the Boolean numbers is *also* given by an equation ax + by + cz = 0, except that now a,b,c are required to be either 0 or 1. So there are eight possible equations of this form. What's more, setting a= b = c = 0 gives an equation (0 = 0) which is satisfied for all x, y, and z, and thus doesn't determine a plane; so in all there are seven planes through the origin in the Boolean 3-space, which means there are seven lines in the Boolean projective plane, just as there should be.

272: **Hamming, a young veteran of the Manhattan Project:** Info on Hamming largely drawn from section 2 of Thomas M. Thompson, *From Error-Correcting Codes Through Sphere Packing to Simple Groups* (Washington, DC: Mathematical Association of America, 1984).

276: **"The Patent Department would not release the thing":** Ibid., 27

276: **Golay published first:** Ibid., 5, 6.

276: **As for the patent:** Ibid., 29.

278: **Ro was an artificial language:** All material on Ro is from the "Dictionary of Ro" at www.sorabji.com/r/ro.

279: **it goes back to the astronomer Johannes Kepler:** Historical material on sphere packing is taken from George Szpiro's book *The Kepler Conjecture* (New York: Wiley, 2003).

281: **Henry Cohn and Abhinav Kumar proved:** Henry Cohn and Abhinav Kumar, "Optimality and Uniqueness of the Leech Lattice Among Lattices," *Annals of Mathematics* 170 (2009): 1003–50.

281: **a single giant roll of paper:** Thompson, *From Error-Correcting Codes*, 121.

285: **was discovered by R. H. F. Denniston:** Ralph H. F. Denniston, "Some New 5-designs," *Bulletin of the London Mathematical Society* 8, no. 3 (1976): 263–67.

289: **"This man spends his life without weariness . . .":** Pascal, *Pensées*, no. 139.

290: **Even businesses that survive:** Information about the "typical entrepreneur" comes from chapter 6 of Scott A. Shane's book *The Illusions of Entrepreneurship: The Costly Myths That Entrepreneurs, Investors, and Policy Makers Live By* (New Haven, CT: Yale University Press, 2010).

Chapter 14: The Triumph of Mediocrity

295: **a widely used statistics textbook:** Horace Secrist, *An Introduction to Statistical Methods: A Textbook for Students, a Manual for Statisticians and Business Executives* (New York: Macmillan, 1917).

296: **"Mediocrity tends to prevail":** Horace Secrist, *The Triumph of Mediocrity in Business* (Chicago: Bureau of Business Research, Northwestern University, 1933), 7.

296: **"The results confront the business man":** Robert Riegel, *Annals of the American Academy of Political and Social Science* 170, no. 1 (Nov. 1933): 179.

297: **"Complete freedom to enter trade":** Secrist, *Triumph of Mediocrity in Business*, 24.

298: **"Pupils of all ages":** Ibid., 25.

298: **"I can cast up any sum at addition":** Karl Pearson, *The Life, Letters and Labours of Francis Galton* (Cambridge, UK: Cambridge University Press, 1930), 66.

298: **"devoured its contents and assimilated them":** Francis Galton, *Memories of My Life* (London: Methuen, 1908), 288. Both Galton's memoir and Pearson's biography are reproduced in full as part of the staggering collection of Galtoniana at galton.org.

299: **as one reviewer complained:** quoted in Emel Aileen Gökyigit, "The Reception of Francis Galton's *Hereditary Genius*," *Journal of the History of Biology* 27, no. 2 (Summer 1994).

300: **"I attempted mathematics":** From Charles Darwin, "Autobiography," in Francis Darwin, ed., *The Life and Letters of Charles Darwin* (New York and London: Appleton, 1911), 40.

304: **"Matt Kemp is off to a blazing start":** Eric Karabell, "Don't Fall for Another Hot April for Ethier," Eric Karabell Blog, Fantasy Baseball, http://insider.espn.go.com/blog/eric-karabell/post/_/id/275/andre-ethier-los-angeles-dodgers-great-start-perfect-sell-high-candidate-fantasy-baseball   (accessed Jan. 14, 2014).

305: **first-half American League home run leaders:** Data about midseason home run totals is drawn from "All-Time Leaders at the All-Star Break," CNN Sports Illustrated, http://sportsillustrated.cnn.com/baseball/mlb/2001/allstar/news/2001/07/04/leaders_break_hr.

306: **famous statistical smackdown:** Harold Hotelling, "Review of *The Triumph of Mediocrity in Business* by Horace Secrist," *Journal of the American Statistical Association* 28, no. 184 (Dec. 1933): 463–65.

306: **Hotelling was a Minnesotan:** Biographical information about Hotelling is drawn from Walter L. Smith, "Harold Hotelling, 1895–1973," *Annals of Statistics* 6, no. 6 (Nov 1978).

307: **Then the hammer drops:** My treatment of the Secrist/Hotelling story owes much to Stephen M. Stigler, "The History of Statistics in 1933," *Statistical Science* 11, no. 3 (1996): 244–52.

308: **"Very few of those biologists":** Walter F. R. Weldon, "Inheritance in Animals and Plants" in *Lectures on the Method of Science* (Oxford: Clarendon Press, 1906). I learned of Weldon's essay from Stephen Stigler.

309: **A 1976 *British Medical Journal* paper:** A. J. M. Broadribb and Daphne M. Humphreys, "Diverticular Disease: Three Studies: Part II: Treatment with Bran," *British Medical Journal* 1, no. 6007 (Feb. 1976): 425–28.

310: **When the program was put through randomized trials:** Anthony Petrosino, Carolyn Turpin-Petrosino, and James O. Finckenauer, "Well-Meaning Programs Can Have Harmful Effects! Lessons from Experiments of Programs Such as Scared Straight," *Crime and Delinquency* 46, no. 3 (2000): 354–79.

Chapter 15: Galton's Ellipse

311: **"I began with a sheet of paper":** Francis Galton, "Kinship and Correlation," *North American Review* 150 (1890), 419–31.

312: **Or at least reinvented it:** All material about the history of the scatterplot is drawn from Michael Friendly and Daniel Denis, "The Early Origins and Development of the Scatterplot," *Journal of the History of the Behavioral Sciences* 41, no. 2 (Spring 2005): 103–30.

318: **[isopleth map]:** Stanley A. Changnon, David Changnon, and Thomas R. Karl, "Temporal and Spatial Characteristics of Snowstorms in the Contiguous United States," Journal of Applied Meteorology and Climatology 45, no. 8 (2006): 1141–55.

318: **the first published isoplethic map:** Information on Halley's isogonal map from Mark Monmonier, *Air Apparent: How Meteorologists Learned to Map, Predict, and Dramatize Weather,* (Chicago: University of Chicago Press, 2000), 24–25.

322: **And here are the 50 U.S. states:** Data and image courtesy of Andrew Gelman.

324: **three of Thomas Pynchon's major novels:** Michael Harris, "An Automorphic Reading of Thomas Pynchon's *Against the Day*" (2008). Available at www.math.jussieu.fr/~harris/Pynchon.pdf (accessed Jan. 14, 2014). See also Roberto Natalini, "David Foster Wallace and the Mathematics of Infinity," in *A Companion to David Foster Wallace Studies* (New York: Palgrave MacMillan, 2013), 43–58, which interprets *Infinite Jest* in a similar way, finding there are not only parabolas and hyperbolas, but the *cycloid,* which is what you get when you subject a parabola to the mathematical operation of inversion.

325: **"The problem may not be difficult to an accomplished mathematician":** Francis Galton, *Natural Inheritance* (New York: Macmillan, 1889), 102.

326: **"The Bertillon cabinet":** Raymond B. Fosdick, "The Passing of the Bertillon System of Identification,"*Journal of the American Institute of Criminal Law and Criminology* 6, no. 3 (1915): 363–69.

326: **height versus "cubit":** Francis Galton, "Co-relations and Their Measurement, Chiefly from Anthropometric Data," *Proceedings of the Royal Society of London* 45 (1888): 135–45; and "Kinship and Correlation," *North American Review* 150 (1890): 419–31. In Galton's own words, from the 1890 paper: "Then a question naturally arose as to the limits of refinement to which M. Bertillon's system could be carried advantageously. An additional *datum* was no doubt obtained through the measurement of each additional limb or other bodily dimension; but what was the corresponding increase of accuracy in the means of identification? The sizes of the various parts of the body of the same person are in some degree related together. A large glove or shoe suggests that the person to whom it belongs is a large man. But the knowledge that a man has a large glove and a large shoe does not give us very much more information than if our knowledge had been confined to only one of the two facts. It would be most incorrect to suppose that the accuracy of the anthropometric method of identification increases with the number of measures in anything like the same marvellous rapidity that the security afforded by the better description of locks increases with the number of wards. The depths of the wards are made to vary quite independently of each other; consequently the addition of each new ward *multiplies* the previous security. But the lengths of the various limbs and bodily dimensions of the same person do not vary independently; so that the addition of each new measure adds to the security of the identification in a constantly-lessening degree."

334: **"As in most other cases of novel views":** Francis Galton, *Memories of My Life,* 310.

340: **a recent Supreme Court oral argument:** *Briscoe v. Virginia,* oral argument, Jan. 11, 2010, available at www.oyez.org/cases/2000-2009/2009/2009_07_11191 (accessed Jan. 14, 2014).

341: **"Like upscale areas everywhere":** David Brooks, "One Nation, Slightly Divisible," *Atlantic,* Dec. 2001.

341: **statistician Andrew Gelman found:** Andrew E. Gelman et al, "Rich State, Poor State, Red State,

Blue State: What's the Matter with Connecticut?" *Quarterly Journal of Political Science* 2, no. 4 (2007): 345–67.

341: **In some states, like Texas and Wisconsin:** See Gelman's book *Rich State, Poor State, Red State, Blue State* (Princeton, NJ: Princeton University Press, 2008), 68–70 for this data.

343: **was halted in 2011:** "NIH Stops Clinical Trial on Combination Cholesterol Treatment," NIH News, May 26, 2011, www.nih.gov/news/health/may2011/nhlbi-26.htm (accessed Jan. 14, 2014).

344: **appeared actually to *increase* the risk:** "NHLBI Stops Trial of Estrogen Plus Progestin Due to In- creased Breast Cancer Risk, Lack of Overall Benefit," NIH press release, July 9, 2002, www.nih .gov/news/pr/jul2002/nhlbi-09.htm (accessed Jan. 14, 2014).

344: **or that estrogen alone:** Philip M. Sarrel et al., "The Mortality Toll of Estrogen Avoidance: An Analysis of Excess Deaths Among Hysterectomized Women Aged 50 to 59 Years," *American Jour- nal of Public Health* 103, no. 9 (2013): 1583–88.

Chapter 16: Does Lung Cancer Make You Smoke Cigarettes?

350: **the relation between smoking and lung cancer:** Material on the early history of the link between smoking and lung cancer from Colin White, "Research on Smoking and Lung Cancer: A Landmark in the History of Chronic Disease Epidemiology," *Yale Journal of Biology and Medicine* 63 (1990): 29–46.

350: **paper of Doll and Hill:** Richard Doll and A. Bradford Hill, "Smoking and Carcinoma of the Lung," *British Medical Journal* 2, no. 4682 (Sept. 30, 1950): 739–48.

351: **"Is it possible then, that lung cancer":** Fisher wrote this in 1958. Quoted in Paul D. Stolley, "When Genius Errs: R. A. Fisher and the Lung Cancer Controversy," *American Journal of Epidemiology* 133, no. 5 (1991).

353: **more recent work has borne out his intuition:** See, e.g., Dorret I. Boomsma, Judith R. Koopmans, Lorenz J. P. Van Doornen, and Jacob F. Orlebeke, "Genetic and Social Influences on Starting to Smoke: A Study of Dutch Adolescent Twins and Their Parents," *Addiction* 89, no. 2 (Feb. 1994): 219–26.

353: **"if only the authors had been on the right side":** Jan P. Vandenbroucke, "Those Who Were Wrong," *American Journal of Epidemiology* 130, no. 1 (1989), 3–5.

353: **"A number of authorities who have examined the same evidence cited by Dr. Burney":** quoted in Jon M. Harkness, "The U.S. Public Health Service and Smoking in the 1950s: The Tale of Two More Statements," *Journal of the History of Medicine and Allied Sciences* 62, no. 2 (Apr. 2007): 171–212.

354: **the remarkable work:** Ibid.

355: **"possible to conceive but impossible to conduct":** Jerome Cornfield, "Statistical Relationships and Proof in Medicine," *American Statistician* 8, no. 5 (1954): 20.

355: **fell well short of disastrous:** For the 2009 pandemic, see Angus Nicoll and Martin McKee, "Mod- erate Pandemic, Not Many Dead—Learning the Right Lessons in Europe from the 2009 Pan- demic," *European Journal of Public Health* 20, no. 5 (2010): 486–88. But note that more recent studies have suggested that the worldwide death toll was much larger than originally estimated, perhaps on the order of 250,000.

358: **"Cancer is a biologic":** Joseph Berkson, "Smoking and Lung Cancer: Some Observations on Two Recent Reports," *Journal of the American Statistical Association* 53, no. 281 (Mar. 1958): 28–38.

358: **"It is as though":** Ibid.

358: **"If 85 to 95 per cent":** Ibid.

Chapter 17: There Is No Such Thing as Public Opinion

365: **In a January 2011 CBS News poll:** "Lowering the Deficit and Making Sacrifices," Jan. 24, 2011, www.cbsnews.com/htdocs/pdf/poll_deficit_011411.pdf (accessed Jan. 14, 2014).

365: **A Pew Research poll from February 2011:** "Fewer Want Spending to Grow, But Most Cuts Remain Unpopular," Feb. 10, 2011, www.people-press.org/files/2011/02/702.pdf.

366: **"The most plausible reading of this data is that the public wants a free lunch":** Bryan Caplan, "Mises and Bastiat on How Democracy Goes Wrong, Part II" (2003), Library of Economics and Liberty, www.econlib.org/library/Columns/y2003/CaplanBastiat.html (accessed Jan. 14, 2014).

366: **"People want spending cut":** Paul Krugman, "Don't Cut You, Don't Cut Me," *New York Times,* Feb. 11, 2011, http://krugman.blogs.nytimes.com/2011/02/11/dont-cut-you-dont-cut-me.

366: **"Many people seem to want to cut down the forest but to keep the trees":** "Cutting Government

Spending May Be Popular but There Is Little Appetite for Cutting Specific Government Programs," Harris Poll, Feb. 16, 2011, www.harrisinteractive.com/NewsRoom/HarrisPolls/tabid/447/mid/1508/articleId/693/ctl/ReadCustom%20Default/Default.aspx (accessed Jan. 14, 2014).

367: **Only 47% of Americans:** These numbers are from the January 2011 CBS poll cited above.

368: **In an October 2010 poll of likely voters:** "The AP-GfK Poll, November 2010," questions HC1 and HC14a, http://surveys.ap.org/data/GfK/AP-GfK%20Poll%20November%20Topline-nonCC.pdf.

371: **"The clause seems to express a great deal of humanity":** *Annals of the Congress of the United States,* Aug. 17, 1789. (Washington, DC: Gales and Seaton, 1834), 782.

373: **"The Court pays lip service":** *Atkins v. Virginia,* 536 US 304 (2002).

373: **Akhil and Vikram Amar:** "Akhil Reed Amar and Vikram David Amar, Eighth Amendment Mathematics (Part One): How the Atkins Justices Divided When Summing," *Writ,* June 28, 2002, writ.news.findlaw.com/amar/20020628.html (accessed Jan. 14, 2014).

375: **Just over six hundred people in all:** Numbers of executions taken from the Death Penalty Information Center, www.deathpenaltyinfo.org/executions-year (accessed Jan. 14, 2014).

376: **you can train a slime mold to run through a maze:** See, e.g., Atsushi Tero, Ryo Kobayashi, and Toshiyuki Nakagaki, "A Mathematical Model for Adaptive Transport Network in Path Finding by True Slime Mold,"*Journal of Theoretical Biology* 244, no. 4 (2007): 553–64.

376: **Tanya Latty and Madeleine Beekman:** Tanya Latty and Madeleine Beekman, "Irrational Decision-Making in an Amoeboid Organism: Transitivity and Context-Dependent Preferences," *Proceedings of the Royal Society B: Biological Sciences* 278, no. 1703 (Jan. 2011): 307–12.

382: **jays, honeybees, and hummingbirds:** Susan C. Edwards and Stephen C. Pratt, "Rationality in Collective Decision-Making by Ant Colonies," *Proceedings of the Royal Society B: Biological Sciences* 276, no. 1673 (2009): 3655–61.

382: **Psychologists Constantine Sedikides, Dan Ariely, and Nils Olsen:** Constantine Sedikides, Dan Ariely, and Nils Olsen, "Contextual and Procedural Determinants of Partner Selection: Of Asymmetric Dominance and Prominence," *Social Cognition* 17, no. 2 (1999): 118–39. But note also Shane Frederick, Leonard Lee, and Ernest Baskin, "The Limits of Attraction" (working paper), which argues that the evidence for the asymmetric domination effect in humans outside artificial lab scenarios is very weak.

384: **"among the very greatest improvements":** John Stuart Mill, *On Liberty and Other Essays* (Oxford: Oxford University Press, 1991), 310.

384: **the 2009 mayoral race in Burlington, Vermont:** The vote totals here are all drawn from "Burlington Vermont IRV Mayor Election," http://rangevoting.org/Burlington.html (accessed Jan. 15, 2014). See also University of Vermont political scientist Anthony Gierzynski's assessment of the election, "Instant Runoff Voting," www.uvm.edu/~vlrs/IRVassessment.pdf (accessed Jan. 15, 2014).

387: **"le mouton enragé":** Ian MacLean and Fiona Hewitt, eds., *Condorcet: Foundations of Social Choice and Political Theory* (Cheltenham, UK: Edward Elgar Publishing, 1994), 7.

388: **"I must act not by what I think reasonable":** From Condorcet's *"Essay on the Applications of Analysis to the Probability of Majority Decisions,"* in Ian MacLean and Fiona Hewitt, Condorcet, 38.

388: **"mathematical charlatan":** Material on Condorcet, Jefferson, and Adams from MacLean and Hewitt, Condorcet, 64.

389: **made an extended visit to Voltaire's house:** The material about the relationship between Voltaire and Condorcet in this section is largely drawn from David Williams, "Signposts to the Secular City: The Voltaire-Condorcet Relationship," in T. D. Hemming, Edward Freeman, and David Meakin, eds., *The Secular City: Studies in the Enlightenment* (Exeter, UK: University of Exeter Press, 1994), 120–33.

389: **Condorcet, like R. A. Fisher after him:** Lorraine Daston, *Classical Probability in the Enlightenment* (Princeton, NJ: Princeton University Press, 1995), 99.

389: **"so beautiful that it was frightening":** recounted in a letter of Mme. Suard of June 3, 1775, quoted in Williams, "Signposts," 128.

Chapter 18: "Out of Nothing I Have Created a Strange New Universe"

395: **"You must not attempt this approach to parallels":** This quote, and much of the historical discussion of Bolyai's work on non-Euclidean geometry, is drawn from Amir Alexander, *Duel at Dawn: Heroes, Martyrs, and the Rise of Modern Mathematics* (Cambridge, MA: Harvard University Press, 2011), part 4.

396: **"To praise it would amount to praising myself":** Steven G. Krantz, *An Episodic History of Mathematics* (Washington, DC: Mathematical Association of America, 2010), 171.

401: **the U.S. Supreme Court said no:** In *Bush v. Gore*, 531 U.S. 98 (2000).
401: **"Long live formalism":** Antonin Scalia, *A Matter of Intepretation: Federal Courts and the Law* (Princeton, NJ: Princeton University Press, 1997), 25.
402: **"The best-umpired game":** Quoted widely, e.g. in Paul Dickson, *Baseball's Greatest Quotations, rev. ed.* (Glasgow: Collins, 2008), 298.
403: **Jeter knew it wasn't a home run:** To be fair, the question "What did Derek Jeter know and when did he know it?" has never been completely settled. In a 2011 interview with Cal Ripken, Jr., he conceded that the Yankees "caught a break" on the play but wasn't willing to go so far as to say he should have been out. But he should have been out.
403: **"Most of the cases the Supreme Court agrees to decide":** from Richard A. Posner, "What's the Biggest Flaw in the Opinions This Term?" *Slate*, June 21, 2013.
404: **what Congress must have meant:** See, e.g., Scalia's concurrence in *Green v. Bock Laundry Machine Co.*, 490 U.S. 504 (1989).
405: **"When we are engaged in investigating":** From the translation of Hilbert's speech by Mary Winston Newson, *Bulletin of the American Mathematical Society*, July 1902, 437–79.
406: **"tables, chairs, and beer mugs":** Reid, *Hilbert*, 57.
410: **"A careful reader":** Hilbert, "Über das unendliche," *Mathematische Annalen* 95 (1926): 161–90; trans. Erna Putnam and Gerald J. Massey, "On the Infinite," in Paul Benacerraf and Hilary Putnam, *Philosophy of Mathematics*, 2^d ed. (Cambridge, UK: Cambridge University Press, 1983).
410: **Terry Tao found a mistake:** If you want to see what it looks like when serious mathematicians go toe-to-toe, you can watch the whole thing play out in real time in the comment section of the math blog *The N-Category Café* from September 27, 2011, "The Inconsistency of Arithmetic," http://golem.ph.utexas.edu/category/2011/09/the_inconsistency_of_arithmeti.html   (accessed Jan. 15, 2014).
411: **"The typical working mathematician":** Phillip J. Davis and Reuben Hersh, *The Mathematical Experience* (Boston: Houghton Mifflin, 1981), 321.
412: **Ramanujan was a prodigy from southern India:** Robert Kanigel's book *The Man Who Knew Infinity* (New York: Scribner, 1991) is a thorough popular account of Ramanujan's life and work, if you want to know more.
412: **that was Hermann Minkowski:** Reid, *Hilbert*, 7.
413: **Psychologists nowadays call it "grit":** See, e.g., the work of Angela Lee Duckworth.
415: **"It takes a thousand men":** From a letter Twain wrote on Mar. 17, 1903, to the young Helen Keller, available as "The Bulk of All Human Utterances Is Plagiarism," Letters of Note, www.lettersof note.com/2012/05/bulk-of-all-human-utterances-is.html (accessed Jan. 15, 2014).
415: **"The popular image":** Terry Tao, "Does One Have to Be a Genius to Do Maths?" http://terrytao .wordpress.com/career-advice/does-one-have-to-be-a-genius-to-do-maths (accessed Jan. 15, 2014).
417: **As to the nature of the contradiction:** The story and the quoted conversation come from "Kurt Gödel and the Institute," Institute for Advanced Study, www.ias.edu/people/godel/institute.
417: **he refused to sign the 1914 Declaration to the Cultural World:** Reid, *Hilbert*, 137.
418: **recounts the conversation:** Constance Reid, *Hilbert* (Berlin: Springer-Verlag, 1970), 210.
418: **"We cannot usually avoid being presented with decisions of this kind":** From "An Election Between Three Candidates," a section of Condorcet's *"Essay on the Applications of Analysis,"* in MacLean and Hewitt, Condorcet.

How to Be Right

424: **went his whole life without lifting a weapon in anger:** Wald did fulfill his compulsory service in the Rumanian army, though, so I can't say for sure he didn't.
424: **"Soonest Mended":** It appears in Ashbery's 1966 book *The Double Dream of Spring*. You can read the poem online at www.poetryfoundation.org/poem/177260 (accessed Jan. 15, 2014).
425: **"Sitting on a fence":** "Sitting on a Fence" appears on the Housemartins's debut record, *London 0 Hull 4.*
426: **I think of Silver as a kind of Kurt Cobain of probability:** Some of this material is adapted from my review of Silver's book *The Signal and the Noise* in the *Boston Globe*, Sept. 29, 2012.
427: **"On September 30, leading into the debates":** Josh Jordan, "Nate Silver's Flawed Model," *National Review Online*, Oct. 22, 2012, www.nationalreview.com/articles/331192/nate-silver-s-flawed-model-josh-jordan (accessed Jan. 15, 2014).
427: **"So should Mitt Romney win on Nov. 6":** Dylan Byers, "Nate Silver: One-Term Celebrity?" *Politico*, Oct. 29, 2012.
428: **states Silver considered potentially competitive:** Nate Silver, "October 25: The State of the States," *New York Times*, Oct. 26, 2012.

429: **"To believe something is to believe that it is true":** Willard Van Orman Quine, *Quiddities: An Intermittently Philosophical Dictionary* (Cambridge, MA: Harvard University Press, 1987), 21.
430: **Obama had a 67% chance:** These aren't Silver's actual numbers, which aren't archived as far as I can tell, just numbers made up to illustrate the kind of predictions he was making before the election.
431: **Sometimes the people speak and they say "I dunno":** The discussion of close elections is adapted from my article "To Resolve Wisconsin's State Supreme Court Election, Flip a Coin," *Washington Post*, Apr. 11, 2011.
432: **"He knew little out of his way":** From *The Autobiography of Benjamin Franklin* (New York: Collier, 1909), www.gutenberg.org/cache/epub/148/pg148.html (accessed Jan. 15, 2014).
433: **their reasoning modules frazzle and halt:** See, e.g., "I, Mudd," Star Trek, air date Nov. 3, 1967.
433: **"The test of a first-rate intelligence":** F. Scott Fitzgerald, "The Crack-Up," *Esquire*, Feb. 1936.
433: **it's said of the topologist R. H. Bing:** For instance, in George G. Szpiro, *Poincaré's Prize: The Hundred-Year Quest to Solve One of Math's Greatest Puzzles* (New York: Dutton, 2007).
435: **"I can't go on, I'll go on":** Samuel Beckett, *The Unnameable* (New York: Grove Press, 1958).
435: **Determined to record and neutralize:** My take on DFW's language is taken from an article I published in *Slate* on Sept. 18, 2008, "Finite Jest: Editors and Writers Remember David Foster Wallace," www.slate.com/articles/arts/culturebox/2008/09/finite_jest_2.html.
436: **"and you go the way of Hippasos":** Samuel Beckett, *Murphy* (London: Routledge, 1938).

INDEX